T0135297

Lecture Notes in Networks and Systems   801

Alvaro Rocha · Hojjat Adeli ·
Gintautas Dzemyda · Fernando Moreira ·
Valentina Colla

Editors

Information Systems and Technologies

WorldCIST 2023, Volume 3

Springer

Editors
Alvaro Rocha
ISEG
Universidade de Lisboa
Lisbon, Cávado, Portugal

Hojjat Adeli
College of Engineering
The Ohio State University
Columbus, OH, USA

Gintautas Dzemyda
Institute of Data Science and Digital
Technologies
Vilnius University
Vilnius, Lithuania

Fernando Moreira
DCT
Universidade Portucalense
Porto, Portugal

Valentina Colla
TeCIP Institute
Scuola Superiore Sant'Anna
Pisa, Italy

ISSN 2367-3370   ISSN 2367-3389 (electronic)
Lecture Notes in Networks and Systems
ISBN 978-3-031-45647-3   ISBN 978-3-031-45648-0 (eBook)
https://doi.org/10.1007/978-3-031-45648-0

This Springer imprint is published by the registered company Springer Nature Switzerland AG
The registered company address is: Gewerbestrasse 11, 6330 Cham, Switzerland

Paper in this product is recyclable.

Preface

This book contains a selection of papers accepted for presentation and discussion at the 2023 World Conference on Information Systems and Technologies (WorldCIST'23). This conference had the scientific support of the Sant'Anna School of Advanced Studies, Pisa, University of Calabria, Information and Technology Management Association (ITMA), IEEE Systems, Man, and Cybernetics Society (IEEE SMC), Iberian Association for Information Systems and Technologies (AISTI), and Global Institute for IT Management (GIIM). It took place in Pisa city, Italy, 4–6 April 2023.

The World Conference on Information Systems and Technologies (WorldCIST) is a global forum for researchers and practitioners to present and discuss recent results and innovations, current trends, professional experiences and challenges of modern Information Systems and Technologies research, technological development, and applications. One of its main aims is to strengthen the drive toward a holistic symbiosis between academy, society, and industry. WorldCIST'23 was built on the successes of: WorldCIST'13 held at Olhão, Algarve, Portugal; WorldCIST'14 held at Funchal, Madeira, Portugal; WorldCIST'15 held at São Miguel, Azores, Portugal; WorldCIST'16 held at Recife, Pernambuco, Brazil; WorldCIST'17 held at Porto Santo, Madeira, Portugal; WorldCIST'18 held at Naples, Italy; WorldCIST'19 held at La Toja, Spain; WorldCIST'20 held at Budva, Montenegro; WorldCIST'21 held at Terceira Island, Portugal; and WorldCIST'22, which took place online at Budva, Montenegro.

The Program Committee of WorldCIST'23 was composed of a multidisciplinary group of 339 experts and those who are intimately concerned with Information Systems and Technologies. They have had the responsibility for evaluating, in a 'blind review' process, and the papers received for each of the main themes proposed for the Conference were: A) Information and Knowledge Management; B) Organizational Models and Information Systems; C) Software and Systems Modeling; D) Software Systems, Architectures, Applications, and Tools; E) Multimedia Systems and Applications; F) Computer Networks, Mobility, and Pervasive Systems; G) Intelligent and Decision Support Systems; H) Big Data Analytics and Applications; I) Human-Computer Interaction; J) Ethics, Computers & Security; K) Health Informatics; L) Information Technologies in Education; M) Information Technologies in Radiocommunications; and N) Technologies for Biomedical Applications.

The conference also included workshop sessions taking place in parallel with the conference ones. Workshop sessions covered themes such as: Novel Computational Paradigms, Methods, and Approaches in Bioinformatics; Artificial Intelligence for Technology Transfer; Blockchain and Distributed Ledger Technology (DLT) in Business; Enabling Software Engineering Practices Via Latest Development's Trends; Information Systems and Technologies for the Steel Sector; Information Systems and Technologies for Digital Cultural Heritage and Tourism; Recent Advances in Deep Learning Methods and Evolutionary Computing for Health Care; Data Mining and Machine Learning in Smart Cities; Digital Marketing and Communication, Technologies, and Applications;

Digital Transformation and Artificial Intelligence; and Open Learning and Inclusive Education Through Information and Communication Technology.

WorldCIST'23 and its workshops received about 400 contributions from 53 countries around the world. The papers accepted for oral presentation and discussion at the conference are published by Springer (this book) in four volumes and will be submitted for indexing by WoS, Scopus, EI-Compendex, DBLP, and/or Google Scholar, among others. Extended versions of selected best papers will be published in special or regular issues of leading and relevant journals, mainly JCR/SCI/SSCI and Scopus/EI-Compendex indexed journals.

We acknowledge all of those that contributed to the staging of WorldCIST'23 (authors, committees, workshop organizers, and sponsors). We deeply appreciate their involvement and support that was crucial for the success of WorldCIST'23.

April 2023

Alvaro Rocha
Hojjat Adeli
Gintautas Dzemyda
Fernando Moreira
Valentina Colla

Organization

Honorary Chair

Hojjat Adeli   The Ohio State University, USA

General Chair

Álvaro Rocha   ISEG, University of Lisbon, Portugal

Co-chairs

Gintautas Dzemyda   Vilnius University, Lithuania
Sandra Costanzo   University of Calabria, Italy

Workshops Chair

Fernando Moreira   Portucalense University, Portugal

Local Organizing Committee

Valentina Cola (Chair)   Scuola Superiore Sant'Anna—TeCIP Institute, Italy
Marco Vannucci   Scuola Superiore Sant'Anna—TeCIP Institute, Italy
Vincenzo Iannino   Scuola Superiore Sant'Anna—TeCIP Institute, Italy
Stefano Dettori   Scuola Superiore Sant'Anna—TeCIP Institute, Italy

Advisory Committee

Ana Maria Correia (Chair)   University of Sheffield, UK
Brandon Randolph-Seng   Texas A&M University, USA

Chris Kimble	KEDGE Business School & MRM, UM2, Montpellier, France
Damian Niwiński	University of Warsaw, Poland
Florin Gheorghe Filip	Romanian Academy, Romania
Janusz Kacprzyk	Polish Academy of Sciences, Poland
João Tavares	University of Porto, Portugal
Jon Hall	The Open University, UK
John MacIntyre	University of Sunderland, UK
Karl Stroetmann	Empirica Communication & Technology Research, Germany
Majed Al-Mashari	King Saud University, Saudi Arabia
Miguel-Angel Sicilia	University of Alcalá, Spain
Mirjana Ivanovic	University of Novi Sad, Serbia
Paulo Novais	University of Minho, Portugal
Wim Van Grembergen	University of Antwerp, Belgium
Mirjana Ivanovic	University of Novi Sad, Serbia
Reza Langari	Texas A&M University, USA
Wim Van Grembergen	University of Antwerp, Belgium

Program Committee

Abderrahmane Ez-Zahout	Mohammed V University, Morocco
Adriana Gradim	University of Aveiro, Portugal
Adriana Peña Pérez Negrón	Universidad de Guadalajara, Mexico
Adriani Besimi	South East European University, Macedonia
Agostinho Sousa Pinto	Polythecnic of Porto, Portugal
Ahmed El Oualkadi	Abdelmalek Essaadi University, Morocco
Akex Rabasa	University Miguel Hernandez, Spain
Alba Córdoba-Cabús	University of Malaga, Spain
Alberto Freitas	FMUP, University of Porto, Portugal
Aleksandra Labus	University of Belgrade, Serbia
Alessio De Santo	HE-ARC, Switzerland
Alexandru Vulpe	University Politehnica of Bucharest, Romania
Ali Idri	ENSIAS, University Mohamed V, Morocco
Alicia García-Holgado	University of Salamanca, Spain
Amélia Badica	Universti of Craiova, Romania
Amélia Cristina Ferreira Silva	Polytechnic of Porto, Portugal
Amit Shelef	Sapir Academic College, Israel
Alanio de Lima	UFC, Brazil
Almir Souza Silva Neto	IFMA, Brazil
Álvaro López-Martín	University of Malaga, Spain

Ana Carla Amaro	Universidade de Aveiro, Portugal
Ana Isabel Martins	University of Aveiro, Portugal
Anabela Tereso	University of Minho, Portugal
Anabela Gomes	University of Coimbra, Portugal
Anacleto Correia	CINAV, Portugal
Andrew Brosnan	University College Cork, Ireland
Andjela Draganic	University of Montenegro, Montenegro
Aneta Polewko-Klim	University of Białystok, Institute of Informatics, Poland
Aneta Poniszewska-Maranda	Lodz University of Technology, Poland
Angeles Quezada	Instituto Tecnologico de Tijuana, Mexico
Anis Tissaoui	University of Jendouba, Tunisia
Ankur Singh Bist	KIET, India
Ann Svensson	University West, Sweden
Anna Gawrońska	Poznański Instytut Technologiczny, Poland
Antoni Oliver	University of the Balearic Islands, Spain
Antonio Jiménez-Martín	Universidad Politécnica de Madrid, Spain
Aroon Abbu	Bell and Howell, USA
Arslan Enikeev	Kazan Federal University, Russia
Beatriz Berrios Aguayo	University of Jaen, Spain
Benedita Malheiro	Polytechnic of Porto, ISEP, Portugal
Bertil Marques	Polytechnic of Porto, ISEP, Portugal
Boris Shishkov	ULSIT/IMI-BAS/IICREST, Bulgaria
Borja Bordel	Universidad Politécnica de Madrid, Spain
Branko Perisic	Faculty of Technical Sciences, Serbia
Carla Pinto	Polytechnic of Porto, ISEP, Portugal
Carlos Balsa	Polythecnic of Bragança, Portugal
Carlos Rompante Cunha	Polytechnic of Bragança, Portugal
Catarina Reis	Polytechnic of Leiria, Portugal
Célio Gonçalo Marques	Polythenic of Tomar, Portugal
Cengiz Acarturk	Middle East Technical University, Turkey
Cesar Collazos	Universidad del Cauca, Colombia
Christine Gruber	K1-MET, Austria
Christophe Guyeux	Universite de Bourgogne Franche Comté, France
Christophe Soares	University Fernando Pessoa, Portugal
Christos Bouras	University of Patras, Greece
Christos Chrysoulas	London South Bank University, UK
Christos Chrysoulas	Edinburgh Napier University, UK
Ciro Martins	University of Aveiro, Portugal
Claudio Sapateiro	Polytechnic of Setúbal, Portugal
Cosmin Striletchi	Technical University of Cluj-Napoca, Romania
Costin Badica	University of Craiova, Romania

Cristian García Bauza	PLADEMA-UNICEN-CONICET, Argentina
Cristina Caridade	Polytechnic of Coimbra, Portugal
David Cortés-Polo	University of Extremadura, Spain
David Kelly	University College London, UK
Daria Bylieva	Peter the Great St.Petersburg Polytechnic University, Russia
Dayana Spagnuelo	Vrije Universiteit Amsterdam, Netherlands
Dhouha Jaziri	University of Sousse, Tunisia
Dmitry Frolov	HSE University, Russia
Dulce Mourato	ISTEC - Higher Advanced Technologies Institute Lisbon, Portugal
Edita Butrime	Lithuanian University of Health Sciences, Lithuania
Edna Dias Canedo	University of Brasilia, Brazil
Egils Ginters	Riga Technical University, Latvia
Ekaterina Isaeva	Perm State University, Russia
Eliana Leite	University of Minho, Portugal
Enrique Pelaez	ESPOL University, Ecuador
Eriks Sneiders	Stockholm University, Sweden
Esperança Amengual	Universitat de les Illes Balears, Spain
Esteban Castellanos	ESPE, Ecuador
Fatima Azzahra Amazal	Ibn Zohr University, Morocco
Fernando Bobillo	University of Zaragoza, Spain
Fernando Molina-Granja	National University of Chimborazo, Ecuador
Fernando Moreira	Portucalense University, Portugal
Fernando Ribeiro	Polytechnic Castelo Branco, Portugal
Filipe Caldeira	Polythecnic of Viseu, Portugal
Filipe Portela	University of Minho, Portugal
Filippo Neri	University of Naples, Italy
Firat Bestepe	Republic of Turkey Ministry of Development, Turkey
Francesco Bianconi	Università degli Studi di Perugia, Italy
Francisco García-Peñalvo	University of Salamanca, Spain
Francisco Valverde	Universidad Central del Ecuador, Ecuador
Frederico Branco	University of Trás-os-Montes e Alto Douro, Portugal
Galim Vakhitov	Kazan Federal University, Russia
Gayo Diallo	University of Bordeaux, France
Gema Bello-Orgaz	Universidad Politecnica de Madrid, Spain
George Suciu	BEIA Consult International, Romania
Ghani Albaali	Princess Sumaya University for Technology, Jordan

Gian Piero Zarri	University Paris-Sorbonne, France
Giovanni Buonanno	University of Calabria, Italy
Gonçalo Paiva Dias	University of Aveiro, Portugal
Goreti Marreiros	ISEP/GECAD, Portugal
Graciela Lara López	University of Guadalajara, Mexico
Habiba Drias	University of Science and Technology Houari Boumediene, Algeria
Hafed Zarzour	University of Souk Ahras, Algeria
Haji Gul	City University of Science and Information Technology, Pakistan
Hakima Benali Mellah	Cerist, Algeria
Hamid Alasadi	Basra University, Iraq
Hatem Ben Sta	University of Tunis at El Manar, Tunisia
Hector Fernando Gomez Alvarado	Universidad Tecnica de Ambato, Ecuador
Hector Menendez	King's College London, UK
Hélder Gomes	University of Aveiro, Portugal
Helia Guerra	University of the Azores, Portugal
Henrique da Mota Silveira	University of Campinas (UNICAMP), Brazil
Henrique S. Mamede	University Aberta, Portugal
Henrique Vicente	University of Évora, Portugal
Hicham Gueddah	University Mohammed V in Rabat, Morocco
Hing Kai Chan	University of Nottingham Ningbo China, China
Igor Aguilar Alonso	Universidad Nacional Tecnológica de Lima Sur, Peru
Inês Domingues	University of Coimbra, Portugal
Isabel Lopes	Polytechnic of Bragança, Portugal
Isabel Pedrosa	Coimbra Business School - ISCAC, Portugal
Isaías Martins	University of Leon, Spain
Issam Moghrabi	Gulf University for Science and Technology, Kuwait
Ivan Armuelles Voinov	University of Panama, Panama
Ivan Dunđer	University of Zabreb, Croatia
Ivone Amorim	University of Porto, Portugal
Jaime Diaz	University of La Frontera, Chile
Jan Egger	IKIM, Germany
Jan Kubicek	Technical University of Ostrava, Czech Republic
Jeimi Cano	Universidad de los Andes, Colombia
Jesús Gallardo Casero	University of Zaragoza, Spain
Jezreel Mejia	CIMAT, Unidad Zacatecas, Mexico
Jikai Li	The College of New Jersey, USA
Jinzhi Lu	KTH-Royal Institute of Technology, Sweden
Joao Carlos Silva	IPCA, Portugal

João Manuel R. S. Tavares	University of Porto, FEUP, Portugal
João Paulo Pereira	Polytechnic of Bragança, Portugal
João Reis	University of Aveiro, Portugal
João Reis	University of Lisbon, Portugal
João Rodrigues	University of the Algarve, Portugal
João Vidal Carvalho	Polythecnic of Coimbra, Portugal
Joaquin Nicolas Ros	University of Murcia, Spain
John W. Castro	University de Atacama, Chile
Jorge Barbosa	Polythecnic of Coimbra, Portugal
Jorge Buele	Technical University of Ambato, Ecuador
Jorge Gomes	University of Lisbon, Portugal
Jorge Oliveira e Sá	University of Minho, Portugal
José Braga de Vasconcelos	Universidade Lusófona, Portugal
Jose M Parente de Oliveira	Aeronautics Institute of Technology, Brazil
José Machado	University of Minho, Portugal
José Paulo Lousado	Polythecnic of Viseu, Portugal
Jose Quiroga	University of Oviedo, Spain
Jose Silvestre Silva	Academia Militar, Portugal
Jose Torres	Universidty Fernando Pessoa, Portugal
Juan M. Santos	University of Vigo, Spain
Juan Manuel Carrillo de Gea	University of Murcia, Spain
Juan Pablo Damato	UNCPBA-CONICET, Argentina
Kalinka Kaloyanova	Sofia University, Bulgaria
Kamran Shaukat	The University of Newcastle, Australia
Karima Moumane	ENSIAS, Morocco
Katerina Zdravkova	University Ss. Cyril and Methodius, North Macedonia
Khawla Tadist	Marocco
Khalid Benali	LORIA—University of Lorraine, France
Khalid Nafil	Mohammed V University in Rabat, Morocco
Korhan Gunel	Adnan Menderes University, Turkey
Krzysztof Wolk	Polish-Japanese Academy of Information Technology, Poland
Kuan Yew Wong	Universiti Teknologi Malaysia (UTM), Malaysia
Kwanghoon Kim	Kyonggi University, South Korea
Laila Cheikhi	Mohammed V University in Rabat, Morocco
Laura Varela-Candamio	Universidade da Coruña, Spain
Laurentiu Boicescu	E.T.T.I. U.P.B., Romania
Lbtissam Abnane	ENSIAS, Morocco
Lia-Anca Hangan	Technical University of Cluj-Napoca, Romania
Ligia Martinez	CECAR, Colombia
Lila Rao-Graham	University of the West Indies, Jamaica

Łukasz Tomczyk	Pedagogical University of Cracow, Poland
Luis Alvarez Sabucedo	University of Vigo, Spain
Luís Filipe Barbosa	University of Trás-os-Montes e Alto Douro
Luis Mendes Gomes	University of the Azores, Portugal
Luis Pinto Ferreira	Polytechnic of Porto, Portugal
Luis Roseiro	Polytechnic of Coimbra, Portugal
Luis Silva Rodrigues	Polythencic of Porto, Portugal
Mahdieh Zakizadeh	MOP, Iran
Maksim Goman	JKU, Austria
Manal el Bajta	ENSIAS, Morocco
Manuel Antonio Fernández-Villacañas Marín	Technical University of Madrid, Spain
Manuel Ignacio Ayala Chauvin	University Indoamerica, Ecuador
Manuel Silva	Polytechnic of Porto and INESC TEC, Portugal
Manuel Tupia	Pontifical Catholic University of Peru, Peru
Manuel Au-Yong-Oliveira	University of Aveiro, Portugal
Marcelo Mendonça Teixeira	Universidade de Pernambuco, Brazil
Marciele Bernardes	University of Minho, Brazil
Marco Ronchetti	Universita' di Trento, Italy
Mareca María PIlar	Universidad Politécnica de Madrid, Spain
Marek Kvet	Zilinska Univerzita v Ziline, Slovakia
Maria João Ferreira	Universidade Portucalense, Portugal
Maria José Sousa	University of Coimbra, Portugal
María Teresa García-Álvarez	University of A Coruna, Spain
Maria Sokhn	University of Applied Sciences of Western Switzerland, Switzerland
Marijana Despotovic-Zrakic	Faculty Organizational Science, Serbia
Marilio Cardoso	Polythecnic of Porto, Portugal
Mário Antunes	Polythecnic of Leiria & CRACS INESC TEC, Portugal
Marisa Maximiano	Polytechnic Institute of Leiria, Portugal
Marisol Garcia-Valls	Polytechnic University of Valencia, Spain
Maristela Holanda	University of Brasilia, Brazil
Marius Vochin	E.T.T.I. U.P.B., Romania
Martin Henkel	Stockholm University, Sweden
Martín López Nores	University of Vigo, Spain
Martin Zelm	INTEROP-VLab, Belgium
Mazyar Zand	MOP, Iran
Mawloud Mosbah	University 20 Août 1955 of Skikda, Algeria
Michal Adamczak	Poznan School of Logistics, Poland
Michal Kvet	University of Zilina, Slovakia
Miguel Garcia	University of Oviedo, Spain

Miguel Melo	INESC TEC, Portugal
Mihai Lungu	University of Craiova, Romania
Mircea Georgescu	Al. I. Cuza University of Iasi, Romania
Mirna Muñoz	Centro de Investigación en Matemáticas A.C., Mexico
Mohamed Hosni	ENSIAS, Morocco
Monica Leba	University of Petrosani, Romania
Nadesda Abbas	UBO, Chile
Narjes Benameur	Laboratory of Biophysics and Medical Technologies of Tunis, Tunisia
Natalia Grafeeva	Saint Petersburg University, Russia
Natalia Miloslavskaya	National Research Nuclear University MEPhI, Russia
Naveed Ahmed	University of Sharjah, United Arab Emirates
Neeraj Gupta	KIET group of institutions Ghaziabad, India
Nelson Rocha	University of Aveiro, Portugal
Nikola S. Nikolov	University of Limerick, Ireland
Nicolas de Araujo Moreira	Federal University of Ceara, Brazil
Nikolai Prokopyev	Kazan Federal University, Russia
Niranjan S. K.	JSS Science and Technology University, India
Noemi Emanuela Cazzaniga	Politecnico di Milano, Italy
Noureddine Kerzazi	Polytechnique Montréal, Canada
Nuno Melão	Polytechnic of Viseu, Portugal
Nuno Octávio Fernandes	Polytechnic of Castelo Branco, Portugal
Nuno Pombo	University of Beira Interior, Portugal
Olga Kurasova	Vilnius University, Lithuania
Olimpiu Stoicuta	University of Petrosani, Romania
Patricia Zachman	Universidad Nacional del Chaco Austral, Argentina
Paula Serdeira Azevedo	University of Algarve, Portugal
Paula Dias	Polytechnic of Guarda, Portugal
Paulo Alejandro Quezada Sarmiento	University of the Basque Country, Spain
Paulo Maio	Polytechnic of Porto, ISEP, Portugal
Paulvanna Nayaki Marimuthu	Kuwait University, Kuwait
Paweł Karczmarek	The John Paul II Catholic University of Lublin, Poland
Pedro Rangel Henriques	University of Minho, Portugal
Pedro Sobral	University Fernando Pessoa, Portugal
Pedro Sousa	University of Minho, Portugal
Philipp Jordan	University of Hawaii at Manoa, USA
Piotr Kulczycki	Systems Research Institute, Polish Academy of Sciences, Poland

Prabhat Mahanti	University of New Brunswick, Canada
Rabia Azzi	Bordeaux University, France
Radu-Emil Precup	Politehnica University of Timisoara, Romania
Rafael Caldeirinha	Polytechnic of Leiria, Portugal
Raghuraman Rangarajan	Sequoia AT, Portugal
Raiani Ali	Hamad Bin Khalifa University, Qatar
Ramadan Elaiess	University of Benghazi, Libya
Ramayah T.	Universiti Sains Malaysia, Malaysia
Ramazy Mahmoudi	University of Monastir, Tunisia
Ramiro Gonçalves	University of Trás-os-Montes e Alto Douro & INESC TEC, Portugal
Ramon Alcarria	Universidad Politécnica de Madrid, Spain
Ramon Fabregat Gesa	University of Girona, Spain
Ramy Rahimi	Chungnam National University, South Korea
Reiko Hishiyama	Waseda University, Japan
Renata Maria Maracho	Federal University of Minas Gerais, Brazil
Renato Toasa	Israel Technological University, Ecuador
Reyes Juárez Ramírez	Universidad Autonoma de Baja California, Mexico
Rocío González-Sánchez	Rey Juan Carlos University, Spain
Rodrigo Franklin Frogeri	University Center of Minas Gerais South, Brazil
Ruben Pereira	ISCTE, Portugal
Rui Alexandre Castanho	WSB University, Poland
Rui S. Moreira	UFP & INESC TEC & LIACC, Portugal
Rustam Burnashev	Kazan Federal University, Russia
Saeed Salah	Al-Quds University, Palestine
Said Achchab	Mohammed V University in Rabat, Morocco
Sajid Anwar	Institute of Management Sciences Peshawar, Pakistan
Sami Habib	Kuwait University, Kuwait
Samuel Sepulveda	University of La Frontera, Chile
Snadra Costanzo	University of Calabria, Italy
Sandra Patricia Cano Mazuera	University of San Buenaventura Cali, Colombia
Sassi Sassi	FSJEGJ, Tunisia
Seppo Sirkemaa	University of Turku, Finland
Shahnawaz Talpur	Mehran University of Engineering & Technology Jamshoro, Pakistan
Silviu Vert	Politehnica University of Timisoara, Romania
Simona Mirela Riurean	University of Petrosani, Romania
Slawomir Zolkiewski	Silesian University of Technology, Poland
Solange Rito Lima	University of Minho, Portugal
Sonia Morgado	ISCPSI, Portugal

Sonia Sobral	Portucalense University, Portugal
Sorin Zoican	Polytechnic University of Bucharest, Romania
Souraya Hamida	Batna 2 University, Algeria
Stalin Figueroa	University of Alcala, Spain
Sümeyya Ilkin	Kocaeli University, Turkey
Syed Asim Ali	University of Karachi, Pakistan
Syed Nasirin	Universiti Malaysia Sabah, Malaysia
Tatiana Antipova	Institute of Certified Specialists, Russia
Tatianna Rosal	Universtiy of Trás-os-Montes e Alto Douro, Portugal
Tero Kokkonen	JAMK University of Applied Sciences, Finland
The Thanh Van	HCMC University of Food Industry, Vietnam
Thomas Weber	EPFL, Switzerland
Timothy Asiedu	TIM Technology Services Ltd., Ghana
Tom Sander	New College of Humanities, Germany
Tomaž Klobučar	Jozef Stefan Institute, Slovenia
Toshihiko Kato	University of Electro-communications, Japan
Tuomo Sipola	Jamk University of Applied Sciences, Finland
Tzung-Pei Hong	National University of Kaohsiung, Taiwan
Valentim Realinho	Polythecnic of Portalegre, Portugal
Valentina Colla	Scuola Superiore Sant'Anna, Italy
Valerio Stallone	ZHAW, Switzerland
Vicenzo Iannino	Scuola Superiore Sant'Anna, Italy
Vitor Gonçalves	Polythecnic of Bragança, Portugal
Victor Alves	University of Minho, Portugal
Victor Georgiev	Kazan Federal University, Russia
Victor Hugo Medina Garcia	Universidad Distrital Francisco José de Caldas, Colombia
Victor Kaptelinin	Umeå University, Sweden
Viktor Medvedev	Vilnius University, Lithuania
Vincenza Carchiolo	University of Catania, Italy
Waqas Bangyal	University of Gujrat, Pakistan
Wolf Zimmermann	Martin Luther University Halle-Wittenberg, Germany
Yadira Quiñonez	Autonomous University of Sinaloa, Mexico
Yair Wiseman	Bar-Ilan University, Israel
Yassine Drias	University of Algiers, Algeria
Yuhua Li	Cardiff University, UK
Yuwei Lin	University of Roehampton, UK
Zbigniew Suraj	University of Rzeszow, Poland
Zorica Bogdanovic	University of Belgrade, Serbia

Contents

Software and Systems Modeling

Organizational Models and Information Systems

Technologies for Biomedical Applications

Software Systems, Architectures, Applications and Tools

Review of Open Software Bug Datasets

Tomas Holek[1] , Miroslav Bures[1(⊠)] , and Tomas Cerny[1,2]

[1] Department of Computer Science, Faculty of Electrical Engineering,
Czech Technical University in Prague, Karlovo namesti 13, 121 35 Prague, Czechia
`miroslav.bures@fel.cvut.cz`
[2] Systems and Industrial Engineering, University of Arizona,
1127 East James E Rogers Way, Tucson, AZ 85721, USA
`tcerny@arizona.edu`
`http://still.felk.cvut.cz`

Abstract. The localisation of the bug position in a source code and the prediction of which specific parts of a source code might be the cause of defects play an important role in maintaining software quality. Both approaches are based on applying information retrieval techniques and machine learning or deep learning methods. The prerequisite for using these approaches is the availability of a consistent bug dataset of sufficient size. This paper presents an overview of available public bug datasets and analyses their specific application areas. The paper also suggests possible future research directions in this field.

Keywords: Software Quality · Software Testing · Software Bugs · Bug Localisation and Prediction · Bug Dataset · Literature Review

1 Introduction

Software bug datasets are useful for a number of purposes, the main ones being the localisation of the bug position in the source code and the prediction of source code parts that might contribute to future defects [1,22]. Additional purposes for which bug datasets can be utilised are automatic bug repair [4] and prediction-related investigations, such as bug triaging, bug-fixing time estimation, and bug information mining. Bug datasets that are publicly available are a great resource for rendering individual studies comparable and reproducible [3].

During the software development lifecycle, various bugs are identified and reported. Bug reports mainly contain a description in a natural language and other metadata. In most cases, bug reports are stored in a database of issue tracking tools. These tools usually do not directly link the reported bug with a place in a source code where this bug was fixed. Such information is available in the source code versioning system, recording all changes to the software source code. A number of techniques have been developed to relate bug reports to the actual bug fixes in the source code, which allows for creation of bug datasets.

Open bug datasets, which are publicly available, are generally helpful in the development of bug localisation techniques, the analysis of software project

A. Rocha et al. (Eds.): WorldCIST 2023, LNNS 801, pp. 3–12, 2024.
https://doi.org/10.1007/978-3-031-45648-0_1

dynamics, defect root cause analysis, and bug prediction. Hence, an overview of available open bug datasets would be beneficial for the software research and engineering community.

As there is currently no comprehensive work providing this overview (only individual aspects or partial summaries were provided), one of the goals of this paper is to provide a summary of existing open bug datasets. This will be based on a defined search methodology, by analysing the available open bug datasets, describing them, and classifying them according to their use, structure, and method of their creation. The overview of open bug datasets provided as an outcome of this study will be a useful resource for researchers, as well as software industry practitioners.

2   Related Work

The first bug datasets were created from databases of bug tracking systems and Code Versioning Systems (CVS) [18,23]. The current trend is to use open-source code-sharing platforms, such as GitHub [21]. As a result of this trend, a number of publicly available datasets were created.

To give some examples, one of the first publicly available datasets is the *NASA Metric Data Program (MDP)* dataset [18], which was used for numerous studies in the past, e.g. [5,10]. Another option, the *Bug prediction dataset* by D'Ambros *et al.* contains data to calculate the established Chidamber & Kemerer (C&K) metrics for class level [1]. An alternative, the *GitHub Bug Dataset* was constructed at class and file-level by Toth *et al.* [21]. *Bugs.jar* by Saha *et al.* is a large and diverse dataset for automatic bug repair [9]. This bug repair is also supported by the *Bears dataset*, by Madeiral *et al.* [7].

From a structural viewpoint, we can distinguish several types of bug datasets: (1) file level, (2) class level, and (3) method level [1,15,21]. The files in the bug datasets vary in formats, the most frequently used are CSV [23], ARFF [18,23] (an input format of Weka environment)[1], XML [17], XLSX [12], or JSON [12]. Some public datasets are downloadable as zip files, e.g. [3,11,12,21], while other datasets are stored in the GitHub repository, e.g. [8,18].

Software bug datasets comprise the bug information and the software metrics information. Two main categories of these metrics are used: (1) process metrics and (2) source code metrics. Process metrics focus on measuring the software quality based on developer activity and software change history [14]. As for the code metrics, bug datasets most frequently contain different object-oriented (OO) metrics that are used to measure the quality of object-oriented software design. It is shown that OO metrics are useful for predicting defect density, and the low code quality indicated by values of OO metrics is significantly associated with defects [19].

When creating a bug dataset, there are two main strategies for locating the source of the bug using the bug reports: (1) Information Retrieval (IR) and (2)

---

[1] Weka ARFF, https://www.cs.waikato.ac.nz/ml/weka/arff.html.

a Hybrid approach consisting of IR and Machine Learning (ML), alternatively Deep Learning (DL) for more recent systems. IR addresses the bug localisation as if it were a document retrieval problem. Many IR-based techniques and tools exist in the literature [11]. In Hybrid approaches, DL and ML methods are used to reduce the lexical gap that exists between source code and bug reports. This approach provides higher accuracy in bug localisation and bug prediction [9].

From available consolidated works, Ferenc *et al.* created the *Unified bug dataset* [3]. In their study, they provide a summarisation of the set of chosen datasets in more detail and investigate each dataset's peculiarities, looking for common characteristics. The paper also provides some basic size statistics on the chosen datasets. All described datasets focus on bugs from the software testing perspective and also support future automatic program repair studies.

In contrast to the study by Ferenc *et al.* [3], our paper analyses a wider range of datasets and only deals with the description of existing bug datasets, not datasets we have created, nor approaches to the data processing of bug reports that we have employed. Moreover, as we did not find any other articles summarising the open bug datasets, we provide this information in the overview within this paper.

3   Research Questions and Search Methodology

In this study, we focus on answering two research questions relating to public software bug datasets:

RQ1: What are the most common application areas of public software bug datasets?

RQ2: Which public software bug datasets are available, and what are their data structure and application areas?

This study follows the methodology recommendations for systematic mapping studies in software engineering, provided by Kitchenham and Charters [6]. We have divided the process of collection and analysis of relevant studies into six stages: (1) *Research scope*, determination and definition of RQs to be answered in the study, (2) *Paper search*, a search for relevant papers, which includes the establishment of a search strategy, (3) *Paper filtering*, identification of truly relevant papers from the initial selection, which includes performing snowball sampling of other relevant studies, (4) *Data extraction* from the remaining papers to allow further detailed analyses, (5) *Paper analysis*, classification of papers and analyses of the extracted data to answer defined RQs, and, (6) *Validity check*, an evaluation and discussion of the possible limitations of the study.

To search for relevant studies, we used the following four established publication databases: Elsevier Science Direct, IEEE Xplore, ACM Digital Library and SpringerLink. After the initial scope settlement and refinement, the search string was established as *(bug OR defect) AND dataset AND software*.

The timespan for selection of the papers was determined to be from 2005 to 2022 (the paper search was conducted in April), and as for article formats used, we included journal papers, book chapters, and conference papers. The number of initially downloaded papers is presented in Table 1, column *Initial sample size.*

Downloaded papers were filtered in two steps. Step one, we selected only conference and journal papers, other media were excluded from the analysis. Step two, we excluded papers not describing any software bug datasets. The number of filtered papers for individual databases is presented in Table 1, column *After filtering.* The next step was the snowball sampling process. We analysed other relevant papers referred to from the filtered papers, which were not already a part of the set of filtered papers. The number of papers found after this step is presented in Table 1, column *After snowball.*

Table 1. Numbers of research papers found in individual stages of the search process.

Source	Initial sample size	After filtering	After snowball
ScienceDirect	5	2	3
IEEE Xplore	111	10	11
ACM Digital Library	3	0	0
SpringerLink	3	0	0
other databases	0	0	8
Total	**119**	**15**	**22**

After the snowballing process, open bug datasets described in the final 22 papers were analysed in detail to answer RQs defined for this study.

To ensure better objectiveness of the paper filtering and selection process, two authors conducted the Paper filtering and Paper analysis phases independently, and their results were compared. In the instance of discrepancies, discussions were held to find a consensus.

4   Results

This section presents the answers to defined RQs.

RQ1: What are the most common application areas of public software bug datasets?

During the analysis, we identified three principal application areas of open bug datasets: (1) bug prediction and localisation, (2) automated code reparation, and (3) investigations and research.

Bug prediction and bug localisation are the oldest and most widespread application areas [1,3,11]. The bug localisation task is locating the existing bug position in the source code. The bug prediction task is identifying the most probable source that causes the defect. Both tasks can be done using IR, ML, and DL through available software tools. These tasks greatly reduce the effort of developers in localising the source of the bug [1,3,11].

The bug prediction can be done on file and class levels, as well as on a method level. A number of algorithms are used for bug prediction. The most frequently used algorithms on a class level are PART, J48, RandomForest, and RandomTree. The most frequently used algorithms on the file level are DecisionTable, RandomForest and Logistic [21].

In recent years, the importance of the second application area, the *automated code repair* area, has increased [7,20]. Several public bug datasets that are used by automatic code reparation systems can be identified [4,7,20]. A variety of tools use open datasets, namely recommender tools, that analyse the code and report errors to the developers so that they can fix it [4].

Another principal application area is *investigation and research* in the field of bug prediction, bug-fixing time estimation, bug-fixing prioritisation and automated assignment to a developer (triaging), and bug information mining for various purposes. Improving bug-fixing time estimation simplifies and optimises software project management. In a bug triaging process, an appropriate developer who could fix the bug is identified. The bug triaging algorithm takes the bug's data as bug title and bug description as the input, and it assigns the bug to the most suitable of the available developers. This algorithm can be formulated as a classification problem [18,20].

RQ2: Which public software bug datasets are available and what are their data structure and application areas?

In a software bug dataset analysis, several aspects or features can be evaluated. One of these features is a dataset application area. Another aspect is whether a dataset is a single software bug dataset or a repository containing multiple software bug datasets. The next important feature is the source code languages of source projects. Other features we examine are the metric used and the level (e.g. method, file, or class). Additionally, an important property of a dataset is the dataset file format (e.g. XML, ARFF, XLSX, CSV, JSON). Open bug datasets identified in this study are summarised in Table 2. In this section, we analyse these datasets in more detail.

PRedictOr Models In Software Engineering (**PROMISE**) [16] is a research data repository in software engineering, promoting predictive models. It is community based, so anybody can donate a new dataset or public tools, which can help other researchers in building predictive models. The repository provides datasets of several different types as code analysis, testing, maintenance, software defects, and others. The **NASA Metrics Data Program (MDP) datasets** is a repository that consists of 13 original datasets. Each original dataset in the repository represents a NASA software system/subsystem, mostly written in C/C++ and Java. The repository was intended for software metrics, and it has been used for bug prediction too. The NASA datasets can be found on GitHub. The **Eclipse bug dataset** maps defects from the bug database (The Bugzilla bug tracking system) of Eclipse 2.0, 2.1, and 3.0 [17,23], and it has been used in many studies. The dataset lists the number of pre-release and post-release defects on the file and package levels. Some features are calculated at a finer granularity by aggregation, taking the average, total, and maximum values of the metrics. The last update of this dataset was done on 25 March 2010. Two formats of the dataset are available: the XML format file contains references to bug reports and transactions that fixed the bugs. The ARRF and CVS format files extend the original data by complexity metrics counted from abstract syntax trees.

The **Bug prediction dataset** contains bugs from five Java projects, which are accompanied by calculated C&K metrics on the class level. CVS, SVN, Bugzilla and Jira were used as sources for this dataset. The dataset content includes pre- and post-release phase defects. D'Ambros et al. also extended the source code metrics with change metrics. This extension improves the performance of the fault prediction methods [1].

The **Defects4J** dataset is a collection of reproducible bugs from 17 open-source projects and a supporting infrastructure. This dataset was first presented by Just *et al.* and its goal is to be used for software engineering research. The Defect4j dataset can be accessed in the GitHub repository (see Table 2). It also provides a command-line interface. The **GitHub bug Dataset** [21] is based on 15 Java projects from GitHub. The fixed bugs are matched with the corresponding elements of source code at the class and file level to calculate a set of software product metrics. This dataset serves for the evaluation of machine learning algorithms for bug prediction [21]. **IntroClassJava** is a collection of 297 small software programs in Java that contain a set of bugs [20]. It is publicly available on Github. The IntroClassJava dataset is very similar to Defects4J. The difference is that IntroClassJava doesn't contain fixing patches that are manually cleaned. This bug dataset can be used for research on defect localisation and automatic repair of Java programs.

Bugs.jar is a dataset for automatic bug repair. Bugs.jar is based on eight open-source Java projects, and it contains 1,158 bugs, which is approximately ten times larger than Defect4J [9]. The **Bears dataset** is a collection of Java program bugs used for automatic repair, which are organised into a benchmark. Individual researchers can extend this dataset [7]. This bug dataset is created from commit reports of open-source projects on GitHub, where program versions containing a bug and its fix are compared. This process is performed through a specially configured Continuous Integration (CI) pipeline.

BuGC is a dataset for research in bug localisation [12]. This dataset is based on 21 C language projects extracted from GitHub. This dataset consists of 36.617 closed issues, out of which 2462 fixed bug reports can be used for bug localisation purposes. The BuGC dataset is available in JSON and XLSX file formats. **BuGL** was presented in 2019 by Muvva *et al.* and is a cross-language bug dataset that comprises more than 10.000 bug reports acquired from open-source projects in C, C++, Java, and Python languages [11]. The **Unified Bug Dataset** serves for bug prediction and focuses on class and file levels. In their work, Ferenc et al. try to unify a set of previous open datasets which contain calculated source code metrics [15]. **NFBugs** is a dataset of non-functional bugs. This work by Radu and Nadi comprises 133 non-functional bugs. These fixes were gathered from a set of 65 open-source projects in Java and Python languages. This dataset can

Table 2. Summary of analysed open bug datasets.

Dataset name (bold), reference, authors		Year[a]	Application area (bold), source (italics), structure
NASA bug datasets [18], Shepperd et al.	[b]	2005, 2016	**bug prediction**, *NASA software projects* Repository of 13 datasets, each representing a NASA software system/subsystem. Available as ARFF files.
Eclipse bug dataset [17], Zimmermann et al.	[c]	2007, 2010	**bug prediction**, *Bugzilla* Dataset contains bug reports and links to their fixes. Available as XML.
Eclipse bug dataset extended [23], Zimmermann et al.	[d]	2007, 2010	**bug prediction**, *Bugzilla* Extension of the Eclipse bug dataset by complexity metrics. Available as ARRF/CVS.
Bug Prediction Dataset, D'Ambros	[e]	2010	**bug prediction**, *CVS, SVN, Bugzilla, Jira* Extracted from 5 Java projects, Fusion and Moose were used to calculate C&K metrics for class level. Contains the numbers of pre- and post-release defects.
Defects4j Dataset [8], Martinez et al.	[f]	2014	**research**, *Github* A large dataset of bugs from real Java projects. Each bug is linked to a test suite and at least one failing test that finds this bug.
GitHub Bug Dataset v 1.0, v 1.1 [21], Ferenc et al.	[g] [h]	2016	**bug prediction**, *Github* Extracted from 15 Java projects selected from GitHub. Known and fixed bugs are linked to corresponding classes and files of the source code.
IntroClassJava [20], Durieux and Monperrus	[i]	2016, 2018	**code repair, research of bug localisation**, *Github* Very similar to Defects4J; however, it does not provide the manually cleaned fixing patches.
bugs.jar [9], Saha and Lyu	[j]	2018	**code repair** *Jira, Git* Diverse dataset of real-world Java code bugs.
Bears [7], Madeiral et al.	[k]	2019	**code repair**, *Github* Dataset including the bugs of the source code, the test suite, and the corresponding fixes of the bugs.
BugHunter Dataset [15], Ferenc et al.	[l]	2020	**bug prediction**, *Github* Automatically constructed bug dataset including a set of code metrics and information about bugs.
BuGC_Dataset [12], Muvva et al.	[m]	2020	**bug localisation**, *Github* The dataset containing more than 36,000 closed issues; available in json and xlsx.
BuGL [11], Muvva et al.	[n]	2020	**bug localisation**, *Github* Cross-language dataset that comprises more than 10.000 bug reports from open-source projects.
Unified Bug Dataset [3], Ferenc et al.	[o]	2020	**bug prediction**, *Github* A dataset of bugs at class and file level, accompanied by a set of unified source code metrics.
NFBugs [13], Radu and Nadi	[p]	2019	**code recommender**, *Github* A dataset of 133 fixes of non-functional bugs from 65 various open-source projects

[a] In case two years are presented, the first one is the year when the dataset has been created and the second one is the year of the last update. If only one year is presented, it is a year of dataset creation.

[b] https://github.com/klainfo/NASADefectDataset/tree/master/OriginalData/MDP.

[c] https://www.st.cs.uni-saarland.de/softevo/bug-data/eclipse/

[d] https://www.st.cs.uni-saarland.de/softevo/bug-data/eclipse/

[e] https://bug.inf.usi.ch/index.php

[f] https://github.com/rjust/defects4j

[g] http://www.inf.u-szeged.hu/~ferenc/papers/GitHubBugDataSet/GitHubBugDataSet.zip

[h] http://www.inf.u-szeged.hu/~ferenc/papers/GitHubBugDataSet/GitHubBugDataSet-1.1.zip

[i] https://github.com/Spirals-Team/IntroClassJava

[j] https://github.com/bugs-dot-jar

[k] https://github.com/bears-bugs/bears-benchmark

[l] https://data.mendeley.com/datasets/8tx7kjbkg4/2/files/b69abdfc-e48b-4744-b36d-a6d41a57445e

[m] https://zenodo.org/record/4153561/files/BuGC_Dataset.zip?download=1

[n] https://zenodo.org/record/3653836/files/BuGL.zip?download=1

[o] https://zenodo.org/record/3693686/files/UnifiedBugDataset-1.2.zip?download=1

[p] https://github.com/ualberta-smr/NFBugs

be utilised in code recommender systems focusing on non-functional properties of the software [13].

5   Discussion and Future Directions

The current trend is creating bug datasets from public projects stored on GitHub and similar platforms. There are alternatives to Github for hosting open-source projects such as GitLab, Bitbucket, Beanstalk, Launchpad, Sourceforge, Phabricator, and GitBucket. These can be considered for the use of creating open bug datasets as well.

The majority of bug datasets are based on bug reports drawn from projects written in the major programming languages, such as C, C++, Java, and Python. Creating other datasets from other programming languages such as Golang, Kotlin, or Javascript may also be considered, as the bugs can be language-specific. Such alternative bug datasets might be considered as one of the possible future directions.

In creating a bug dataset, it is helpful to accompany it with appropriate code metrics. The situation is not unified at this point; different bug datasets use different code metric suites. Despite the fact that there is an effort to create the public unified bug dataset, which contains selected publicly available ones accompanied by a set of metrics in a unified format [3], such efforts are rather isolated. This field represents one of the possible future directions.

In general, the importance and use of bug datasets are increasing and will increase as data processing technologies using machine learning techniques (e.g. word embedding techniques, such as word2vec) have evolved in recent years [9]. This automated processing to create and analyse open bug datasets represents other future directions.

The usefulness of bug datasets can be evaluated using the *F-measure* metric when testing different bug localisation or prediction models [2]. However, in this summary paper, we did not compare the F-measure of individual bug datasets. Moreover, bug datasets are extracted from a different number of projects and different numbers of bug reports; and additionally, the projects differ in a number of lines of code. It would be worth finding out how these parameters affect F-measure, which is another possible further direction to research.

6   Conclusion

This paper summarises available open bug datasets, their analysis and identification of trends and possible future research directions in this field. This area was under researched, as the majority of the papers on this topic focus on the description of their own datasets; in the instances that they do mention other bug datasets, the information was incomplete.

In our analysis, we identified three major application areas of software bug datasets: bug prediction and localisation, automated code reparation, and investigations and research. In the set of analysed papers, we identified 14 open bug datasets, for which we presented the main characteristics.

In the analysis of these datasets, we discussed how the use of bug datasets, their structure, and the resources used to create them have evolved over time. Furthermore, we suggested some possible future research directions, such as the creation of other open datasets for less frequent programming languages, unification of metrics used in the datasets, automated processing to create and analyse open bug datasets, and a deeper examination of F-measure in the datasets.

References

1. D'Ambros, M., Lanza, M., Robbes, R.: An extensive comparison of bug prediction approaches. In: 2010 7th IEEE Working Conference on Mining Software Repositories (MSR 2010), pp. 31–41 (2010)
2. Ferenc, R., Gyimesi, P., Gyimesi, G., Tóth, Z., Gyimóthy, T.: An automatically created novel bug dataset and its validation in bug prediction. J. Syst. Softw. **169**, 110691 (2020)
3. Ferenc, R., Tóth, Z., Ladányi, G., Siket, I., Gyimóthy, T.: A public unified bug dataset for java and its assessment regarding metrics and bug prediction, March 2020
4. Goues, C., Forrest, S., Weimer, W.: Current challenges in automatic software repair. Software Qual. J. **21**, 421–443 (2013)
5. Gray, D., Bowes, D., Davey, N., Sun, Y., Christianson, B.: Reflections on the NASA MDP data sets. IET Softw. **6**, 549–558 (2012)
6. Kitchenham, B., Charters, S.: Guidelines for performing systematic literature reviews in software engineering (2017)
7. Madeiral, F., Urli, S., Maia, M., Monperrus, M.: Bears: an extensible java bug benchmark for automatic program repair studies. In: 2019 IEEE 26th International Conference on Software Analysis, Evolution and Reengineering (SANER), February 2019
8. Martinez, M., Durieux, T., Sommerard, R., Xuan, J., Monperrus, M.: Automatic repair of real bugs in java: a large-scale experiment on the Defects4J dataset. Empir. Softw. Eng. **22**(4), 1936–1964 (2017)
9. Matias, M., Thomas, D., Romain, S., Jifeng, X., Martin, M.: Proceedings of the 15th International Conference on Mining Software Repositories, MSR 2018, pp. 10–13 (2018)
10. Murillo-Morera, J., Quesada-López, C., Castro-Herrera, C., Jenkins, M.: An empirical evaluation of nasa-mdp data sets using a genetic defect-proneness prediction framework. In: 2016 IEEE 36th Central American and Panama Convention (CONCAPAN XXXVI), pp. 1–6 (2016)
11. Muvva, S., Rao, A.E., Chimalakonda, S.: BuGL–a cross-language dataset for bug localization. arXiv preprint arXiv:2004.08846 (2020)
12. Muvva, S., Sangle, S., Chimalakonda, S.: BuGC: C dataset for bug localization. Zenodo, October 2020
13. Radu, A., Nadi, S.: A dataset of non-functional bugs. In: 2019 IEEE/ACM 16th International Conference on Mining Software Repositories (MSR), pp. 399–403 (2019)
14. Ramadhina, S., Bahaweres, R., Hermadi, I., Suroso, A., Rodoni, A., Arkeman, Y.: Software defect prediction using process metrics systematic literature review: dataset and granularity level, pp. 1–7, September 2021

15. Rudolf, F., Péter, G., Gábor, G., Zoltán, T., Tibor, G.: An automatically created novel bug dataset and its validation in bug prediction. J. Syst. Softw. **169**, 110691 (2020)
16. Sayyad Shirabad, J., Menzies, T.: The PROMISE Repository of Software Engineering Databases. University of Ottawa, Canada, School of Information Technology and Engineering (2005)
17. Schröter, A., Zimmermann, T., Premraj, R., Zeller, A.: If your bug database could talk. In: Proceedings of the 5th International Symposium on Empirical Software Engineering, pp. 18–20 (2006)
18. Shepperd, M., Song, Q., Sun, Z., Mair, C.: NASA MDP software defects data sets (2018)
19. Thapaliyal, D., Verma, G.: Software defects and object oriented metrics - an empirical analysis. Int. J. Comput. Appl. **9**, 41–44 (2010)
20. Thomas, D., Martin, M.: IntroClassJava: a benchmark of 297 small and buggy Java programs, pp. 10–13. Universite Lille 1 (2016)
21. Tóth, Z., Gyimesi, P., Ferenc, R.: A public bug database of github projects and its application in bug prediction. In: Gervasi, O., et al. (eds.) ICCSA 2016. LNCS, vol. 9789, pp. 625–638. Springer, Cham (2016). https://doi.org/10.1007/978-3-319-42089-9_44
22. Zhou, J., Zhang, H., Lo, D.: Where should the bugs be fixed? More accurate information retrieval-based bug localization based on bug reports. In: Proceedings - International Conference on Software Engineering, pp. 14–24, June 2012
23. Zimmermann, T., Premraj, R., Zeller, A.: Predicting defects for eclipse. In: Third International Workshop on Predictor Models in Software Engineering (PROMISE 2007: ICSE Workshops 2007), p. 9 (2007)

Web Cloud Services with Distribution Modules (SaaS) and Java

Nelson Salgado Reyes$^{(\boxtimes)}$ and Henry N. Roa

Pontificia Universidad Católica del Ecuador, 12 de Octubre, 1076 Quito, Ecuador
{nesalgado,hnroa}@puce.edu.ec

Abstract. Currently, companies seek to automate their processes. It means the ability of technology to perform many of the tasks that humans perform daily in their workplace. This automatic function also controls, modifies, and clarifies the operational nature of the work and activities. This research presents a model for software development, which allows the rapid construction of corporate applications in the cloud (SaaS), using Java EE 11 technology, and based on agile development methodologies. For the design of this model, we consider appropriate processes and techniques on the software development stages. Particularly, on the analysis, construction, test, and deployment stages. For the development of this research, we studied several agile methodologies, but we only considered those that have the largest number of bibliographic sources such as SCRUM, XP (Extreme Programming), and other methodologies derived from Lean Development. Additionally, in each methodology, we analyze the principle of the agile manifesto to verify to what extent they comply with it. We also review the best practices and applicable design patterns in each methodology. In the end, we find common points between the different methodologies, which promote standardization in conjunction with the Java EE platform. As result, we present a solution based on three pillars for rapid development: (a) the automatic generation of initial code, (b) the use of predefined templates, and (c) the execution of automatic scripts. These three pillars are the basis for rapid development, with which the company may be able to support the development of business software in the cloud using Java EE technology, and with this, optimize costs, time, and resources.

Keywords: Cloud Computing · TIC · Software as a Service (SaaS)

1 Introduction

Today's world has taken and continues to take significant steps in terms of technology. Through technology, people can remotely carry out various activities worldwide.

Companies, professionals, and users increasingly use cloud services (Cloud Computing). This model is based on different layers and handles new terminology such as IaaS, PaaS, SaaS, etc., which lend themselves to being confused with each other. Thanks to these services, companies are avoiding making large in-vestments in both software

A. Rocha et al. (Eds.): WorldCIST 2023, LNNS 801, pp. 13–22, 2024.
https://doi.org/10.1007/978-3-031-45648-0_2

and hardware. In addition, they obtain multiple advantages from having all their applications in the cloud, allowing their employees to access them from any device, anywhere, and at any time.

This research involved the study of various sources of information. Mainly about:

1. Systems prepared for Cloud Computing.
2. Java EE platform as an alternative for development in the Cloud.
3. Agile development methodologies.
4. Tools available for code generation that increase developer productivity.
5. Different types of management applications.
6. Upcoming cloud trends in terms of software applications.

The demand for services provided by SaaS is continually increasing. The reason for this increase is that companies have realized that the future is today. Accordingly, they must be by it. Thus, if they do not evolve, they run the risk of not being able to compete with other companies and disappear.

Nowadays, acquiring a service is very easy and reliable if it is developed under an open-source language such as Java. This tool emerges as a powerful, flexible, and scalable approach to solving computational problems in the construction of a model for the development software, which allows the rapid construction of corporate applications in the cloud (SaaS), using Java EE 11 technology and using agile development methodologies.

2 Web Cloud Services (SaaS) with Distribution Modules

To improve the understanding of this article, we proceed to detail the distribution model.

2.1 Cloud Computing

It is a way of describing a technology that has renewed the world today. This technology allows access to a service through the Internet.

The operation is effortless: the client is not interested in the infrastructure and the requirements for the service to work, the only thing that matters to him is that he can access and consume the service. For this, the service is implemented in the cloud in an agile, timely, and secure way (Cárdenas Sánchez 2022).

Cloud Infrastructure Classes. The National Institute of Standards and Technology (NIST) classifies clouds into four types:

- Public
- Private
- Community
- Hybrid

Public Cloud. Refers to when several entities share the same infrastructure because it is given for any type of client. (Molina 2019) (Fig. 1).

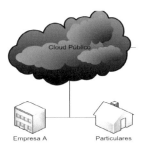

Fig. 1.   Public cloud.

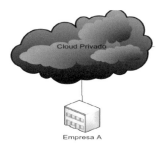

Fig. 2.   Private cloud.

Private Cloud. It is given under the creation guidelines under the company resources and is conducted by companies that are specialists in this field (Fig. 2).

Community Cloud. It is when several companies come together to solve an infrastructure problem-oriented to similar objectives and with a private security framework (Fig. 3).

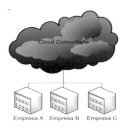

Fig. 3.   Community cloud.

Hybrid Cloud. It is the union of any of the previous three. They are separate entities but united by the same standardized technology (Fig. 4).

Cloud Computing Benefits. The benefits it gives to customers are:

- Access to information from anywhere in the world and at any time without time restrictions and with high availability.

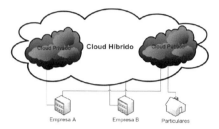

Fig. 4. Hybrid cloud.

- The user does not worry about the preventive and corrective maintenance that oversees the service providers.
- Information backups and the good performance of the service are linked to the service provider.
- Economic savings, since the initial investment of installation, maintenance, and administration costs are large in traditional systems.

Cloud Computing Disadvantages. The most relevant disadvantages are mentioned to follow:

- The criterion of several analysts is the risk of the information as it is not in the custody of the company. The information could suffer some type of leak or loss, which tends to be detrimental to the entity.
- Another problem that can be highlighted is that customers now depend on the operating policies of service providers and their policies, which may be in the functional area of connection and security.
- Users are linked to an Internet connection because it is the bridge that unites them with the services providers provide.

Distributed Model. In the distributed model, the main task is to provide or manage a service based on the company business logic. When applying the model, several variables are required that, when modified, are oriented to the results that are required (Fig. 5).

Fig. 5. Distributed model.

The main task of this model is to provide and manage a service based on the company business logic. Applying a distributed model requires several variables that, when modified, are oriented to the results that are required.

Servers TIC are the pillars required today to boost the capacity and improvement of the company and much more when they obtain Cloud Computing services. These resources are necessary to carry out data management (Guillermo Stuyck González 2020).

Software on Demand. This type of software provides users with services hosted in the cloud. For this reason, the cost is for the client's application use.

Unlike Cloud Computing, total access control is given since the company that provides the service allows it. The process that this type of service entails is very easy and of great importance, because no investment is required, which is very beneficial for the client.

2.2   SaaS Solutions

SaaS (software as a service) solutions are a new trend in the market. It is a way of consuming technological resources that meets a latent need of the environment, primarily in the business area. The solutions are leased through internet payments. With this, the institutions that use the solution do not invest in infrastructure, licenses, and maintenance of the solutions. Simply with authentication in their internet browser they access the services, at available prices and depending on the number of users (Candel 2021).

In contrast, Platforms as a Service (PaaS) is an environment managed by a service provider that enables program developers to host and run their programs without the complications of infrastructure implementation and deployment (Molina 2020).

Saas Advantages. Among the most relevant advantages, we have that the SaaS model eliminates the high costs of purchasing licenses, maintenance, and installation costs. Another advantage is that most applications run with the Internet service. Additionally, when a new version is generated for the end user, it is totally transparent since it does not require any intervention from the client.

The user or client only requires a stable connection to the Internet service to be able to access the application without worrying about the geographical situation in which they are located (Fig. 6).

SaaS Disadvantages. The user does not have its data locally; instead, its data is in an external server owned by the provider.

The user is vulnerable to the service providers' policies since, if they do not have a good encryption, the user's data could be modified or, in the worst case, deleted.

SaaS applications have problems when these applications need to be modified and adapted to each client. This happens because the application already has its own policies and rules.

Saas Impact
The solutions described below use a SaaS distribution model (Fig. 7).

Fig. 6. SaaS Advantages.

Fig. 7. Tools that use the SaaS model.

- Back-office solutions (ERPs).
- Messaging solutions (E-mail management, SPAM treatment, antivirus protection).
- CRM applications.
- Application integration solutions

2.3 Java Scripting Tools

Apache Maven reuses mechanisms to simplify the build process and improves upon the capabilities of the popular Ant tool.

This new tool (along with a fresh approach) can automate the Java build process. Maven has all the functionality required to build projects (clean, compile, copy resources, etc.). In other words, we do not need to create any build scripts for standard builds.

Maven uses a Project Object Model (POM), which describes a project in the form of an XML file, project.xml. Specifically, it represents the project's directory structure, JAR file dependencies, and other project control details.

Tool Box. Java 11 has two tools, jdeprscan and jdeps, which help detect potential problems. These tools can run on existing JAR or class files. Java 11 can assess the effort required for the transition without recompiling. Jdeprscan checks if deprecated or removed APIs are in use. The use of deprecated APIs is not a problem that causes a crash, but it is something that should be examined before a crash occurs. Jdeps is a Java class dependency analyzer. When used with the --jdk-internals option, jdeps indicates which class depends on which internal API (Oracle Corporation 2019).

Java Enterprise Application Architecture. The technologies discussed in the previous section make it possible to configure different types of Java EE application architectures. The following figure shows some of the possible combinations. Web services are delivered through web servers (Fig. 8).

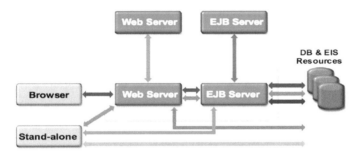

Fig. 8. Java Enterprise Application Architecture.

Servers and Containers. The Java EE platform specifies the operation of distinct types of containers: Web containers and EJB containers (Fig. 9).

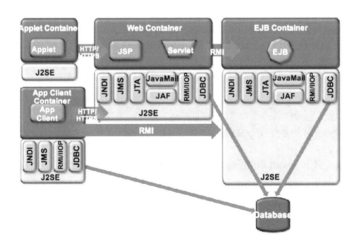

Fig. 9. Web and EJB containers.

These containers are included in servers that also incorporate added services (clustering, resource management, APIs, etc.). They are Web servers and application servers (Fig. 10).

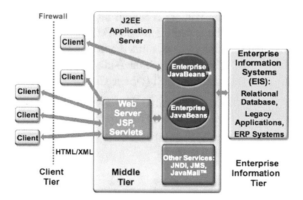

Fig. 10. Web Servers and Application Servers.

Java EE web servers are usually free and open-source platforms (Tomcat, Resin, Jetty, Rimfaxe, Jigsaw). Application servers are usually paid (Bea Weblogic, IBM Web-Sphere, etc.), although free versions are becoming popular (JBoss, Glassfish, EasyBeans, OpenEJB).

3   Methodology

3.1   Agile Methodologies

The main objective of agile methodologies is to avoid overloads and optimize the way of working of each one of the teams. This allows you to develop software quickly and respond to changes that might arise throughout a project.

The Agile Manifesto is a document that summarizes the essence of agile methodologies, and a series of postulates that mainly values the following aspects (Martin Fowler 2016):

1. More attention is paid to individual and development team interactions than to processes and tools.
2. The main goal is to develop software that works, rather than to get good documentation.
3. At all times, it is about collaboration with the client more than the negotiation of a contract.
4. Respond to change, rather than strictly following a plan.

3.2   Proposed Methodology

The methodology is composed of several aspects that can be taken independently. In this way, aspects can be gradually implemented by continuously monitoring the processes. We emphasize the use of 3 pillars for rapid development that originate from the investigation of existing problems (Fig. 11).

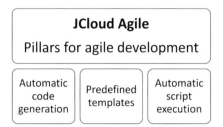

Fig. 11. Pillars for agile development

3.3 SaaS Implemented with Java

In SaaS solutions we can have two configuration levels:

- *User interface level:* the ability to configure the user interface implies being able to adapt the "look and feel" according to the needs of the user and the tenant in question, such as icons, colors, fonts, titles, etc.
- *Functional level:* the ability to configure the functional level means that the behavior of the application may vary for each tenant, where certain system functions may have different business logic for other tenants or even wholly different workflows. In addition, the variability of certain functionality can cause the interface also to vary.

However, for both levels, it is necessary to store the configuration data to implement this feature in the application. Configuration data is the heart of a SaaS application. Thus, we must design how to storage and retrieval configuration data of users (Rodríguez 2016).

In these cases, we use Java for business applications; with this edition, we can create web applications, creation of JSP (Java Server Pages), Servlets, JSF (Java Server Faces), Beans, Web Services (SOA and Rest), Web Sockets, JSON-Processing.

4   Conclusions

In this type of model, it has been possible to determine that:

The focus on development processes using the Java EE standard for its execution in the cloud is proposed to work with the services of the SaaS model with the distribution module, supported by agile methodologies.

An essential aspect of the created model is that it is based on the generation of work tools that constitute the three pillars of rapid development and support the entire main stage of development in its distinct phases.

Through this SaaS model and with the help of Java applications, companies could improve their application development and be technology leaders.

Taking advantage of the fact that the application is in the cloud and the software distributor has complete control of the platform, it is easy for them to manage versions or updates.

The investment cost of using SaaS compared to the traditional development method is lower, so it is a real benefit.

References

Oracle Corporation: JDK-8198756: Asignación diferida de subprocesos del compilador, 29 de octubre de 2018 (en línea). https://bugs.java.com/bugdatabase/view_bug.do?bug_id=8198756. (Último acceso el 20 de octubre de 2022)
Oracle Corporation: JEP 193: Identificadores de variables, 17 de agosto de 2017 (en línea). https://openjdk.java.net/jeps/193. (Último acceso el 20 de octubre de 2022)
Oracle Corporation: JEP 269: Convenience Factory Methods for Collections, 26 de junio de 2017 (en línea). https://openjdk.java.net/jeps/269. (Último acceso el 20 de octubre de 2022)
Oracle Corporation: JEP 285: Spin-Wait Hints, 20 de agosto de 2017 (en línea). https://openjdk.java.net/jeps/285. (Último acceso el 20 de octubre de 2022)
Oracle Corporation: JEP 321: Cliente HTTP (Estándar), 27 de septiembre de 2018 (en línea). https://openjdk.java.net/jeps/321. (Último acceso el 20 de octubre de 2022)
González, G.S.: Consultoría previa, auditoría informática y asesoramiento para una empresa real (2020)
Global IT Solution: Computación en la nube (2018). http://globalitss.com/services/cloud/
Inteco-cert: RIESGOS Y AMENAZAS EN CLOUD COMPUTING (2018)
Molina, S.G.R.: Metodologías Agiles enfocadas al modelo de requerimientos. Universidad Nacional de la Patagonia Austral, Argentina (2019)
González, I.: Las ventajas de un modelo de seguridad SaaS. http://www.aunclicdelastic.com/las-ventajas-de-un-modelo-de-seguridad-saas/. (Último acceso el 20 de octubre de 2022)
Evaluando ERP.com Protección legal de información en los modelos SaaS y Cloud Computing. http://www.evaluandoerp.com/nota-1086-Proteccion-legal-de-informacion-en-los-modelos-SaaS-y-Cloud-Computing-.html. (Último acceso el 20 de octubre de 2022)
Trend Micro Soluciones de seguridad de software como servicio (SaaS). http://es.trendmicro.com/imperia/md/content/es/products/datasheets/datasheet_saas_es.pdf. (Último acceso el 20 de octubre de 2022)
Hernán, S.M.: Diseño de una Metodología Ágil de Desarrollo de Software, Fiuba, Argentina (2004)
Molina, J.J.: Jornadas de Ingeniería del Software y Bases de Datos. Ediciones Universitarias de Salamanca, Madrid, 2016 (2020)

Virtual Reality Training Platform: A Proposal for Heavy Machinery Operators in Immersive Environments

Manuel Pinto[1(✉)], Ricardo Rodrigues[1], Rui Machado[1], Miguel Melo[2], Luís Barbosa[1,2], and Maximino Bessa[1,2]

[1] University of Trás-os-Montes e Alto Douro, Vila Real, Portugal
armandop@utad.pt
[2] INESC TEC, Porto, Portugal

Abstract. Training in a virtual environment can augment the current methods of professional's training, preparing them better for possible situations in the field of work while taking advantage of Virtual Reality (VR) benefits. This paper proposes a cost-effective immersive VR platform designed in real-context usage, consisting of an authoring tool that permits the creation and manipulation of training courses and the execution of these courses in an immersive environment. Accomplishing a good training experience in an immersive simulation requires an equilibrium between the simulator performance and the virtual world aesthetics quality. Thus, in addition to presenting the development of the proposed training platform based on Unity technologies, this paper describes an objective performance evaluation of a virtual training scene using the different render pipelines and across immersive and non-immersive setups. Results confirmed the platform's viability and revealed that the rendering pipeline should be defined according to the display device used.

Keywords: virtual reality · training · authoring tool

1 Introduction

Immersive virtual reality (VR), a technology for creating computer-generated worlds, allows the simulation of the thinkable and the creation of the unthinkable. These worlds have been used successfully in the most diverse areas of applications, such as entertainment, education, or training of professionals in areas ranging from first responder [1], medical staff [2], industrial workers [3], and military. In this last case of application, the real-world training courses are augmented with digital ones, where the most dangerous, complicated, or even life-threatening situations are recreated in a controlled environment without placing the trainee in harm [4]; as stated by [5], VR is a promising training tool for industrial workers, as long the simulation meets high levels of realism and is immersive as possible without compromising the learning, turning into a valid training method.

A. Rocha et al. (Eds.): WorldCIST 2023, LNNS 801, pp. 23–32, 2024.
https://doi.org/10.1007/978-3-031-45648-0_3

The more the industry specializes, the more it is critical that its professionals have proper qualifications to execute their tasks, the current methods of training that utilise simulations are dominated by a non-immersive interaction. However, only recently, it started to be incorporated with immersive technologies to achieve better professional training [6,7], but it is still at an early stage due to the associated costs of creating and maintaining this type of technology, creating a barrier to its adoption since companies have to allocate resources to guarantee that the trainees get proper learning [8,9], resulting in a low adherence to these technologies [10]; when compared with the current training methods, was proved to be a high-success solution in the training of technicians [11]. This method offers companies the advantage of not needing to pause production or allocate trainers since the exercises are permanently available, granting greater control over them and an increase in safety [8,12].

To reduce the costs of the development of VR training programs, authoring tools (AT) can be created, making available to the trainers the tools for the creation and maintenance of the training courses. These can be combined with visual program representation to build interactive experiences [13], with modular architecture and visual scripting tools, enhancing the visualization and speeding up the creation of content [14]. However, these tools do not offer the possibility to reuse or import new assets limiting is life-cycle of operability [15].

The AT developed in [16] offer a fully immersive interaction, where the users interact with the courses without leaving the virtual world, import and prepare new 3D models to be used and create the training steps by recording the trainer choreography while he performs training procedure. However, the models imported could not be modified inside the tool, requiring the trainer to repeat the process to fix the model.

Training to operate heavy machinery can be expensive due to the maintenance of the equipment and dangerous when they are wrongly handled. To alleviate this, [17] developed a VR training simulator about the operation of forklifts, where through the usage of controllers similar to the real ones, the users could learn how to operate them safely inside of a controlled environment since it was proven to be an effective training method by a certified specialist operator.

This paper tackles the adoption barrier by proposing a VR training platform with an AT for anyone to create personalized virtual training scenarios without requiring specialized knowledge. The proposal platform and its development will increase the availability of training, making it safer to execute and improving employee resilience/readiness for different real-world situations.

In the remaining paper, Sect. 2 presents the proposal for an immersive VR-based training platform for the industry, complemented by an AT for creating and managing immersive VR courses. Section 3 sets forth the proposal development by testing different rendering solutions. Ending with the taken decisions and future work for the training platform development.

2   Immersive VR Training Platform

Training platforms require careful planning before their development as training programs involve different dimensions of a corporation ranging from the managing perspective of who supervises the training to the individual particularities of each collaborator that undergoes the corporate training. This section proposes an immersive VR training platform designed based on real-context usage, with the long-term vision of creating a fully modular VR training platform that can be adopted by any stakeholder willing to adopt a VR training platform for corporate training.

2.1   Case Study Specifications

The case study was designed together with company CUTPLANT Solutions, S. A. [18], a company that is devoted to the design, development and sale of machinery and equipment for the agricultural and forestry sectors. The purpose of this training platform accents in the preparation of the respective professionals for real-life situations, permitting an improvement in the repair time of the forest machines by reducing the number of breakdowns. To achieve it, the trainees will face a series of tasks in the immersive simulation, where the corresponding performance is matched with the previously determined key performance indicators.

The proposed training platform will focus on immersive VR training with forest machines, where it was established that it will have two types of users: trainers and trainees. It is expected that trainers can create and manipulate the training courses, the environments and the forest machines, and re-purpose training courses with new heavy equipment supported through a model importation system. It is expected also that trainers can monitor the trainees during the execution of the training courses, as well as consult past executed training courses to augment the evaluation of the trainee. The trainees, as the name suggests, will be able to undergo the training courses, perform the required tasks to fulfil the simulation, and learn about the operation and maintenance of the respective heavy equipment. The previously stated objectives serve as guidelines for the proposed training platform development; in conjunction with the analysed related work, the following requirements can be defined where the platform must allow:

- The creation and storage of the training courses.
- The creation of the training environments where the course will take place.
- The definition of the environment terrain orography, creating deformations, inclinations, or even flattening the land.
- The scattering of different types of the trees through the environment.
- The importation of 3D models of the harvester head.
- The definition of the period of the day for the training courses, indicating the time of the day where the operation will take place.
- To load the previously created and stored training courses.

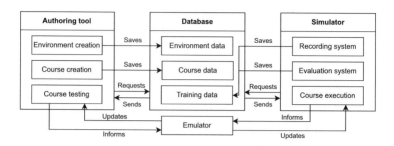

Fig. 1. Training platform integration of the development modules.

- The execution of the training courses in an immersive mode with 90fps.
- The representation of the forest machine functionality in the virtual world, simulating the rotation and oscillation when it is operated.
- The simulation of the forest machine physics during his locomotion and rotation, producing the corresponding sounds.
- The simulation of the harvest head functionality with the physics and sounds.
- To record the training courses execution for a post evaluation of the user.
- Be generic and scalable to newly developed products.

Based on the above-identified requirements, the platform was designed to be divided into two modules: authoring tool and simulator. The first one will emphasize the creation and management of the immersive training courses, and the second one the execution of these courses while evaluating the performance of the trainee (Fig. 1).

2.2   Authoring Tool

To achieve the previous stated requirements, the development of the AT is going to be divided into several modules to fulfil each one of them. This way, dependencies between modules are not created and could be developed independently granting modularity to the process (Fig. 2). The Environments module will focus on the development of tools and features to accommodate the need to design the environments for the immersive training courses; the Asset import will allow the trainer to import new assets of the forest machine, preparing it to be operated on the training course. The features of the Courses module focus on the creation and maintenance of courses, where the trainer defines the environment, forest machine, and the success criteria, among others; the trainer could verify the credibility of the course on the Course testing module, simulating the functionality of the forest machine on the corresponding course.

In an industrial environment, the companies produce 3D models of the respective products, allowing the modification, analysis, and optimization of the same products before even being created. These models can be designed in Computer-aided design (CAD), becoming a challenge when used for the training of professionals since they provide an inefficient performance in a virtual environment, or

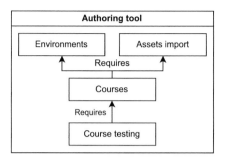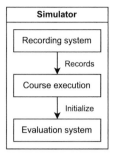

Fig. 2. Integration of the Authoring tool and Simulator modules.

the constituted parts are not split for the user interaction. To solve these complications, a middleware is required to allow the trainers to transform the already existing 3D models into models that can be imported by the AT. Since CAD remains an industry standard, if the immersive training platforms can accommodate the respective models without modifying the existing systems, it will encourage the adoption of these training technologies.

2.3   Simulator

The development of the simulator will be divided into smaller parts to guarantee its modularity and independent development (Fig. 2), where the module Course execution and the Course testing module from the AT will be developed at the same time since they share the concept for the execution of the course. In these, the information from the course will be loaded into the virtual world, presenting to the user the virtual environment, the forest machine, and the defined tasks, among others. During these simulations the virtual forest machines will react based on the instructions given by the user through the controllers, producing visual, auditory, and haptic feedback. Another module will focus on the evaluation of the trainee at the end of the training experience, where a training report is created containing the respective course, the time that took to complete the tasks, the outcome for each of the tasks, as well the trainee grades for that course. To help the trainer with the evaluation of the trainee, the training session will be recorded and stored, then played back when needed.

3   Development of the VR Training Platform

The proposed VR training platform was developed using the game engine Unity since it allows a feasible development of immersive VR experiences. With this engine the authors design and implement the base structure for the AT and the Simulator, connecting them to a database for sharing the training data.

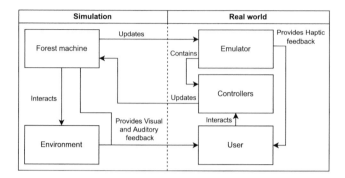

Fig. 3. Integration of the Simulator with the Emulator.

Fig. 4. Temporary terrain orography interface on the authoring tool.

3.1   Authoring Tool

In this, a trainer has access to the module of Environments, where he can manage the existing environments or create new ones; with the help of resizable brushes, he can modify the terrain orography, creating deformations and inclinations, as well as flatten the same land (Fig. 4); to increase the environment realism, forest elements can be scattered through the terrain randomly, with resizable brushes or placed one by one, some of this elements serve as targets for the interaction during the Simulation.

Within the AT, the trainer it is permitted to import new 3D models of harvest heads attachments for forest machines. The process starts when the desired head attachment CAD file was chosen and processed through a middleware system to one of the extensions .fbx or .obj, allowing the trainer to import it to the AT. Once loaded the trainer has to identify each of the harvest head parts with the respective tag from the available ones, this permits the tool to apply the corresponding functionality to the harvest part; the trainer can adjust the functionality of each harvest part by modifying the available parameters. At this stage, the visual aspect of the head attachment can be modified by changing the corresponding material of the several harvest parts, and the auditory feedback by adding the desired audio to the several harvest parts. When it meets the desired requirements, the trainer can save it and use it on the simulator.

With the previous components available on the AT, the trainer can create new courses or manage the existing ones, indicating the corresponding title and description, associating the desired environment and harvest head, and defining the hour of the day that the course will start.

3.2   Simulator

On the Simulator, is loaded the corresponding environment and the selected harvest head attachment preparing them for the immersive experience. Once ready, the forest machine and the harvesting head will react based on the user input, creating visual and auditory feedback on the different parts and the interaction between the forest machines with the environment and its elements. As represented in the (Fig. 3), the user interacts with the controllers presented on the emulator; these are composed of a combination of joysticks, pedals, and buttons that recreate the cabin of a forest machine. The presence of these controller types in the simulated training course will permit the training experience to be more realistic and multisensorial for the user.

3.3   Performance Evaluation of Rendering Pipelines Across Different VR Setups

Achieving the best training experience in the immersive simulation requires an equilibrium between the Simulator performance and the virtual world aesthetic. When it becomes imbalanced, the simulation can become sluggish, impacting the overall experience and the trainee's knowledge retention or even inflicting cybersickness on the trainee. For the proposed training platform the aesthetics tuning was found through a study of a custom scene in different render pipelines available in the Unity engine: the High Definition Render Pipeline (HDRP) provides tools to create applications in high-fidelity graphics on high-end platforms (Fig. 5c), and the Universal Render Pipeline (URP) offers artist-friendly workflows that allow a quickly and easily creation of optimized graphics across a wide range of platforms (Fig. 5b), and the Built-in Render Pipeline that has limited options for customization since it is the general-purpose render pipeline for the engine (Fig. 5a). To assure fairness and reliability of the study results, each one of the custom scenes was executed for a total duration of one minute, in a computer with the following specifications, Intel Core i7-9700k CPU, NVIDIA GeForce RTX 2080 Ti, 16 GB of RAM. Since the Simulator application is expected to allow immersive execution, the custom scenes were tested using three different display interfaces, two immersive and the other non-immersive, represented respectively by the head-mounted display (HMD) HTC VIVE, the HMD Oculus Quest 2 and the conventional computer display LG E2242T-BN 22″. The proposed training platform it is targeted to achieve 90 frames per second (fps) in the simulation so for this study, the HMD Oculus Quest 2 refresh rate was set to 90 Hz, and was tested in the recommended resolution (RR) of 3712×1872 and in the lowest resolution (LR) of 2432×1216.

(a) Built-In   (b) URP   (c) HDRP

Fig. 5. Visual aesthetics of the different pipelines from the user perspective

Table 1. Display performance for the different render pipelines in fps

Display	Render pipeline								
	Built-In			URP			HDRP		
	avg	1%	0.1%	avg	1%	0.1%	avg	1%	0.1%
Monitor	468	329	147	472	257	101	323	264	102
VIVE	89	82	70	89	85	67	89	83	65
Quest RR	89	81	70	56	40	39	66	42	36
Quest LR	89	82	71	90	70	46	89	81	65

To foment the choice for the pipeline, during these tests was collected for each one of the conditions the average (avg.) frame rate, the worst 1% of frame rate, and the worst 0.1% of frame rate. Each HMD was placed standing still on a platform to avoid human intervention during the tests, granting the same visual perspective of the virtual environment.

Based on the obtained results presented in Table 1, it is possible to conclude that the Built-In pipeline offers a stable performance independently of the immersive display while keeping a pleasant virtual environment. The Universal and High Definition render pipeline provide new tools to create a virtual environment but require higher tailoring of the different settings to achieve the same level of balance of Built-In; the results also showed that the resolution of the immersive displays affects the performance. Consequently, results suggest that built-in render pipeline should be adopted as default, but when more computational power is available, URP or HDRP can and should be privileged.

4   Conclusions

The current paper presents a training platform targeted to optimize the productivity of the trainees by improving the efficacy of the operators and maintenance technicians, reducing the number of breakdowns and improving the time needed for repair. Based on a real case study of a company in the agricultural and forestry sectors, the development of the training platform was divided into Authoring tool and Simulator components, where the users could create and execute the training simulations.

The present study allowed the authors to define the Built-In as the default render pipeline of choice for the proposed training platform development since it performed stable across the tested immersive displays while keeping the aesthetics pleasant. Nevertheless, depending on the computational power of the system, users can adopt a different rendering pipeline.

Based on the provided feedback in the interactions with company CUTPLANT SOLUTIONS, S.A., the Authoring tool component from the training platform must allow for registration of the key performance indicators and the success criteria of the courses. The Simulator component needs to be connected with a digital twin system, abstracting the interactions between the Emulator and the simulation and modifying the simulation with new data, such as possible breakdowns. Other sources of sensory feedback will be added to the Emulator, improving the training simulation realism and the user training experience. Such as haptic feedback based on the virtual environment interaction and placing the emulator on a rig with 6 degrees of freedom, allowing it to represent the locomotion of the virtual forest machine through the virtual terrain.

As for future work, the authors are committed to the evolution of the proposed platform by modifying the simulations with data from a digital twin to improve training realism. Validating it in usability, performance and training augmentation with a real-context case study created in partnership with company CUTPLANT SOLUTIONS, S.A., with the professional operators.

Acknowledgments. This work is co-financed by the ERDF - European Regional Development Fund through the Operational Programme for Competitiveness and Internationalisation - COMPETE 2020 under the PORTUGAL 2020 Partnership Agreement, and through the Portuguese National Innovation Agency (ANI) as a part of project "SMARTCUT - Diagnóstico e Manutenção Remota e Simuladores para Formação de operação e manutenção de Máquinas Florestais: POCI-01-0247-FEDER-048183"

References

1. Narciso, D., Melo, M., Vasconcelos-Raposo, J., Paulo Cunha, J., Bessa, M.: Virtual reality in training: an experimental study with firefighters. Multimedia Tools Appl. **79**(9-10), 6227–6245 (2020). https://doi.org/10.1007/s11042-019-08323-4. http://link.springer.com/10.1007/s11042-019-08323-4
2. The virtual reality training platform for nursing (2022). https://www.ubisimvr.com/. Accessed 30 Mar 2022
3. Virtual assembly line training (2022). https://www.seriousgames.net/en/portfolio/opel-virtual-assembly-line-training/. Accessed 30 Mar 2022
4. Narciso, D., et al.: A systematic review on the use of immersive virtual reality to train professionals. Multimedia Tools Appl. **80**(9), 13, 195–13, 214 (2021). https://doi.org/10.1007/s11042-020-10454-y. https://link.springer.com/10.1007/s11042-020-10454-y
5. Radhakrishnan, U., Koumaditis, K., Chinello, F.: A systematic review of immersive virtual reality for industrial skills training. Behav. Inf. Technol. **40**(12), 1310–1339 (2021). https://doi.org/10.1080/0144929X.2021.1954693

6. Yildiz, E., Melo, M., Moller, C., Bessa, M.: Designing collaborative and coordinated virtual reality training integrated with virtual and physical factories. In: 2019 International Conference on Graphics and Interaction (ICGI), Faro, Portugal, pp. 48–55. IEEE (2019). https://doi.org/10.1109/ICGI47575.2019.8955033. https://ieeexplore.ieee.org/document/8955033/

7. Pérez, L., Rodríguez-Jiménez, S., Rodríguez, N., Usamentiaga, R., García, D.F.: Digital twin and virtual reality based methodology for multi-robot manufacturing cell commissioning. Appl. Sci. **10**(10) (2020). https://doi.org/10.3390/app10103633. https://www.mdpi.com/2076-3417/10/10/3633

8. Ayala García, A., Galván Bobadilla, I., Arroyo Figueroa, G., Pérez Ramírez, M., Muñoz Román, J.: Virtual reality training system for maintenance and operation of high-voltage overhead power lines. Virtual Reality **20**(1), 27–40 (2016). https://doi.org/10.1007/s10055-015-0280-6

9. Oi, W.: The Fixed Employment Costs of Specialized Labor, chap. 2, pp. 63–122. University of Chicago Press (1983). http://www.nber.org/chapters/c7374

10. Morozova, A.: How much does a virtual reality app cost? (2019). https://jasoren.com/how-much-does-a-virtual-reality-app-cost-in-2018/

11. Bertram, J., Moskaliuk, J., Cress, U., Bertram, J., Moskaliuk, J., Cress, U.: Virtual training: making reality work? Comput. Hum. Behav. **43**, 284–292 (2015). https://doi.org/10.15496/publikation-5845

12. Visser, H., Watson, M.O., Salvado, O., Passenger, J.D.: Progress in virtual reality simulators for surgical training and certification. Med. J. Aust. **194**(S4) (2011). https://doi.org/10.5694/j.1326-5377.2011.tb02942.x

13. Zhang, L., Oney, S.: FlowMatic: An Immersive Authoring Tool for Creating Interactive Scenes in Virtual Reality, chap. 6A, pp. 342–353. Association for Computing Machinery, New York (2020). https://doi.org/10.1145/3379337.3415824

14. Zikas, P., et al.: Immersive visual scripting based on VR software design patterns for experiential training. Vis. Comput. **36**(10), 1965–1977 (2020). https://doi.org/10.1007/s00371-020-01919-0

15. Coelho, H., Monteiro, P., Gonçalves, G., Melo, M., Bessa, M.: Authoring tools for virtual reality experiences: a systematic review. Multimedia Tools Appl. **81**(19), 28037–28060 (2022). https://doi.org/10.1007/s11042-022-12829-9

16. Cassola, F., et al.: Design and evaluation of a choreography-based virtual reality authoring tool for experiential learning in industrial training. IEEE Trans. Learn. Technol., 1 (2022). https://doi.org/10.1109/TLT.2022.3157065

17. Lustosa, E.B.S., de Macedo, D.V., Formico Rodrigues, M.A.: Virtual simulator for forklift training. In: 2018 20th Symposium on Virtual and Augmented Reality (SVR), pp. 18–26 (2018). https://doi.org/10.1109/SVR.2018.00016

18. Cutplant Solutions S.A.: VICORT, Cutplant Solutions, S.A. (2022). https://www.vicort.com/. Accessed 18 Mar 2022

Digital Twin Technologies for Immersive Virtual Reality Training Environments

Ricardo Rodrigues[1]([✉]), Rui Machado[1], Pedro Monteiro[1], Miguel Melo[2],
Luís Barbosa[1,2], and Maximino Bessa[1,2]

[1] University of Trás-os-Montes e Alto Douro, Vila Real, Portugal
`rrodrigues1999.rr@gmail.com`
[2] INESC TEC, Porto, Portugal

Abstract. With industry evolution and the development of Industry 4.0, manufacturers are trying to leverage it and find a way to increase productivity. Digital Twins (DT) technologies allow them to achieve this objective and revolutionize Product Life-cycle Management as they provide real-time information and insights for companies, allowing real-time product monitoring. Virtual Reality (VR) is a technology that permits users to interact with virtual objects in immersive environments; even under constant development, VR has proven efficient and effective in enhancing training. DT integration into immersive VR environments is constantly developing, with many challenges ahead. This study aims the development of an immersive virtual world for training integrated with DT technologies to handle all users' input using the simulator. Those were subject to a performance evaluation to understand how the application handles different input types, which confirmed the viability and reliability of this integration.

Keywords: digital twin · virtual reality · immersive environments

1 Introduction

With the evolution of technologies for the Industry and the development of Industry 4.0, manufacturers have incorporated several types of sensors into their equipment that allow them to collect data that leads to obtaining relevant information about the machine's operation status or measuring the machine's past, present, or future performances. As new sensors for equipment development grows, it is also necessary to increase the capacity to communicate the data obtained to central systems, preferably in real-time [1]. Digital Twin (DT) technologies have revolutionized Product Lifecycle Management (PLM) as they provide live, or near real-time, information and insights for manufacturers that allow real-time monitoring and design of all PLM-related processes [2].

DT data is real-time data that reflects the physical object state. The DT integrates and converges real-time data and uses previously processed data to provide more helpful information to the system. This technology represents real entities synchronized through various sources such as sensors and continuous

A. Rocha et al. (Eds.): WorldCIST 2023, LNNS 801, pp. 33–42, 2024.
https://doi.org/10.1007/978-3-031-45648-0_4

collection of information, allowing real-time representation of their state [3]. This data is processed and synchronized with the DT applications to provide more helpful information to the system and can be used to understand how the physical object works, detect anomalies and prevent malfunctions through simulations.

Along the DT, there are two other ways of representing reality in a digital system, the Digital Model and the Digital Shadow. Figure 1 represents the three models, the flow of information, and the direction of communications between the real and virtual objects [4].

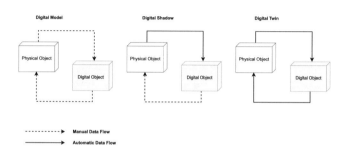

Fig. 1. Digital Twin Models

While Digital Model is used to present a concept and to create production and construction documentation, where data flow between the physical and the digital model as to be made manually, the Digital Shadow concept refers to the representation of the physical model on a digital model, where data flows automatically in a single direction, from physical to digital model, unlike the DT model where communication is bi-directional [5].

VR is the representation of a virtual environment and 3D models that can be interacted by the users with various input sources, such as using hand controllers, tracking gloves, or with Head-Mount Displays (HMDs). In these environments, various senses, such as vision, hearing, touch, and even smell, can be stimulated to give the user a greater sense of presence in the virtual environment [6]. Therefore, this technology can improve training and be used in several areas like medicine, marketing, entertainment, and others [7]. Different authors have defined VR according to their views on this technology. Some define "virtual" as something false or not real, and "reality" as a reference to the real world, which the conjunction might be interpreted as something like an imitation of reality [8]. A definition proposed by Steve Bryson is "The use of computer technology to create the effect of an interactive three-dimensional world in which the objects have a sense of spatial presence." [9], that is, simulation of a 3D environment generated through a computer, where the user can interact, move and become immersed in this virtual world.

In a VR system, all human senses can be stimulated, which creates different impressions and sensations in the user inside the virtual world, increasing the user's sense of presence and immersion when using the VR application. Thus, the quality and immersion levels are determined by how involved and engaged the user is with the virtual environment during the VR experience and how these impression types stimulate all user's sensations [10,11]. The VR systems can be divided by the immersion that users feel when using the application in three main levels [11]: immersive, semi-immersive, and non-immersive. In particular, immersive systems allow users to feel fully immersed in the virtual environment, using HMDs responsible for tracking head movements and the user's position in the virtual world. In these systems, and to increase the level of immersion, different systems can stimulate different sensations [11].

VR technologies can benefit from other technologies, such as DT, though most DTs, depending on the data's characteristics, are presented as text or graphics in non-immersive environments. Through the integration of DT and VR, users may have a more in-depth understanding of the dynamics of the represented model by experiencing deep immersion and a sense of reality [12].

One example of an application combining DT and VR is "Virtual Assembly Line Training". This application is a car assembly line simulator created to reduce critical errors, assembly costs and training time. This software allows customization of the assembly line to each scenario by analyzing the DT data [13].

Another application that uses VR and DT is Flexsim. This solution permits the design and development of 3D factory simulations. FlexSim includes a standard models library that can replicate real-world operations functionalities so that users can build different scenarios. Flexsim is used in different areas such as industrial manufacturing, healthcare, and robotics, among others [14].

This work focuses on the integration of the DT in immersive virtual environments, having as its basis the SMARTCUT project, which intends to develop a solution that envisages making the most of these two technologies by simulating forestry and agricultural equipment based on CAN technology to allow the execution of telediagnostic tasks, remote maintenance and, monitoring of the forest. This project will use information from an existing DT, which will be processed and used in immersive VR environments. These tasks will be complemented with the use of the DT. A performance evaluation will be made to understand how the application handles different input types to validate and evaluate the proposed solution.

The next sections present the proposal and integration of DT technologies in a training VR application for training.

2   Digital Twin Integration in a VR Application for Training

The main objective of this work is to integrate a DT into an immersive virtual environment (simulator), mainly to give visual feedback to the user. The DT reads inputs made at a physical emulator and sends them to the simulator that applies them to the corresponding virtual elements. The emulator replicates a real forestry machine with all associated controls, such as pedals, joysticks, and buttons.

In this work, the case study focuses on the operation of forest machines and harvester heads, where users control all those components. Users can use these 3D components to complete tasks associated with individual training sessions in the immersive environment. The training session is customized for each user and different scenarios by modifying the environment props, terrain, lighting, harvester model, and tasks. Upon completion of all tasks, the application generates results for each completed task and compares them with those of other users.

Through DT technology, several tasks can be accomplished during the training. They include the representation of the operation of the entire forestry machine and a harvesting head representing its operation and malfunctions that can be associated with each component. Those training courses consist of a group of tasks, including the operation and maintenance of forest machines.

The simulator and all virtual worlds were developed with the UNITY 3D graphics engine.

2.1   Communication Architecture

As shown in Fig. 2, the emulator and simulator communicate through the DT with a broker to emulate all machine and harvesting head movements.

There are two types of users: the trainee and the trainer. When using the application, the trainee does not interact directly with the simulator but instead utilizes all available controls in the emulator that sends all necessary information to DT. The simulator uses this information to update all virtual models. A trainer panel is built into the simulator and is used to interact directly with the application by changing the environment variables, such as time of day and weather, and affecting forest machine and harvesting head functionality.

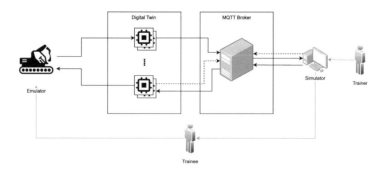

Fig. 2. MQTT Architecture

Upon initialization, the simulator subscribes to multiple topics (represented by dashed lines) and receives values from each (incoming solid lines). The simulator can also send data to Broker (outgoing solid line) to notify and synchronize the DT about what is happening with the virtual machine and harvesting head.

The DT is composed of several controllers, some responsible for storing and sending to the server the different states of each emulator component and others that will subscribe to multiple topics and are responsible for receiving information sent by the simulator and acting on the emulator. An MQTT broker transmits messages using the MQTT Protocol, a lightweight messaging protocol for IoT devices. By implementing the Observer pattern, the Broker effectively exchanges every message between the DT and simulator [15].

According to Listing 1.1, MQTT messages are JSON messages, composed of a topic that identifies the component, along with three values: (a) the current component value, the actual input value of the component at the emulator, (b) the default component value, the input value of component at emulator when not being used, and (c) the maximum component value, the maximum input value of component at the emulator. This message structure allows us to easily handle all needed values to apply to the virtual model. This data is continuously sent to the MQTT Broker, sending the message to all listeners who have subscribed to the current topic.

Listing 1.1. MQTT Message Json Schema

```
{
  "$schema": "http://json-schema.org/draft-04/schema#",
  "type": "object",
  "properties": {
    "atualValue": {
      "type": "integer"
    },
    "defaultValue": {
      "type": "integer"
    },
    "maxValue": {
      "type": "integer"
    }
  },
  "required": [
    "atualValue",
    "defaultValue",
    "maxValue"
  ]
}
```

2.2   Application Flowchart

To better understand and interpret the entire operation of starting and executing a training session in the application, a flowchart (Fig. 3) was designed that shows how the entire process will work.

When starting the application, the user is presented with an initial menu with several options, one of which is the training menu. When entering this menu, if there are already courses created, these are listed, and the user can choose one of them and start the train. Otherwise, it is possible to create training using the training creation menu, where the user chooses the environment and the desired harvesting head for the training execution. When starting the training, the trainee interacts with the machine and the harvesting head through the emulator until all the tasks required for the training are completed. At the end of the training, the obtained results are generated and saved, and the user is redirected back to the initial menu.

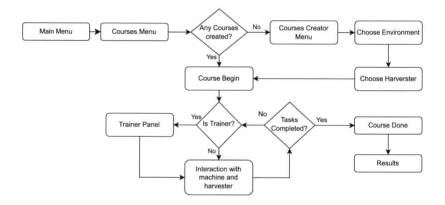

Fig. 3. Simulator's Flowchart

2.3   Integration Development

One of the purposes of the tool is to serve as a base for the operation of forest machines and harvesting heads in virtual environments, integrated into a DT system that will process all the control inputs and send the result to the simulator that will apply the proper effect to the forest machine.

This subsection will focus on how the communication between the DT and the simulator was implemented and how the messages are processed by the simulator and applied to the machine and the harvesting head to perform the training tasks.

To integrate Digital Twin with Unity 3D, a client was created to connect to the MQTT Broker and receive all messages for each topic using an external library.

Before starting the training, the user can access the options menu (Fig. 4) and choose, in the "Input Mode" option, to use the Digital Twin as input. When using Digital Twin as input, two other options are presented to define if the Broker Server is local, that is, if the server is executed on the machine that is running the simulator or if you want the Broker Server to be remote and executed on a remote machine and accessed through the IP Address defined in the "IP Broker" option. If you choose a local server, the simulator starts the Broker without any action from the user. All these options are saved in an external configuration file in JSON format, which allows these options to be set automatically whenever the application is started.

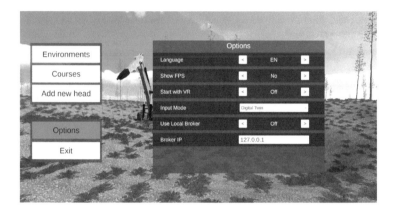

Fig. 4. Application options menu

When starting the training scene, the application connects to the Broker Server and subscribes to all topics of the *"digital_twin/controlo/"* type. After these steps, the application waits until the DT connects to the Broker. When connected to the Broker, the DT starts sending messages to the topics associated with each control that contains all the information necessary for the application to work correctly. These topics, as well as all associated values, are defined in the DT on each controller. Each of these controllers can be associated with an extra controller, called an error block, which can change the original value of the input to another value. These blocks are helpful to replicate malfunctions and problems in the operation of some of the components of the forest machine, as well as in the harvesting head.

After receiving the values sent by the DT, the client processes the MQTT message and associates an event to each topic responsible for processing and applying the due effect to the intended component. These events have, as input parameters, the values sent in the MQTT message (Current Value, Default Value, and Maximum Value).

3   Performance Evaluation

To test the viability and reliability of the integration of the DT and the interactive immersive virtual environment, a protocol was designed with all the actions possible in the simulator, which is the basis of the performance evaluation. The main purpose of this evaluation is to assess how a DT-based approach performs against a conventional approach and validate its viability in terms of latency and responsiveness.

3.1   Procedure

To ensure that all protocol steps were executed uniformly, a script was developed to simulate user inputs through a direct input to the virtual environment or via a DT. In addition, each test was executed five times in two different environments for consistency purposes.

For evaluation purposes, two environments were created: one empty scene only with the forest machine (baseline) and the harvesting head attached, and the other was a highly populated scene with many trees, grass, and rocks (representative of a real case scenario). In addition, the input mode was also considered as a variable, direct input (representative of the conventional solutions). This resulted in eight different test conditions:

– Keyboard Inputs + Non VR mode + Empty Scene
– Keyboard Inputs + Non-VR mode + High Populated Scene
– Keyboard Inputs + VR mode + Empty Scene
– Keyboard Inputs + VR mode + High Populated Scene
– Digital Twin Inputs + Non-VR mode + Empty Scene
– Digital Twin Inputs + Non-VR mode + High Populated Scene
– Digital Twin Inputs + VR mode + Empty Scene
– Digital Twin Inputs + VR mode + High Populated Scene

The considered performance variables were Input Latency, FPS, RAM Usage, CPU Usage Percentage, CPU Time, GPU Usage Percentage, and GPU Time. The performance tests were executed on a Desktop Computer with the following specifications: Intel(R) Core(TM) i7-6700K CPU, NVIDIA GeForce GTX 1080 GPU, and 32 GB of RAM.

3.2   Results

The obtained results are shown in Table 1. The best performance was achieved when non-immersive displays and testing the application in an empty scene (ES), reaching an average of more than 300 FPS, and the worst when using VR displays and a High Populated scene (HPS), which was not able to reach 20 FPS in most of the cases. Looking at CPU Usage Percentage (CPU %) and CPU Time, we can see that they are slightly higher for the same scene type and display mode combinations when the DT makes application inputs. This was also expected since the application must process all MQTT messages. In terms of latency, the values are lower when using the keyboard compared with DT when executing the application in an empty scene, but this changes when we run the application in a complex scene and when FPS are lower, where DT inputs are more responsive. This mainly happens because, even though the application has to process all MQTT messages, this type of input allows us to transport input processing from the main application to the DT application.

Table 1. All average variables values of each test

| | Digital Twin | | | | Keyboard | | | |
| | VR | | NON VR | | VR | | NON VR | |
	ES	HPS	ES	HPS	ES	HPS	ES	HPS
Latency (ms)	11.53	31.72	6.55	17.31	4.03	47.48	1.52	22.81
FPS	72	16	309	32	72	18	334	37
RAM (Mb)	738	10771	724	10619	781	11540	750	11157
CPU %	10.53	25.32	21.11	34.21	8.97	24.24	19.09	32.28
CPU TIME (ms)	14.01	64.76	3.40	31.42	13.92	55.14	3.10	27.14
GPU %	58.82	78.75	83.29	84.21	63.21	98.90	89.56	98.03
GPU TIME (ms)	3.62	23.92	0.66	17.24	1.92	33.27	0.77	6.84

4 Conclusions

The study aimed to integrate an immersive VR application for training with DT and evaluate and validate it by analyzing its performance and comparing it across different scenarios, including conventional simulation (without DT) and across VR setups (non-immersive vs immersive).

According to the results obtained, we can conclude that creating an application with the primary purpose of rendering a simulation with complex scenes that leverages DT potential is feasible. In addition, DT approaches helped to improve application stability and reliability by handling input processing since CPU time and usage percentage were lower.

The developed tool allows trainers to create different training scenarios and change, whenever they want, the type of inputs and controllers without reprogramming the simulator's entire Input System. Also, the MQTT Broker can be instantiated on a dedicated server, which allows users to use both the training application for trainees and the control of the application by the trainers remotely without having to be in the same place where the simulator is being executed.

Acknowledgments. This work is co-financed by the ERDF - European Regional Development Fund through the Operational Programme for Competitiveness and Internationalisation - COMPETE 2020 under the PORTUGAL 2020 Partnership Agreement, and through the Portuguese National Innovation Agency (ANI) as a part of project "SMARTCUT - Diagnóstico e Manutenção Remota e Simuladores para Formação de operação e manutenção de Máquinas Florestais: POCI-01-0247-FEDER-048183"

References

1. Melesse, T.Y., Di Pasquale, V., Riemma, S.: Digital twin models in industrial operations: state-of-the-art and future research directions. IET Collab. Intell. Manuf. **3**(1), 37–47 (2021). https://doi.org/10.1049/cim2.12010
2. Tao, F., Cheng, J., Qi, Q., Zhang, M., Zhang, H., Sui, F.: Digital twin-driven product design, manufacturing and service with big data. Int. J. Adv. Manuf. Technol. **94**(9), 3563–3576 (2018). https://doi.org/10.1007/s00170-017-0233-1
3. Bergs, T., Gierlings, S., Auerbach, T., Klink, A., Schraknepper, D., Augspurger, T.: The concept of digital twin and digital shadow in manufacturing. Procedia CIRP **101**, 81–84 (2021). https://doi.org/10.1016/j.procir.2021.02.010
4. Seppälä, L.: Data-driven shipbuilding. https://www.cadmatic.com/en/resources/publications-and-brochures/cadmatic-data-driven-shipbuilding-article-collection.pdf
5. Sepasgozar, S.: Differentiating digital twin from digital shadow: elucidating a paradigm shift to expedite a smart, sustainable built environment. Buildings **11**, 151 (2021). https://doi.org/10.3390/buildings11040151
6. Wheeler, A.: Understanding Virtual Reality Headsets (2016). https://www.engineering.com/story/understanding-virtual-reality-headsets
7. Sherman, W.R., Craig, A.B.: Understanding virtual reality-interface, application, and design. Presence **12**(4), 441–442 (2003). https://doi.org/10.1162/105474603322391668
8. Muhanna, M.A.: Virtual reality and the CAVE: taxonomy, interaction challenges and research directions. J. King Saud Univ. Comput. Inf. Sci. **27**(3), 344–361 (2015). https://doi.org/10.1016/j.jksuci.2014.03.023
9. Virtual reality: Definition and requirements (2022). https://www.nas.nasa.gov/Software/VWT/vr.html. Accessed 19 Feb 2022
10. Radianti, J., Majchrzak, T.A., Fromm, J., Wohlgenannt, I.: A systematic review of immersive virtual reality applications for higher education: design elements, lessons learned, and research agenda. Comput. Educ. **147**, 103,778 (2020). https://doi.org/10.1016/j.compedu.2019.103778
11. Mandal, S.: Brief introduction of virtual reality and its challenges. Int. J. Sci. Eng. Res. **4**(4), 304–309 (2013)
12. Zhu, Z., Liu, C., Xu, X.: Visualisation of the digital twin data in manufacturing by using augmented reality. Procedia CIRP **81**, 898–903 (2019). https://doi.org/10.1016/j.procir.2019.03.223
13. Virtual assembly line training (2022). https://www.seriousgames.net/en/portfolio/opel-virtual-assembly-line-training/. Accessed 13 July 2022
14. Flexsim (2022). https://www.flexsim.com/. Accessed 13 July 2022
15. Martin, R.C.: Design principles and design patterns. Object Mentor **1**(34), 597 (2000)

A Prototype of the Crowdsensing System for Pollution Monitoring in a Smart City Based on Data Streaming

Aleksa Miletić[1] , Marijana Despotović-Zrakić[1] , Zorica Bogdanović[1]([⊠]) ,
Miloš Radenković[2] , and Tamara Naumović[1]

[1] Faculty of Organizational Sciences, University of Belgrade, Belgrade, Serbia
{aleksa.miletic,maja,zorica,tamara}@elab.rs
[2] School of Computing, Union University, Belgrade, Serbia
mradenkovic@raf.rs

Abstract. This paper proposes a prototype of the crowdsensing system for pollution monitoring in a smart city based on data streaming. The first part of the paper analyses concepts, characteristics, and platforms for data streaming. The Apache Kafka solution for data flow management is described. The paper proposes infrastructure for data streaming from the IoT crowdsensing systems for monitoring pollution in smart cities. Crowdsensing services included in this system enable the monitoring of pollution parameters in smart cities. Collected pollution data in the smart city (traffic vibrations, noise, allergens, and air pollution) can be conducted using the Internet of Things (microcomputers, microcontrollers, sensors, etc.) and mobile devices, sent to the Apache Kafka cluster using the MQTT protocol, and then data can be streamed via the web application to end users. Active parts in collecting pollution data have citizens. All collected data can be processed, analyzed, and streamed using the proposed data streaming infrastructure for smart city crowdsensing systems.

Keywords: Internet of Things · Crowdsensing · Data Streaming

1 Introduction

Data streaming is used to integrate, process, filter, analyze and react to data that is being collected in real-time and can be useful for industries that have to deal with big data. Data can be updated frequently using streams and users can get notifications for changes of interest. Data stream refers to data in an ordered sequence that is continuously updated and unbounded. Usually, it refers to the continuous transmission of large amounts of small data, from data producers to data consumers. Data is processed incrementally and sequentially, in small batches or one at a time, using various algorithms.

Many applications use data streaming, such as recommender services, fraud detection, online shopping, IoT systems, etc. The Internet of Things (hereinafter: IoT) enables the connection of intelligent devices to the Internet and data exchange between devices [1–3]. Data from smart environments can also be collected through mobile devices with embedded sensors, using the crowdsensing method. Many crowdsensing mobile applications allow users to collect information from the environment. This data can be streamed to other users in real time via the web or mobile applications.

There are crowdsensing applications that allow monitoring of pollution in the environment such as air pollution, the presence of allergens, noise, etc. Users can subscribe to use these apps and get real-time data. In this paper, we present a data streaming infrastructure for a crowdsensing system for monitoring pollution in smart cities. The proposed crowdsensing system enables the measurement of pollution in the smart city using IoT and mobile devices. All collected data will be sent to Apache Kafka and stored in the private cloud. As a part of the pilot project, an implementation of smart city crowdsensing services for pollution monitoring will be presented.

2   Data Stream Management

A data stream is an abstraction that represents an unbounded set of data. The data set is unlimited because new records are constantly arriving [4]. Data streams refer to data that is frequently generated, in large volumes and speed [5]. Typical sources of data streams are systems are: IoT sensors, server logs, and clickstreams from applications and websites. To manage data streams, specific software components can be used.

Data stream management has become a very important process in big data systems. Some of the advantages of data stream management are real-time data processing, the ability to manage endless streams of data, observing patterns in data dynamics, and scalability. Additional properties of data streams are [4]: event streams are arranged, but it is difficult to predict which event will be executed next; events are immutable, and once they have occurred, they cannot be changed; event streams can be replayed.

Stream processing is common in companies dealing with IoT, machine learning, and data flow analysis. Also, larger companies have apps, websites, and IoT devices that generate large amounts of data that require processing in real-time. Therefore, it is necessary to process that data with data streams. A stream processing system refers to the manipulation, processing, and combination of data before it is stored [6]. This system is built on multiple elements called stream processing elements [6]. Common examples of streaming data sources are [7]: IoT sensors, server and security records, and real-time advertising platforms. The basic concepts of the stream processing system are:

- Real-time systems with batch data processing have a large amount of input data that needs to be processed and produce output data.
- Data processing is continuous.
- An event is generated by a producer once and processed by many subscribers.
- Related events are generally grouped into one topic.

For the smooth management of data streams, it is necessary to integrate the following components [8]: message broker, ETL tools, analytics tools, and storage. A message broker is software that stands in between different applications, systems, and services and allows them to communicate and transfer information [9]. It can be realized in two forms [10]: Point-to-point messaging, used when message queues have a one-to-one relationship between the sender and receiver; and Publish/subscribe messaging, which is a broadcast-style distribution. ETL (extract, transform, and load) tools reduce data storage, prepare data for quick and easy access and processing [11]. A storage system stores incoming data streams, as well as the results of queries. A message broker is different from a traditional database because it stores events temporarily until an application requests that data. After that, it keeps data for a while, and then deletes it. Traditional databases and data lakes work differently because data is persistent [9, 12, 13].

Apache Kafka is a data streaming platform capable of fast processing numerous requests [14]. It is used for storing and analyzing large amounts of data as well as streaming data in real-time. Unlike the AMQ or MQTT protocols, Apache Kafka allows messages not to be forwarded to the user until the user deems the message [15]. Apache Kafka aims to unify offline and online processing by providing a mechanism for parallel loading in Hadoop systems [16]. Kafka accepts data from multiple producers and sends it to multiple subscribers [17]. Data inside topics are stored in offsets where they stay for a certain time, and then they get deleted [15, 18]. Basic Kafka components are:

- **Producer**. The producer writes data to Kafka topics and it represents the data source. This data source can be sensors, laptops, IoT devices, records, etc.
- **Consumer**. Consumers subscribe to topics from the Kafka cluster, and they automatically know from which broker to read data.
- **Connectors**. Connectors are used to connect topics and applications/databases. The Source Connector is responsible for feeding data into Kafka, while the Sink Connector is used for extracting data from Kafka. Kafka topics connect to relational databases, Hadoop, and Amazon services through standard interfaces [19].
- **Stream handlers**. Kafka streams are a library for creating applications and microservices. A stream allows data to be processed in real time, with the ability to filter, merge, aggregate, and perform various transformations on the data.
- **Broker**. Brokers are servers within a Kafka cluster. A broker represents a program running on a Java virtual machine. Kafka topics are nested within the brokers to ensure that data is not lost if a broker "goes down", there is replication to keep data across multiple brokers.
- **Topic**. Kafka topics are used to store messages coming from producers. Each topic must have a unique name. Within one topic we can have multiple partitions.
- **Zookeeper**. Zookeeper monitors Kafka brokers and routes messages arriving at the cluster to the appropriate broker. It is also in charge of metadata management.

3   A Crowdsensing System Based on Data Streaming

Crowdsensing is an approach that includes the e-participation of smart city residents who use mobile devices to collect data from the environment [20]. Using mobile devices with embedded sensors, it is possible to monitor various parameters of the environment. One of the most critical problems in big cities is the problem of air pollution, the existence of allergens, noise, and vibrations produced in the traffic, which can significantly affect the health of citizens [21].

Crowdsensing healthcare mobile applications based on smart devices (microcomputers, microcontrollers, and sensors) usually enable citizens to monitor parameters from the environment related to the microclimate conditions, air pollution, and allergens (temperature, humidity, flammable gases, carbon monoxide, harmful gases, allergens particles, etc.) [22, 23]. This implies that citizens actively participate in the monitoring of environmental pollution at a specific micro-location using IoT systems and crowdsensing mobile applications. Furthermore, citizens can use crowdsensing mobile applications for mapping locations with evident air pollution or if they recognize some the allergens such as ambrosia. In this way, they enable other citizens to receive information about these locations.

Noise pollution and vibrations in the city traffic can be measured using crowdsensing mobile applications or IoT systems. Intelligent devices can be placed in specific micro-locations where the traffic intensity is high, or in public transportation. Citizens can participate in measuring these parameters using crowdsensing applications for smartphones. All collected data from smartphones are stored in the cloud and can be browsed using a web platform.

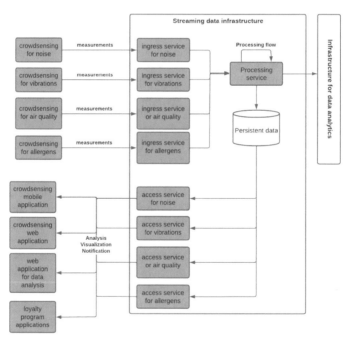

Fig. 1. Infrastructure for data streaming from the IoT crowdsensing systems

To deliver the collected data, it is necessary to have an infrastructure for data streaming that can integrate multiple crowdsensing systems. Figure 1 shows the infrastructure for data streaming from the IoT crowdsensing systems. In this example, IoT crowdsensing systems refer to the monitoring of pollution parameters in a smart city, such as noise and vibration measurement, air quality, and allergen measurement.

The main elements of the proposed system are IoT devices to which sensors are attached. The collected data is transformed on the IoT device before being sent further. The data streaming infrastructure consists of three components:

1. Input services that receive data from mobile phones and IoT devices;
2. A system that processes flows, and stores the necessary data in the database;
3. Output services that send data to the user upon request from a specific application.

Machine learning and other prediction models are good for implementation in real-time applications [21]. User applications include mobile and web applications that require crowd-sensed data. The data is presented in a visual form, and the visualization of notifications is also enabled.

4 Prototype of the Crowdsensing System for Pollution Monitoring in a Smart City Based on Data Streaming

The idea of the prototype is to create a system based on event processing, which will show users real-time data on the noise and vibrations, air pollution, and the presence of allergenic particles in the city. The data will be gathered using the crowdsensing method. This means that part of the data can come from the users themselves, by collecting parameters from their environment. Another type of data collection is based on static IoT stations that can be placed to preselected locations. Such a system will greatly contribute to the quality of life by enabling citizens to receive personalized data they have subscribed to.

The microservice architecture will be used for implementing the system, with main components implemented in Node.js. Event processing will be done using the KafkaJS library, which is open-source. On the client side, the Angular framework will be used. The architecture of the prototype of the crowdsensing system for pollution monitoring in a smart city based on data streaming is shown in Fig. 2.

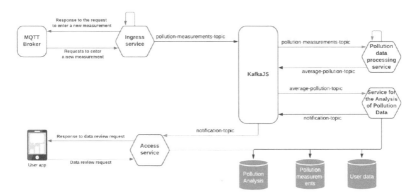

Fig. 2. Prototype of the crowdsensing system for pollution monitoring in a smart city

Collected data is sent to the cloud and accepted using the ingress service. After data has passed the ingress service, it is processed, analyzed, stored, and delivered to users through different delivery models (pull, push) and through different applications [15]. User applications are connected to the system through the access service API. Data from the MQTT broker arrives at the Ingress service, where it is checked. Feedback is sent to the broker if everything went well. After the first check, the data is sent to a KafkaJS topic. KafkaJS then forwards the data to all subscribers to this topic. A subscriber to this topic is the Pollution Data Processing Service where data is processed and forwarded to average-pollution-topic. The Service for the Analysis of Pollution Data is subscribed to this topic, where a more detailed analysis is carried out and data is stored in the database. Data is then forwarded to the notification topic. Access service that sends notifications to users.

5   Conclusion

The primary goal of this paper is to present a prototype of the crowdsensing system for pollution monitoring in a smart city based on data streaming. The proposed system can enable the monitoring of pollution in the smart city using citizens' mobile devices or various intelligent devices and sensors, sending data to the Apache Kafka cluster using the MQTT protocol, and storing it in the private cloud. The real-time data will be processed with Kafka and sent to the users over web applications.

Future directions of research will refer to the implementation of the proposed prototype and the additional features within all the system components. The proposed prototype will be implemented in the scope of the cloud, big data, and IoT-based infrastructure of the Department of E-Business at the Faculty of Organizational Sciences, University of Belgrade. Furthermore, the proposed crowdsensing system should become a part of the loyalty programs within smart cities where all citizens participate in collecting environmental data. This smart city loyalty program can be implemented using blockchain technologies.

References

1. Ashton, K.: That 'Internet of Things' Thing. That "Internet Things" Thing-RFID J. (2010). http://www.rfidjournal.com/article/print/4986. Accessed 13 Nov 2022
2. Gubbi, J., Buyya, R., Marusic, S., Palaniswamia, M.: Internet of Things (IoT): a vision, architectural elements, and future directions. Future Gener. Comput. Syst. **29**(7), 1645–1660 (2013)
3. Jezdović, I., Popović, S., Radenković, M., Labus, A., Bogdanović, Z.: A crowdsensing platform for real-time monitoring and analysis of noise pollution in smart cities. Sustain. Comput. Informatics Syst. **31** (2021). https://doi.org/10.1016/J.SUSCOM.2021.100588
4. Narkhede, N., Shapira, G., Palino, T.: Kafka the Definitive Guide (2017). https://www.oreilly.com/library/view/kafka-the-definitive/9781491936153/. Accessed 27 Aug 2022
5. Akidau, T., Chernyak, S., Lax, R.: Streaming systems - the what, where, when, and how of large-scale data processing, vol. 134, no. 4 (2018). https://www.oreilly.com/library/view/streaming-systems/9781491983867/. Accessed 27 Aug 2022
6. Hiraman, B.R., Viresh, M.C., Abhijeet, C.K.: A study of apache kafka in big data stream processing. In: 2018 International Conference on Information Technology and Communication Engineering, ICICET 2018, November 2018. https://doi.org/10.1109/ICICET.2018.8533771

7. Levy, E.: 4 Key Components of a Streaming Data Architecture (with Examples) | Upsolver, 6 January 2022. https://www.upsolver.com/blog/streaming-data-architecture-key-components. Accessed 28 Aug 2022

8. Radenković, B., Despotović-Zrakić, M., Bogdanović, Z., Barać, D., Labus, A.: Materijali sa predavanja i vežbi, Beograd (2020)

9. IBM Cloud Education: Message Brokers, 23 January 2020

10. Chappell, D.A., Richards, M., Monson-Haefel, R.: Java Message Services, 2nd edn. (2009)

11. Majchrzak, T., Jansen, T., Kuchen, H.: Efficiency evaluation of open source ETL tools. In: Proceedings 2011 ACM Symposium on Applied Computing, pp. 287–294 (2011)

12. What is a data lake? https://aws.amazon.com/big-data/datalakes-and-analytics/what-is-a-data-lake/. Accessed 19 Nov 2022

13. Nargesian, F., Zhu, E., Miller, R.J., Pu, K.Q., Arocena, P.C.: Data lake management: challenges and opportunities. Proc. VLDB Endow., 1986–1989 (2019)

14. Kuduz, N., Ozren, Ć., Slaven, P.: Modelovanje sistema za upravljanje tokovima poruka primjenom Apache Kafka. In: 18th International Symposium Infoteh-Jahorina, March 2019. https://infoteh.etf.ues.rs.ba/zbornik/2019/radovi/KST-2/KST-2-1.pdf. Accessed 3 Jan 2023

15. Miletić, A., Lukovac, P., Jovanić, B., Radenković, B.: Designing a data streaming infrastructure for a smart city crowdsensing platform. In: E-Business Technologies Conference Proceedings, pp. 61–64 (2022)

16. Garg, N.: Apache Kafka. Packt Publishing, Birmingham, UK (2013)

17. Miletić, A., Lukovac, P., Radenković, B., Jovanić, B.: Designing a data streaming infrastructure for a smart city crowdsensing platform. In: E-Business Technologies Conference Proceedings, vol. 2, no. 1, pp. 61–64, June 2022. https://ebt.rs/journals/index.php/conf-proc/article/view/128. Accessed 19 Nov 2022

18. Zelenin, A., Kropp, A.: Apache Kafka, pp. I–XVII (2021). https://doi.org/10.3139/9783446470460.fm

19. Sun, A., Zhong, Z., Jeong, H., Yang, Q.: Building complex event processing capability for intelligent environmental monitoring. Environ Model Softw. **116**, 1–6 (2019)

20. Staletić, N., Labus, A., Bogdanović, Z., Despotović-Zrakić, M., Radenković, B.: Citizens' readiness to crowdsource smart city services: a developing country perspective. Cities **107** (2020). https://doi.org/10.1016/J.CITIES.2020.102883

21. Labus, A., Radenković, M., Nešković, S., Popović, S., Mitrović, S.: A smart city IoT crowdsensing system based on data streaming architecture. Mark. Smart Technol. **279**, 319–328 (2022). https://doi.org/10.1007/978-981-16-9268-0_26

22. Stefanović, S., Nešković, S., Rodić, B., Bjelica, A., Jovanić, B., Labus, A.: Development of a crowdsensing IoT system for tracking air quality. E-bus. Technol. Conf. Proc. **1**(1), 182–184 (2021). https://doi.org/10.1109/RISE.2017.8378212

23. Labus, A., Radenković, M., Despotović-Zrakić, M., Bogdanović, Z., Barać, D.: Crowdsensing system for smart cities. In: International Scientific Conference on Digital Economy—DIEC, pp. 27–42 (2021)

Transcolar Rural New Route Cost Calculation Software and Cost Analysis Tool

Marcelo F. Porto$^{(\boxtimes)}$, Mateus A. Silveira, Lucas V. R. Alves,
Renata'M. A. Baracho, and Nilson T. R. Nunes

Federal University of Minas Gerais, Belo Horizonte, Brazil
{marceloporto,aranha1004,lucasvra,renatabaracho}@ufmg.br,
nilson@etg.ufmg.br

Abstract. In Brazil, 4.2 million children depend on rural school transport. Due to the lack of resources, it is mandatory to look for options to minimize costs. The Transcolar Rural project proposes a solution to optimize school transportation in the most undeveloped regions of Brazil's rural areas. A problem detected in the country's education is the high rate of school dropout that may be due to school transport. The project aims to reduce evasion and proposes to implement efficient methods of routing optimization and cost calculation. The project consists of creating a large database of geographical coordinates for the students, schools, and roads, creating and optimizing routes, and calculating the approximate cost of each new route generated. The new route cost model is an update that makes it more transparent, personalized, and adaptable. To identify these errors, a cost analysis methodology that uses statistical process control and normal approximations was developed. A program that applies the methodology to each parameter of the cost model was created in order to evaluate the final cost of each route. This allows the final value of all routes in a city to be compared with others, finding those with the highest probability of error and proposing the analysis to solve the problem.

Keywords: scholar transportation · cost calculation · statistical control · information modeling

1 Introduction

The Brazilian Federal Constitution of 1988 guarantees equal conditions of access and permanence in education to all citizens [1]. According to the National Institute of Statistics and Research INEP, more than 6 million students live or attend schools in rural areas and approximately 70% need school transport to get to the classroom [2]. The rural exodus contributes to the decrease of schools in these areas, from $103,328$ in 2003 to $70,816$ in 2013 [3] increasing the distance between students and schools.

   The Transcolar Rural project [4] was founded in 2012 at the School of Engineering of the Federal University of Minas Gerais seeking to create a Geographic

A. Rocha et al. (Eds.): WorldCIST 2023, LNNS 801, pp. 50–59, 2024.
https://doi.org/10.1007/978-3-031-45648-0_6

Information System [5] for rural scholar transportation in Brazil. The Transcolar Rural project implements a geographic information system for students, schools, and roads to generate optimized routes. Then, for each route, the cost calculation is made considering expenses such as fuel consumption, driver payment, vehicle maintenance, and depreciation [6].

The new route costing software is an upgrade to the existing one with the aim of increasing transparency and customization. Allowing the municipal manager to change more than 60 parameters used in price accounting, keeping the final result more appropriate to the local reality. To achieve this objective, a graphical interface was implemented in which the local manager can create and edit parameters. The parameter database can be audited by chief managers.

Editing cost parameters increase possible data entry errors by users that invalidate the final result. The need to create a program to evaluate the results and detect possible errors arose. This study adapts statistical process control (SPC) methods, specifically Shewhart control charts. To analyze the cost calculation, a group of results is compared with a copy of it, artificially generated with normally distributed parameters. The statistical measures generated for cost highlight parameters that are likely to be errors so that they can be manually revised over successive iterations to achieve the desired accuracy.

2   Literature Survey

After researching scientific databases, only a few articles discussing the problems related to rural scholar transportation in Brazil were found. The Transcolar Rural project already targets some of these problems, such as the absence of accurate mapping of these areas, high costs caused by non-optimized routes, and inaccurate pricing caused by the lack of technical knowledge of the local managers [4,6]. The new update aims to adapt the program to Brazil's continental size, with different states having considerable differences in geographical and economical aspects with volatile pricing across time and space [7,8] that justify the need to develop adaptable real-time systems. The theoretical fundamentals used to develop the adaptable route costing calculation software and the cost analysis tool, which is based on statistical process control will be presented.

2.1   Routing Cost Calculation

Routing cost calculation is very documented in modern literature [6,9–11]. The approach presented in [11] was used in Transcolar's original program [6], the cost of transportation can be divided into two categories: variable and fixed costs. Variable costs are those proportional to the distance traveled such as fuel, oil, maintenance, and tire wear. Fixed costs can be described as the costs to keep the vehicles stopped, such as taxes, drivers' and monitors' salaries, and vehicular depreciation. Fixed costs are divided among all the routes that use the same vehicle. The optimization process consists of best-allocating vehicles to reduce costs, therefore focusing on reducing fixed costs.

In [6], the final cost of a route is a function of some parameters unique to the route, called travel parameters, municipal and state parameters common among all routes of a city/state, and vehicular parameters particular to the type of vehicle used in each route.

2.2   Shewhart's Control Charts

Shewhart's control charts were first introduced in the early 1930s by Walter A. Shewhart as a tool of Statistic Process Control (SPC) [12]. Control charts' main concept relies on plotting a central control line (CL) referring to the mean value of a sample, an upper control line (UCL), and a lower control line (LCL) both determined by the maximum variation permitted for the process as shown in Fig. 4. Placing the measurement of a variable as points on the chart, it's possible to imply that the process is in a statistical control state if the points are within the limits or if it should be revised [13]. In this chart, these limits are determined by the standard deviation(σ) of the sample population, usually, the lines are placed 3σ above and below the average line (CL). The value of 3σ is used because, in a normal distribution, a random value has a 99.7% chance of being drawn inside these lines. Uses of SPC outside a manufacturing context have been tested increasingly since the 1990s. MacCarthy's work [14] evaluates the use of SPC for this context and possible problems with this adaptation if the basic assumptions of [12] are not followed. MacCarthy [14] groups a variety of non-standard applications accordingly to the final objective: "Group 1 - monitoring non-manufacturing processes", in which the cost calculation fits the best, tends to use conventional Shewhart charts if considering a normal distribution of independent data and determined control states [15].

3   Procedure Structure and Methodology

The Transcolar Rural project aims at planning and managing rural school transport in Brazil. A set of routes optimized for transporting students to schools are generated for each Brazilian municipality. The cost of each route is calculated and then reports are generated for the municipal administration. The Transcolar Rural project can be described in three main steps: (i) geo-localization and road and street mapping, (ii) route optimization, and (iii) route cost calculation. The framework of the process is presented in Fig. 1.

The first stage consists of collecting geographic data on students and schools obtained from the state department of education (SEDUC) and its municipal departments of education (SEMED). The consumer unit code referring to the energy bill of students' homes and schools is also collected. Electricity companies provide the geographic coordinates of all consumer units. By joining this information, it is possible to georeference students and schools. A mobile application was developed that allows students to update their location themselves. To generate student transport routes, the municipal road network is required. The creation of the road network and initially made with data from the collaborative mapping OpenStreetMaps [16] and completed with manual digitization using QGIS [17], open-source software that allows editing geographical information.

The second step consists of making a route optimization request for a municipality on the project's website. In this request, it is necessary to define some restrictions that interfere with the final routes such as the time a student can remain on the bus, how far each type of student can walk to a boarding point, and who must be picked up at the door.

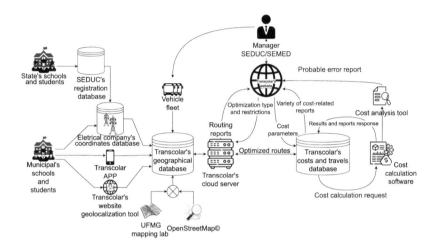

Fig. 1. Transcolar Rural's process framework

The third step, which will be detailed in Sect. 3.1, lies in calculating the cost of every route generated in step two. To achieve this goal the software uses more than 60 cost parameters defined by the municipalities managers on Transcolar's website such as fuel price, driver's salary, and taxes, among others. These parameters are combined with travel parameters obtained in the optimization step, such as distance traveled, vehicle type model, year of fabrication, amount of state and city students and monitors in each vehicle, type of fuel used, and dirt road percentage. Parameters filled, the software calculates the estimated price for each route from the required city and returns the results in reports that can be accessed from the project's website.

3.1   New Cost Calculation Software

The new cost calculation software follows the same methodology as [6] breaking apart the cost parameters into three primary categories: county, state, and vehicular type parameters which are filled in a graphical interface available on the project's website respecting the type of parameter, that is, all municipalities from a state must use the same state parameters and only a state manager can change them. One advantage of allowing this degree of customization is that the final result will be personalized to each municipality's reality and parameters that often change won't require changing the program's source code, just accessing the website and updating the values.

Another new feature is that the software is now highly adaptable to new types of vehicles and fuel. It currently operates with four types of vehicles and fuel: gasoline, diesel, natural gas, and ethanol, but if it were to add another one, for example, electrical vehicles or hydrogen fuel, the cost program would not need to be updated, only a new type of vehicular parameters have to be inserted on the website.

To achieve these goals a graphical interface was made using Microsoft's web-developing framework ASP.NET using C# as the back-end language and HTML combined with JavaScript for the front-end connecting with Transcolar's Postgres [18] cloud database.

With the interface ready, local and state managers can add, edit and delete parameters that are directly loaded into the C# cost program class *LoadParameters*.

LoadParameters. To load the parameters that will be used in the cost calculation, a class named *LoadParameters* is initially called, loading the three parameters categories: *county_parameters*, *state_parameters*, *county_vehicular_parameters* into a new object that will handle all input and output values called *output_costs*. Since all the parameters are kept in a cloud database, this "preload" is important to optimize the code considering that all routes will use the same parameters, and downloading these values for every route would be exceedingly time-consuming. With the values loaded, the next step is to call the class *CalculateCosts* for every route.

CalculateCosts. This class is called to do all the calculations following the formulas defined in [6] that will be combined into the final cost of each route. In the new route cost program, all variables used in these formulas are obtained from *output_costs* and all direct comparisons were removed in order to make the program robust and adaptable to changes outside its source code.

3.2   Cost Analysis Software

The robustness given to the cost program creates a new problem: by allowing several municipalities to alter various parameters, the chance of human error increases greatly. Since the software is designed to work with almost all editable values, the final results are compromised, and considering the volume of participating cities, it's not feasible to manually analyze all of them, hence the necessity of developing a cost result analysis tool.

To use Shewhart's control charts [12] the first step would be to determine which final result would be compared. Since the objective of the analysis is to determine possible errors in the value of cost parameters the results should be independent of all travel parameters (such as distance traveled, vehicle type model, year of fabrication, amount of state and city students and monitors in each vehicle, type of fuel used, and dirt road percentage). Even though the variable cost already is separated from these parameters, the fixed cost is heavily impaired by those considering that it's divided among all the routes that use the same vehicle, and the vehicle-route distribution is decided on external parameters.

To work around this problem, a reference route was defined, keeping all the travel parameters constant, therefore, analyzing only the cost parameters, and for each municipality a new reference route and cost are determined. Following Shewhart's methodology [12], the next step would be defining the average and the standard deviation of the sample to plot the graphical lines.

CalculateReferences Method. This method involves calculating a new route using fixed predetermined route parameters and recovering the cost parameters used in the original calculation from the original *output_costs* object from the database. A new route is created, with the same fixed route parameters between all cities and the final cost per kilometer of this route will be used in the analysis. This step is essential, seeing that two cities can have two very different end results using the same cost parameters if the travel parameters are distinct enough, and the premises required to use a non-standard Shewhart chart considered in Sect. 2.2 would not be followed.

CalculateShewhartLines Method. In parallel, this method generates the standard deviation and average lines required to plot Shewhart's charts [12]. In the context of Transcolar Rural, simply making the σ (standard deviation) and average values of *CalculateReferences* can be problematic, considering that the type of errors that can occur can alter radically the final value, and using a finite sample population the lines can be heavily displaced, obscuring possible errors. The solution was to use a Monte Carlo random generation [19], consisting of creating a large number of cost calculations and, for each one, randomizing all cost parameters following a normal distribution of the original values of all cities and calculating the average and standard deviation of this artificial population.

Plotting the Graphs. The final step is to combine all the methods into a single chart for each type of vehicle and mark those outside the 3σ lines as possible errors. These graphs were plotted on Transcolar's website using the Plotly open-source JavaScript library [20] (Fig. 2).

4 Results

The new route cost program is already implemented in the Transcolar Project, used by more than 450 municipalities to transport more than 180,000 students. Cost parameters are constantly updated by local managers prior to requesting a cost calculation. The cost analysis tool is also available and has been shown capable of detecting errors in specific cities while analyzing an entire state.

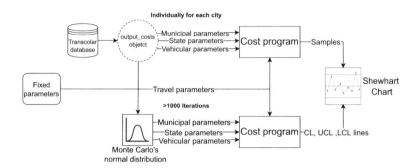

Fig. 2. Cost analysis framework

4.1   New Cost Program

Original Versus New Cost Program. Transcolar Rural's new route cost program is already implemented in several cities and it is possible to compare the results with the old model implemented using the same parameters in [6] as it's shown in Table 1. Considering that the two models follow the same base methodology, the differences are caused by adjustments and changes in the model.

Table 1. Cost program results for determined city [a]

Value	Old model	New model	Variation
Distance traveled daily	4, 841.34	4, 841.34	0.00%
Annual total cost	$ 4, 960, 795.34	$ 5, 136, 267.52	3.54%
Annual fixed cost	$ 1, 515, 810.42	$ 1, 612, 079.80	6.35%
Average cost per km	$ 4.76	$ 4.86	2.07%

[a] Values considered in Brazilian's currency.

Changing Parameters on the New Route Cost Program. Considering that the main objective of the new version of the route cost program was to make the cost parameters accessible, adaptable, and editable to customize the different values between different cities and through time, an assessment of the impact of this parameter variation is shown in Fig. 3 in which five parameters were increased 20% each time and compared to a reference.

The results obtained in Sect. 4.1 show that for that city a 20% variation in the driver's salary can increase the final cost by 4% equivalent to $ 444, 935.00 in a year. A common error that can occur is the misplacement of the decimal separator, altering a value in at least 1, 000% compromising completely the final result. Figure 4 show the cost analyst tool applied to two cities with decimal separator errors, the first one altering the driver's salary by a factor of 10 times, and the second by 100 times. Because a Monte Carlo random generation is used for the control and limit lines, the farther point doesn't alter the average and standard deviation hiding another potential error as shown in Fig. 5.

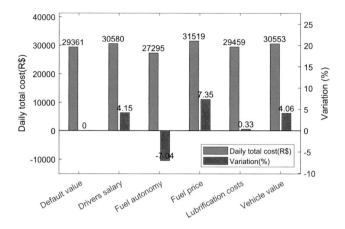

Fig. 3. Results for a 20% variation in each parameter

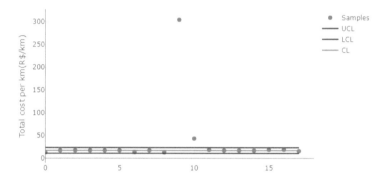

Fig. 4. Shewhart Chart with Monte Carlo random number generation

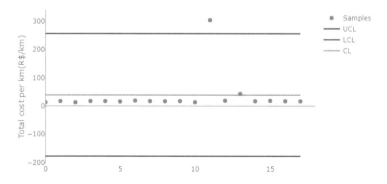

Fig. 5. Shewhart Chart without Monte Carlo random number generation

5 Conclusion

The main purpose of the cost program is to provide an accurate approximation of the cost of a route after the optimization step was reached. With the implementation of the new methodology, it was observed an increase in the values compared to the old model as shown in Table 1. This result is important because the accuracy of the model is essential to make the cost program closer to reality making the public bidding process more transparent and reliable.

Considering Brazil's continental size some prices may variate more than 20% (e.g., fuel prices [7,8]) across different cities and states and through time influences the final cost. The adaptability present in the update is important to keep the program precise throughout time and in different locations. The results in Fig. 3 show that a 20% variation can result in a substantial difference in the final cost depending on what parameter is changed thus the importance of the program to be real-time.

One of the main theoretical contributions of this paper and of the Transcolar Rural Project is to fill the scientific gap in Brazil's rural transportation since researching the subject does not show up-to-date publications.

Acknowledgements. The Transcolar Rural Project of the Engineering School of the Federal University of Minas Gerais (EE/UFMG); the State Research Foundation (FAPEMIG), the Brazilian National Council (CNPq); Coordination for the Improvement of Higher Education Personnel (CAPES).

References

1. Do Brasil, S.F.: Constituição da república federativa do Brasil. Brasília: Senado Federal, Centro Gráfico (1988)
2. Instituto Nacional de Estudos e. Pesquisas Educacionais Anísio Teixeira (Inep): Censo escolar da educação básica 2014: notas estatísticas (2014)
3. Instituto Nacional de Estudos e. Pesquisas Educacionais Anísio Teixeira (Inep): Censo escolar da educação básica 2013: notas estatísticas (2013)
4. Porto, M.F., Sarubbi, J.F.M., Thiéry, S., da Silva, C.M., Nunes, N.T.R., de Carvalho, I.R.V.: Developing a GIS for rural school transportation in minas Gerais, Brazil. J. Syst. Cybern. Inf. **13**, 1–6 (2015)
5. Marble, D.F.: Geographic information systems: an overview. In: Introductory Readings in Geographic Information Systems, vol. 3, no. (4), p. 8 (1990)
6. Porto, M.F., Machado, A.A., Nunes, N.T.R., Alves, L.V.R., Tavares, R.H.C., Porto, P.B.: Developing an environment for cost calculation of rural school transportation. In: Proceedings of The 22nd World Multi-Conference on Systemics, Cybernetics, and Informatics (WMSCI 2018) (2018)
7. Gás Natural e Biocombustíveis. Agência Nacional do Petróleo: Levantamento de preços de combustíveis (últimas semanas pesquisadas) (2022)
8. Raeder, F.T., Rodrigues, N., Losekann, L.D.: Asymmetry in gasoline price transmission: how do fuel pricing strategy and the ethanol addition mandate affect consumers? Int. J. Energy Econ. Policy **12**(4), 517–527 (2022)

9. Souza, A.A.D., Avelar, E.A., Boina, T.M.: Análise da Gestão de Custos no Transporte Público Urbano: um estudo de caso. Revista Mineira de Contabilidade **1**(37) (2010)
10. Gomes da Silva Sant'Ana, I., Soares Regattieri Oliveira, R., Roberto Vallim, C.: Gestão de custos no transporte urbano por aplicativo: Uma análise comparativa na visão do prestador de serviços. Anais do Congresso Brasileiro de Custos - ABC, November 2021
11. Neto, E.L.: Proposta de metodologia de cálculo do custo operacional para o transporte escolar rural: estudo de caso do estado do espírito santo. Master's thesis, Universidade Federal de Minas Gerais (2016)
12. Shewhart, W.A., Deming, W.E.: Statistical Method from the Viewpoint of Quality Control. Courier Corporation (1986)
13. Qiu, P.: Introduction to Statistical Process Control. CRC Press (2013)
14. MacCarthy, B., Wasusri, T.: A review of non-standard applications of statistical process control (SPC) charts. Int. J. Qual. Reliab. Manag. **19**, 295–320 (2002)
15. Batista, L.T., Franco, J.R.Q., Fakury, R.H., Porto, M.F., Braga, C.M.P.: Methodology for determining sustainable water consumption indicators for buildings. Sustainability **14**(9) (2022)
16. Bennett, J.: OpenStreetMap. Packt Publishing Ltd. (2010)
17. Graser, A.: Learning QGIS 2.0. Packt Publishing Ltd. (2013)
18. PostgreSQL: Postgresql (1996). http://www.PostgreSQL.org/about
19. Gentle, J.E.: Random Number Generation and Monte Carlo Methods, vol. 381. Springer, New York (2003). https://doi.org/10.1007/978-1-4757-2960-3
20. Sievert, C.: Interactive Web-Based Data Visualization with R, Plotly, and Shiny. CRC Press (2020)

Agent Team Management Using Distributed Ledger Technology

Kurt Geihs[1]([⊠]) and Alexander Jahl[2]

[1] ITeG Kassel, Pfannkuchstrasse 1, 34121 Kassel, Germany
geihs@uni-kassel.de
[2] University of Kassel, Wilhelmshoeher Allee 73, 34121 Kassel, Germany

Abstract. The notion of Multi-Agent System has turned out to be an appropriate modeling paradigm in dynamic application domains where adaptation and reconfiguration is required at runtime due to changing execution contexts. We present a comprehensive solution for the management of agent teams in highly dynamic environments. The main contribution is to demonstrate how Distributed Ledger technology can effectively support the organizational team management in a decentralized and self-organized manner. Our implementation builds on an extended Hyperledger Sawtooth framework. Our work has shown that the concepts of Multi-Agent System and Distributed Ledger represent an ideal combination.

Keywords: Multi-Agent System · Management · Blockchain · Hyperledger

1 Introduction

Multi-Agent Systems consist of multiple autonomous agents that interact in a shared environment to achieve common or individual tasks. Although in the literature there is no clear definition, the notion of *autonomous agent* is used generally as a model for a reactive component that makes decisions and acts based on its current knowledge and sensed perception of its environment [1]. For example, an autonomous agent can represent a robot, computer program, or human actor. Recent advances in artificial intelligence techniques are key enablers for agent applications. Thus, autonomous agents have become the tool of choice in numerous application domains, particularly in dynamic environments with goal-oriented collaboration. A multi-agent system (MAS) [1] is able to perform tasks that exceed the capabilities of a single agent, in terms of workload sharing and functionality. Just like a team of human beings can achieve more than a single individual, the teamwork of autonomous agents provides opportunities to accomplish tasks that a single agent cannot do alone.

In this paper we focus on the organization of agent teamwork. Specifically, we present a solution for the management of hierarchical teams of autonomous agents that operate in dynamic application environments where adaptation and reconfiguration of the team organization may be necessary due to frequently changing runtime contexts. The main contribution of the presented research is to demonstrate how Distributed Ledger (DL) technology can effectively support the organizational team management in a MAS that operates in such application environments.

The remainder of this paper is organized as follows. In Sect. 2 we discuss the goals of our work and the research-guiding requirements particularly focusing on the combination of MAS and DL. Section 3 gives a brief overview of the main technical building blocks of our solution. Section 4 shows how we accomplish the team management by combining MAS and DL technology. In Sect. 5 we describe a prototype implementation and show results of evaluation experiments. Section 6 discusses related work. Section 7 concludes the paper and points to future work.

2 Requirements and Problem Statement

MAS research covers a very broad field. In this section we narrow down the objectives of our research.

2.1 Requirements

With the ongoing automation and digitalization, more and more loosely coupled, heterogeneous distributed systems are constructed that consist of autonomous sub-systems operating in dynamic environments in a decentralized, self-organized fashion. Typical application domains are service robotics, warehouse logistics, search & rescue missions, autonomous driving, and many more. The MAS paradigm has turned out to be an appropriate modelling abstraction for such teamwork scenarios, where teams are dynamically created for certain tasks and team membership may change over time because individual agents may join and leave the team due to new arrivals, agent failures, loss of communication, etc.

The management of a dynamically changing MAS needs to address a variety of technical challenges, such as heterogeneity of agents, agent communication, team building, knowledge management and decision making, task planning and assignment, secure collaboration and more.

In this paper, we focus specifically on the team management aspects. Thus, we observe the following central requirements:

R1: Design an architecture for the decentralized, self-organized management of hierarchical agent teams including teams of teams.

R2: Provide a decentralized storage facility for management information.

R3: Integrate mechanisms for the monitoring and debugging of agent team activities.

2.2 Problem Statement

Based on the above requirements we address the following research challenges. Agent team configuration and membership are not defined at design time. Instead, a team is formed dynamically at runtime according to the tasks at hand and the properties and skills of the available agents. Such a team is disbanded after task completion or when a state is reached that makes completion impossible.

Dynamic team building demands a middleware that connects the agents and manages the team membership. Agents can join or leave the system during the execution of a task. There is no restriction on the maximum number of participants. Decentralization, scalability and robustness are important requirements for the middleware.

Moreover, team management requires a system-wide monitoring and traceability of the team building process and team interactions. Since we aim at a fully decentralized system, it may contain a potentially large number of processes working in parallel. Parallel processes are difficult to observe and analyze, especially if they interact heavily. Therefore, the integration of an adequate monitoring system is necessary to understand the behavior of individual components as well as the overall system.

3 Foundations

Before we present our solution, we give a brief overview of the key concepts and techniques whereupon our solution is based.

3.1 Self-organization

The concept of self-organization has been elaborated in numerous scientific domains, but no single consistent definition of self-organization exists. In [2] self-organization is defined as a mechanism that enables a system to modify its organization or structure during runtime without explicit instructions from the outside. Components of a self-organizing system are able to manipulate and reorganize other components in order to maintain the system functionality without external control if the context changes.

Usually, self-organizing systems are characterized as dynamic, non-deterministic, and self-modifying. Properties such as autonomy, emergent behavior and adaptation are commonly associated with self-organizing systems [3]. They dynamically evolve over time and space [4].

3.2 Multi-agent Organization

There is a variety of organizational models for MAS [5], that resemble organizational structures in business and society. These organizational models demand different management approaches. Some examples:

- Flat: All agents are equal without a particular leader.
- Hierarchical: Agents are arranged in a tree-like structure. Parent agents control their leaf agents, which in turn can be parents of other agents.
- Team: A team is formed by a group of agents who strive towards a common goal. Team goals can change over time. Agents in a team usually adopt specific roles and authority. If the team goal changes, these roles may change. A team may have an internal hierarchical structure.
- Holonic: Holonic organizations contain layers of so-called holons that encapsulate groups of holons. At the lowest level, holons are represented by individual agents.

- Coalition: Agents form a coalition dynamically to achieve a certain common goal. If the goal is reached, the coalition is dissolved. The internal organization usually is flat but can be hierarchical.

Obviously, there is considerable overlap between these forms of organization. This reflects the large diversity of models, views and application areas for MAS. In our work we focus on teams of agents that may be composed of a hierarchy of teams.

3.3 Distributed Ledger

Several definitions of DL exist in the literature. Essentially, the core concept of a DL is a concatenation of cryptographically linked blocks of data. A fully decentralized network of nodes collectively manages the resulting global data structure [6].

A DL can be viewed as a distributed data store without the need for a central administrator. Thus, nodes in the network have to store a local replicated version of the ledger. The concatenation of blocks in a DL enables the integration of untrusted nodes by using a protected ledger. Hence, management by a central trusted third party is not required. Instead of a central server that ensures trust between the participants, the DL maintains the trust [7]. In general, all DL architectures have the following four ingredients:

- The DL stores all transactions created by the collaborating nodes in the network.
- Public-key cryptography is applied to ensure a secure identity for all participants, which is required to enforce rights over the ownership of data stored in the DL.
- A peer-to-peer-based network structure allows the network to grow without causing interruptions and prevents single point-of-failures.
- Consensus protocols are required to maintain replicated, shared, and synchronized data identical across all nodes and to protect against manipulation [7]. Thus, it is ensured that all involved participants in the network agree on a single version of the data. A trusted third party is not required [8].

Hyperledger Sawtooth. Our agent management solution applies the DL framework called Hyperledger Sawtooth [9, 10]. Validator nodes represent the main components in the Sawtooth network. Such a node essentially consists of the *Validator* component, a consensus module termed *Consensus Engine*, and one or more *Transaction Processors*. Transactions sent by a client to a Validator node are distributed to the corresponding Transaction Processors and possibly to other Validator nodes. The Transaction Processors contain the business logic. They apply the incoming transaction and accordingly modify the data in the blockchain.

4   Solution

Our research has shown that the combination of MAS and DL is an ideal foundation for the management of teams of autonomous agents. Due to space constraints we cannot discuss all the details here. The reader is referred to [11] for further information.

4.1   Team Building

Agents have access to a joint DL. The process of new agents joining a team is controlled using the DL node component of the involved agents. In the following, the term *applicant* refers to an agent that wants to join the Sawtooth network or a team. By *member* we denote an agent that is already part of the network or team. Furthermore, messages are encrypted such that participants outside of a team cannot read team messages recorded in the DL. The process of joining is divided into two steps. In step 1, applicants register at the DL. In step 2, applicants apply for access to a new or an existing team.

Joining the Distributed Ledger Network

In order to connect to the network, applicants need to register their public keys, i.e. their identity in the DL network. New members that want to participate need to have their public key linked to a permission role. Each member that has registered a valid key and owns the permission role *allowed_keys* can perform the key assignment. Hence, only members with this property can manage the joining of new applicants to the network. For listening to join requests via a predefined address, a dedicated member is authorized to manage the connections of members and the DL network. Applicants must have a suitable public-private key pair. Subsequently, the applicant starts the joining process by sending a *join_request* to the predefined address. The request contains the public key of the applicant and the address for the response. This may optionally be extended with additional information for processing the request. If the authentication is successful, the member starts a voting process, as follows.

Voting Mechanism

The voting process sends a voting request to all participants and collects the votes. The defined time limit in which the voting must occur is important here. Hence, partial voting must be possible where not all votes of the participants are required. If a defined quorum is not achieved, the result is a failed voting. In our system a replaceable voting strategy enables different majority voting methods such as simple and qualified majority. The voting strategy can be chosen according to the desired security level. Furthermore, it can be set individually for each team when the team is created. The chosen strategy is then the same for all members of the team.

In addition to the voting strategy, the *voting maintainer* and the number of voting participants are used in the voting algorithm. The voting maintainer represents the member of the DL network that received the request. The voting maintainer sends a voting request indicating the candidate via broadcast to all participating members. After all voting participants have returned their answers, or the voting timeout is reached, the maintainer validates the votes according to the voting strategy. If access is granted, the maintainer adds the candidate as a new member of the network.

In order to avoid unnecessary communication, a selection strategy is applied to limit the number of members participating in the voting. Again, this can be specified for each team individually. No further network communication is required to arrange the voting. Only the result is sent back by each voting member. The selection of participants in the voting depends only on the total number of members and the defined number of voters.

Joining a Team

Members form teams by exchanging a secret key to encrypt messages. Only members that have this key can decrypt the team messages. Each member can create a new team and can request entry into an existing team. A member who is already part of the requested team is required to process the join request. By default, this is an automatic process performed by searching an entry in the list of team members. The highest entry in the list, i.e. the team member that joined the team at last, will be requested to handle the join operation. A list of team members able to handle the join request can be queried directly through the DL network. Alternatively, before sending the join request, an agent can ask a known team member directly to obtain the public key.

Team members communicate their encrypted messages via broadcast. Members can decrypt the messages if they own the secret key. The first node in the team defines a secret key. Afterwards, it generates a transaction including all team information and stores it in the DL network. Moreover, the node sets itself as a team voting maintainer and registers itself for all incoming team events. Further teams, as well as sub-teams, are created in the same way.

5   Implementation and Evaluation

In our prototypical implementation we have extended Hyperledger Sawtooth by the components described in the previous chapter.

5.1   Hyperledger Sawtooth Extension

The modular structure of Hyperledger Sawtooth facilitates the integration of extensions. Implementing the team management functionality, our Hyperledger Sawtooth Extension[1] supports the exchange of team keys via a flexible voting mechanism. Thus, a team member can safely share its secret keys with another peer.

The *Receipt Transaction Family*[2] serves multiple purposes. First, it records the exchange of secret keys and the corresponding voting results. Thus, it is traceable which peers are involved in the key exchange. Second, it enables the tracking of team IDs in order to guarantee the uniqueness of a team ID. All existing team IDs are contained in the state of the DL. Each peer can thus query the existence of a team, for example, to join the team in the next step. Moreover, a peer can be determined to handle a join request using the list of team members managed by the Transaction Family.

The *Transaction Processor* responsible for processing the Receipt Transaction Family only accepts transactions from the same peer that created the receipt. If the transaction header does not match the member public key, the transaction is marked as invalid. The Transaction Processor additionally searches for an address derived from the name of the team when the receipt is of type *join_team*. Subsequently, the Transaction Processor stores a list of public keys of the team members at this address. The public key of the

---

[1] The Hyperledger Sawtooth Extension is available on GitHub 18.

[2] A Transaction Family defines a data model to record and store data as well as an associated Transaction Processor for the business logic of the application.

requester is entered in the list, except if a transaction is marked as invalid. It is the case whenever the list is not empty and does not contain a member public key, since it is not determinable whether the peer who shared the secret team key is indeed a member of the team.

In order to improve responsiveness and to be able to run processes in parallel, additional threads are integrated into the Hyperledger Sawtooth Extension. For example, long running operations are executed in a separate thread. Thus, one thread is provided for preparing and submitting batches of transactions to the Validator because it may take some time to process a batch depending on the workload of the Validator.

A receipt transaction of type *join_team* is submitted when a peer creates a new team. In this case, the member public key and the applicant public key are equal and provided by the creating peer. Thus, the Transaction Processor can determine that the peer has just created the team. Therefore, the Transaction Processor creates a new list corresponding to the address, including the public key of the peer.

A receipt transaction of type *leave_team* is sent if a peer deletes the secret key from its keystore and thus removes itself from a team. It results in the removal of the public key of the peer from the list of public keys. If the peer is the last member of the team, the team itself is deleted.

The *Voting Manager* implements the voting mechanism. In total, it comprises four different processes running in separate threads:

- The *Voting Handler* manages the voting after a *join_request* is received. It determines the voting members and starts the voting process. After the voting process is completed, the Voting Handler checks the signature of each vote to ensure that the votes come from the desired participants.
- The *Voting Evaluator* computes the vote result based on the defined Voting Strategy. Furthermore, it supports a whitelist for applicants that do not require a voting.
- The *Voting Completer* informs the applicant of the voting result. If the result is positive, the key exchange is performed. In a separate thread, the Voting Completer continues the processing of new join requests to maintain its responsiveness.
- In a voting team member, the *Voting Caster* responds to voting requests received via the DL by returning the vote in a Vote Object.

5.2 Evaluation

In order to support the evaluation, an additional monitoring component is implemented. It serves to display information about individual agents during team formation. The monitoring component is linked to a Hyperledger Sawtooth peer for collecting information about the network.

In the experiments[3], simple majority is used as Voting Strategy. All voting casts are counted, and incomplete votes are allowed, which means that not all requested voters have to cast their votes. The voting process has a timeout of 5000 ms. To evaluate the different functionalities of our Hyperledger Sawtooth Extension, they are tested in two

---

[3] Tests were performed on a desktop PC with Intel six-core CPU i7-9850H 2.6 Mhz and 32 GB RAM; Sawtooth peers running in virtual machines, each with Ubuntu 18.04 kernel version 5.4.0-45-generic x86_64, Java Runtime Environment 1.8.0_271, and Docker version 18.09.7.

different scenarios. The evaluation focuses on team creation and the associated voting process. Different team sizes and numbers of participants are considered.

Let us first look at the time taken by the system when a candidate joins a team of a certain size. The initial process of creating the team is not part of the measurement. The experiment is divided into two parts. In the first part, the system selects all members in the team for voting. In the second part, a voter threshold of five is used. The Selection Strategy randomly selects five team members if the team size is at least five.

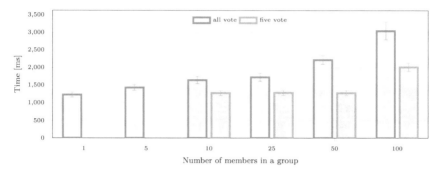

Fig. 1. Time to join a team depending on the team size and number of voters

Looking at Fig. 1, it is obvious that the team size has a significant influence on the duration of the join process when all members of the team are involved in the voting process. The increase in the duration of the joining process results from resource limitations as all instances run on a single computer. The voting request is pre-processed by all team members, even if only five members are involved in the final voting. However, the diagram demonstrates that a reduced number of voters leads to a significant reduction in the duration of the joining process.

The second experiment explores the time needed to form multiple teams, each consisting of pairs of agents. Here, the time for creating a team is contained in the measurements. Each agent is executed as a separate process in a virtual machine, as in the previous experiment. Figure 2 shows the results.

Combining the results of the previous scenario where joining an existing small-size team requires approx. 1100 ms, and these results for the creation of teams of size two, we conclude that the team creation itself also takes approx. 1000 ms.

6 Related Work

There are only a few comparable approaches where the management of a MAS is supported by a DL. The authors in [12] propose a distributed system where vehicles record vehicle, maintenance, sensor and context information in a blockchain, which however only stores the hash of the data. The data itself is stored at the corresponding owners, i.e. car manufacturers, insurance companies, and maintenance service providers.

In [13, 14], the authors consider the ecosystem of smart vehicles modeled as a composition of IoT (Internet of Things) devices. A Hyperledger Sawtooth framework

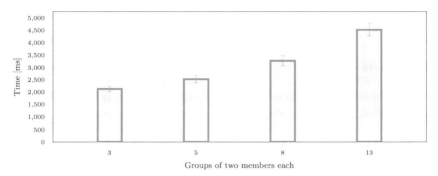

Fig. 2. Time to create teams of two agents each

is combined with an InterPlanetary File System (IPFS), which provides the distributed storage system. It is used as an additional distributed storage to reduce the amount of data in the blockchain and to implement an IoT authentication protocol. The Hyperledger Sawtooth nodes do not run on their assigned IoT devices, but on spatially separated servers. The approach is extended in [15] with Rancher and Kubernetes as container orchestration to enable Cloud-based scenarios.

The authors of [16] introduce a smart grid controller blockchain platform based on Hyperledger Fabric. The platform simulates virtual power plant communication together with load profiles and data of smart meter measurements. The communication require-ments are similar to the ones described in this paper. However, in a Hyperledger Fabric network different node types are required to maintain the network which adds to the complexity of the network interactions. In contrast, Hyperledger Sawtooth has a simpler architecture with only one type of network node.

7   Conclusions

The main contribution of this paper is a demonstration that DL technology can effec-tively support the management of a highly dynamic MAS. Using a DL has many advan-tages, such as no central authority needed, immutability of the replicated shared data, and automation of operations by means of Smart Contracts. Our solution builds on an extended Hyperledger Sawtooth system. The management framework guarantees neces-sary access protection and communication restrictions for teams of agents. Furthermore, a monitoring extension applies the Hyperledger Sawtooth framework to store relevant information about the team communication, and thus to support the debugging of MAS operations. The presented management framework is part of a comprehensive MAS platform providing a fully decentralized and self-organized collaboration support for autonomous agents in highly dynamic environments [11].

Further improvements of the Hyperledger Sawtooth integration are conceivable, particularly aiming at improved performance and fault-tolerance. We expect that an extension of Sawtooth by IOTA Tangle mechanisms [17] would shorten the response time substantially. Furthermore, the integration of a Proof of Learning (PoL) consensus algorithm as proposed in [18, 19] could increase the fault tolerance significantly. In [18], outlier detection techniques are proposed to verify the compatibility of incoming data and subsequently discard suspect data. In [19], consensus nodes train neural network models using training and test data shared across the blockchain network. In case of consensus, the neural network model is valid, and a new block is attached to the blockchain.

Acknowledgement. The research was supported by the emergenCITY project as part of the LOEWE program of the state Hessen in Germany.

References

1. Russell, S., Norvig, P.: Artificial Intelligence: A Modern Approach. Pearson (2002)
2. Banzhaf, W.: Self-organizing systems. In: Meyers, R.A. (ed.) Encyclopedia of Complexity and Systems Science, pp. 8040–8050. Springer, New York (2009). https://doi.org/10.1007/978-0-387-30440-3_475
3. Gorodetskii, V.I.: Self-organization and multiagent systems: I. Models of multiagent self-organization. J. Comput. Syst. Sci. Int. **51**(2), 256–281 (2012)
4. Bonabeau, E., Dorigo, M., Théraulaz, G.: Swarm Intelligence: From Natural to Artificial Systems. Oxford University Press (1999)
5. Bryan Horling, B., Lesser, V.: A survey of multi-agent organizational paradigms. In: The Knowledge Engineering Review, vol. 19, no. 4, pp. 281–316. Cambridge University Press (2004)
6. Rauchs, M., et al.: Distributed ledger technology systems: a conceptual framework. In: EnergyRN: Other Energy Engineering (Topic) (2018). SSRNs eLibrary 3230013 (2008)
7. Atlam, H., Wills, G.: Intersections between IoT and distributed ledger. In: Advances in Computers, vol. 115, pp. 73–113. Elsevier (2019)
8. El Ioini, N., Pahl, C.: A review of distributed ledger technologies. In: Panetto, H., Debruyne, C., Proper, H.A., Ardagna, C.A., Roman, D., Meersman, R. (eds.) On the Move to Meaningful Internet Systems. OTM 2018 Conferences. LNCS, vol. 11230, pp. 277–288. Springer, Cham (2018). https://doi.org/10.1007/978-3-030-02671-4_16
9. Blummer, T., et al.: An Introduction to Hyperledger. The Linux Foundation (2018)
10. Olson, K., et al.: Sawtooth: An Introduction. The Linux Foundation (2018)
11. Jahl, A.: Situative teams in cooperative autonomous systems. Ph.D thesis. University of Kassel, Faculty of Electrical Engineering and Computer Science (2023)
12. Cebe, M., et al.: Block4Forensic: an integrated lightweight blockchain framework for forensics applications of connected vehicles. IEEE Commun. Mag. **56**(10), 50–57 (2018)
13. Gerrits, L., Kromes, R., Verdier, F.: A true decentralized implementation based on IoT and blockchain: a vehicle accident use case. In: International Conference on Omni-Layer Intelligent Systems (COINS 2020) (2020)
14. Kromes, R., Gerrits, L., Verdier, F.: Adaptation of an embedded architecture to run hyperledger sawtooth application. In: IEEE 10th Annual Information Technology, Electronics and Mobile Communication Conference (IEMCON 2019), pp. 0409–0415. IEEE (2019)
15. Gerrits, L., et al.: A blockchain cloud architecture deployment for an industrial IoT use case. In: International Conference on Omni-Layer Intelligent System (COINS 2021), pp. 1–6. IEEE (2021)
16. Goranović, A., et al.: Hyperledger fabric smart grid communication testbed on raspberry PI ARM architecture. In: 15th IEEE International Workshop on Factory Communication Systems (WFCS 2019), pp. 1–4. IEEE (2019)

17. Wellington, F., Roderval, M.: Iota tangle: a cryptocurrency to communicate internet-of-things data. Future Gener. Comput. Syst. **112**, 307–319 (2020)
18. Lan, Y., Liu, Y., Li, B., Miao, C.: Proof of Learning (PoLe): empowering machine learning with consensus building on blockchains, vol. 35, no. 18, pp. 16063–16066 (2021)
19. Salimitari, M., Joneidi, M., Chatterjee, M.: AI-enabled blockchain: an outlier-aware consensus protocol for blockchain-based IoT networks. In: 2019 IEEE Global Communications Conference (GLOBECOM), pp. 1–6. IEEE (2019)

AirVA - Indoor Air Quality Monitoring and Control with Occupants Alerting System

Agostinho Ramos, Vagner Bom Jesus, Celestino Gonçalves, Filipe Caetano, and Clara Silveira[✉]

Instituto Politécnico da Guarda, Escola Superior de Tecnologia e Gestão, Guarda, Portugal
{celestin,caetano,mclara}@ipg.pt

Abstract. The objective of this work is to detail the development of an IoT system to monitor the indoor air quality index and record the entry and exit of occupants of the same space. Data is collected, processed, and sent to a data server through the MQTT communication protocol. This server is responsible for storing and displaying information on the ThingsBoard platform. To evaluate the solution, the air quality index and people count were monitored in some classrooms of Polytechnic Institute of Guarda. The results obtained show that the prototype meets all the requirements proposed for the system.

Keywords: Indoor Air Quality Monitoring · People Counting · IoT · Zigbee · ThingsBoard

1 Introduction

Disease surveillance is essential for the control of respiratory infectious diseases including the novel coronavirus (COVID-19). The lack of good air quality, both inside and outside buildings, has become a global concern for the World Health Organization (WHO), and another fact is that the general population spends approximately 90% of their lifetime inside a building [1, 2]. The degradation of air quality in this type of spaces is a known problem that results from natural environmental conditions, such as temperature and humidity variations, as well as from several existing processes [1, 3], like, for example, the several computer devices or electronic circuits that generate different types of pollutant gases that interfere with the air quality in a negative way, which can compromise the health and well-being of the people who work in those spaces, as well as cause the degradation of materials, components, and equipment [1, 2].

In this context, this paper aims to present AirVA - Indoor Air Quality Monitoring and Control with Occupants Alerting System solution, to monitor the air quality index continuously, also recording the entry and exit of people in the indoor space. Having in mind the objective of the Indoor Air Monitoring and Occupant Counting System, it was decided to choose the agile development methodology, namely the Scrum methodology [4], because it has an approach and perspective that fits with the system to be developed. To evaluate the prototype, tests were performed in the development environment and in some of the classrooms of the Polytechnic Institute of Guarda, in order to verify the validity and efficiency of the system.

The paper is organized into five sections. After the Introduction, comes Sect. 2 that presents the related works, some of them having similar characteristics to the proposed system. Section 3 presents the requirements, development architecture, the components used in the system and their protocols, as well as a description of the performed data management and processing, with emphasis on the ThingsBoard platform. In Sect. 4 some tests and results are presented, and Sect. 5 ends the article with the conclusions and the work that is intended to be done in the future.

2   State of the Art

There are currently several systems developed for air quality monitoring and occupant counting. The following sections describes some of them, considering examples with similar characteristics to the proposed system.

2.1   Indoor Air Quality Monitoring

Good air quality is of paramount importance whenever a room is occupied. High air purity, large air flows, efficient air filtration, and thermal comfort are required to achieve good indoor environmental quality.

The authors of [5] developed a system for monitoring and controlling Indoor Air Quality (IAQ) in school building offices. The system is capable of monitoring CO_2, CO, relative humidity, and temperature concentrations, integrating a dynamic ventilation mechanism capable of regulating IAQ levels according to CO and/or CO_2 levels, also offering the possibility of sending in real time the collected data to a web server, allowing remote access to all the information. The system has an alert mechanism that sends SMS messages to the user's cell phone and/or warnings to the web page. CitiSense [6] is an air quality monitoring system that allows individuals to identify when and where they are exposed to harmful indoor and outdoor air in real time. The system is characterized as a distributed system and has a web server that provides a customized daily pollution map. According to Adochiei [1], the proposed Indoor Air Quality Monitoring System (IAQMS) can perform real-time measurements of a wide range of ambient air parameters. The data generated by the 8 detectors in the system is processed by an Arduino Mega 2560 microcontroller, sent via the ESP-01S Wi-Fi module to the Blynk server, and then displayed on the Blynk iOS/Android mobile app using three visualization techniques: line graphs, meters, and values. The authors of [3] have developed an assistive technology with mobile SMS and e-mail notification that can be used in real time to indoor environment and provide Air Quality Index (AQI) information. The environmental information measured by the sensors are sent to the occupants of the building via mobile SMS and e-mail. The study [7] presents an implementation of MQTT based air quality monitoring system. The air quality measuring device uses a ESP8266 Node MCU that connects to the different sensors used. Table 1 presents a comparison of the systems mentioned above with our AirVA.

Comparison of the AirVA system with other systems

System	Distributed	Autonomous	Smart	Scalability	Notification
AirVA	X	X	X	X	X
[1]	X	--	X	X	--
[3]	X	--	--	--	X
[5]	X	X	--	--	X
[6]	X	X	--	--	--
[7]	X	X	--	X	X

Several technologies can be used in different and efficient ways for monitoring the Air Quality Index (AQI), including the use of different sensors. It is intended to build a system for monitoring and controlling indoor air quality in school building offices, to have a technical, reliable, and economical solution to assess AQI.

2.2 Occupants Count System

Most counting systems developed are based on different types of technologies, such as mobile devices, video cameras, and infrared sensors, which produce results with both advantages and disadvantages.

Nasir et al. [9] proposed an Automatic Passenger Counting (APC) system using image processing based on a skin color detection approach. The lighting conditions demonstrate great performance, with counting accuracy of 90.64%. However, the system has been shown to be unsuitable for application if images are captured under high illumination conditions, such as most outdoor scenarios. The paper [10] provides a case study for counting people using sensors that record the carbon dioxide concentration and estimate indoor occupancy. Satisfactory accuracy was achieved. Myrvoll et al. [11] developed a method for counting public transport passengers using Wi-Fi signatures of mobile devices carried inside a transport vehicle. The results are promising. However, there are some challenges that are difficult to predict, such as the correlation between the number of mobile devices that can be detected and the number of passengers on board. Kalikova et al. [12] proposed a method for counting the number of mobile devices present in smart buildings using mechanisms that capture network packets sent by mobile devices. The measurements differed for each captured mobile device, impacting the accuracy of the results. Barbosa et al. [13] developed a method to count the number of passengers boarding and alighting in an urban transport system. The results proved the viability of the method in the urban transport system. However, the device has no location data recorded in each case of entry or exit of people, and it is not easy to install.

3   AirVA Implementation

This section presents the requirements, the development architecture, the protocols and the components that the AirVA system is composed of.

3.1 Requirements Specification

For the problem of poor indoor air quality, it is intended to create a system that includes the following functional requirements (User Story format):

- As a user (student, teacher) I want to know the indoor air quality index by means of a color code (green, yellow and red), to know if I can enter the space;
- As a user (student, teacher) I want to check the total number of occupants inside the space, to know if more people can enter;
- As a user, I want to be visually informed about the air quality index:

  o Situation 1 - Good Air Quality (Green Traffic Light) - the system reports the number of occupants present in the space;
  o Situation 2 - Critical Air Quality (Yellow Traffic Light) - the system alerts you to open the window/door to improve the air quality and advises against occupants entering the space;
  o Situation 3 - Bad Air Quality (Red Traffic Light) - the system acts in the air quality correction (air extractor, dehumidifier, air purifier, among others), issues an alert so that people can leave the room, keeping the alert for opening the windows and door;

- As an AirVA manager, I want to collect the data from the outside environment (Temperature and Humidity) to compare with the inside data;
- As an AirVA manager, I want to determine the opening or closing status of the windows;
- As an AirVA manager, I want to configure the parameters that allow to determine the air quality index;
- As an AirVA manager, I want to receive a message when the air quality index is critical or bad (yellow or red traffic light).

As non-functional requirements of the AirVA system, we can consider the following:

- The system must have high availability, even without an Internet connection;
- The system must run on any Linux server;
- The system must use programming oriented under the ThingsBoard platform;
- The system must not identify users;
- The system must meet the legal standards of General Data Protection Regulation (GDPR) of European Union.

It should be noted that the AirVA system integrates several sensors to collect data from the physical environment (temperature, humidity, CO_2, dust, smoke, and other gases) and thus determine the air quality index.

3.2 System Architecture

This section describes the hardware and communication protocols used to implement AirVA - Indoor Air Monitoring and Occupant Counting System.

The solution consists of prototypes responsible for measuring gases with ammonia, nitric oxide, alcohol, benzene, carbon dioxide, and smoke and dust, using the MQ-135,

TELAIRE T6615, Grove Dust (PPD42NS) sensors, and for recording the entry of people the Sharp gp2y0a21yk0f sensor was used. The data is collected and processed by an ESP-8266 prototyping board and an Arduino Uno and sent to a data server via MQTT. Table 2 describes the components used for the developed prototype.

Table 2. AirVA system components

Components	Description
Lolin D1 Mini Pro	ESP8266 microcontroller with Wi-Fi. It is used to control and process the information from the sensors and send the data to the Raspberry Pi
OLED 0.66 Shield	64×48 pixels display. Used to visualize the number of occupants inside the room
LED RGB W S2812B	Additional board for the Wemos platform. Used as a traffic light to signal the status of air quality
Li-lon 18650	Rechargeable battery with a capacity of 2200 mAh - 3.7 V, input power DC 5.0 V, output power DC 5.0 V. Used to power the microcontrollers
Telaire T6615	Dual channel CO_2 sensor, for applications up to 50,000 ppm. Used to measure CO_2 concentration
SONOFF SNZB-04	Wireless ZigBee door or window status notification sensor. Used to know if the door or window is open
ZigBee s26 r2zb	Smart socket to turn on and off a device. Used to attach an air purifier (example: fan)
SONOFF SNZB-02	ZigBee temperature and humidity sensor. Used to measure the temperature and humidity of the indoor and outdoor environment
ZigBee CC2531	Provides PC interface for IEEE 802.15.4/ZigBee applications. Used to establish the connection between ZigBee and the Raspberry Pi 4
USB-AC51	Network adapter capable of connecting to the home network via Wi-Fi. Used to establish Wi-Fi connection between microcontrollers and Raspberry Pi
PPD42NS	Pollution sensor responsible for measuring the level of particles in the air. Used to measure the particles in the air
MQ-135	Sensor for gases such as ammonia, carbon dioxide, benzene, nitric oxide, and alcohol. Used to detect these gases in the interior space
Sharp GP2Y0 A21YK	Infrared proximity sensor, analog output ranging from 3.2 V at 10 cm to 0.4 V at 80 cm. Used to count occupants entering and leaving the space
Raspberry Pi 4	Minicomputer that connects to the microcontrollers. Used as a server and gateway to process all information that arrives from the sensors
Arduino Uno	ATmega328 microcontroller, 32 KB flash memory and 2 KB SRAM. Used to send number of occupants to the ESP8266 by serial communication

The components were chosen to have a large range and control with efficiency on the air quality index, allowing the system to be scalable, distributed, and robust. Figure 1 illustrates the final architecture of the developed prototype.

The AirVA system consists of several microcontrollers that connect to the Raspberry Pi via Wi-Fi signal provided by it, allowing two-way communication even in offline mode. Each microcontroller processes all information before publishing it to the Raspberry Pi thread via the MQTT protocol. These microcontrollers publish data in JSON format that is validated and recorded in the Raspberry Pi database in real time. With this metadata, the system can generate an air quality index to determine if the air quality is "GOOD", "CRITICAL" or "BAD".

The AirVA system can communicate with each microcontroller individually, providing different dashboards with different accesses for different users, through the HTTP/TCP protocol for the devices inside the same network, but also outside the network, as long as the Raspberry Pi has access to the Internet. The architecture of the AirVA system was designed to allow the integration of new devices with the parameters known in the API documentation.

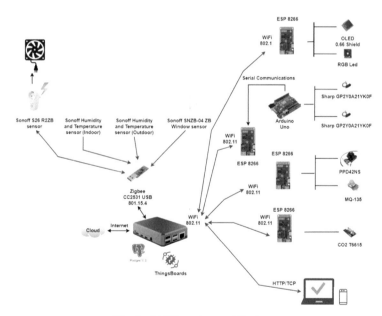

Fig. 1. AirVA system architecture

3.3 System Functionality

As the goal is to perform continuous monitoring of the indoor air quality index and registration of people in each space, it was necessary to add the solution to a system capable of storing and presenting the collected data called ThingsBoard.

ThingsBoard is an open source IoT platform for collecting, processing, visualizing, and managing data devices. It enables device connectivity via industry-standard IoT protocols - MQTT, CoAP and HTTP, supporting both cloud and on-premises deployments [14]. ThingsBoard combines scalability, fault tolerance and performance and can securely provide, monitor and control IoT devices using advanced server-side API [14].

Figure 2 illustrates the Activity Diagram under which the AirVA project was developed. Initially, the AirVA system reads the IQA values and the number of occupants present in the room. When the number of occupants is less than the maximum limit and the IQA is classified as "GOOD", the system will turn on the green traffic light, allowing other occupants to enter. If the number of occupants is equal to the maximum and the IQA is "GOOD" or "CRITICAL", the system will turn on the red light of the traffic light. If the number of occupants is less than the maximum limit and the IQA is "CRITICAL", the system will turn on the yellow light, otherwise, for any number of occupants and for an IQA of "BAD", the system will turn on the red light, automatically activating the air purifier (Fan). As soon as the IQA returns to the "GOOD" state it turns off the actuator and continues the whole process normally. If the temperature value reaches the maximum limit set, the system will act by turning on a fan and will only turn off when the temperature value is 5° below of the defined limit.

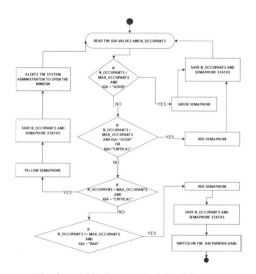

Fig. 2. AirVA System Activity Diagram

4   Simulation and Testing

To test the operation of the system, evaluating the behavior both in monitoring the indoor air quality index and in controlling the entry of people in the space and the respective alerts, three different tests were performed in three classrooms of School of Technology and Management (classrooms 45, 49 and 54) with differences in volume, location and number of occupants. It should be noted that the tests were carried out in a period with weather conditions characteristic of the summer months in the region of Guarda, Portugal.

The first test took place in classroom 54 on June-28-2022, from 13:00 to 17:00. The classroom has 90 seats and the results were as expected, but there was a problem with the database records that led to the loss of the sensor data record. Still, it was important to validate the system configuration and sensors calibration.

The second test took place in classroom 45 on July-2-2022, from 9:00 am until 6:00 pm, and consisted in monitoring the IAQ parameters, without using any ventilation, other than that resulting from the normal opening of doors whenever there was an entrance or exit of occupants. The indoor air temperature and humidity did not vary significantly during the experiment. At 15:23 the fluid gas from a lighter was released into the interior space to verify if it was detected. It was verified that in response, the ventilation was automatically activated to correct the situation. The CO_2 concentration did not suffer critical variations because the maximum number of occupants present in the space was 5, and the space has a maximum capacity of 25 occupants. Unfortunately, it was not possible to register the occupants data.

A third test was performed on July-12-2022, from 8:00 am to 12:00 pm, in a computer room with a maximum occupancy of 21. Figure 3 illustrates the evolution of the values of the air quality monitoring sensors and the number of occupants of AirVA system. As a result of the actual test, the temporal evolution of CO_2 values, number of occupants, temperature and humidity inside the classroom are presented.

At the beginning of the test, data collection was done with the room closed and without ventilation. For covid-19 reasons, the windows and door had to be opened to allow better air circulation. The CO_2 levels of the indoor space oscillate between 602 and 1285 ppm, obtaining an average value of 858 ppm. This reduction is due to the opening of the room's windows and door allowing cross ventilation that provides adequate air renewal. CO_2 concentrations above 1250 ppm indicate the existence of insufficient ventilation and, in addition, may have adverse effects on human health, at the level of physical and psychological performance [15–17], this situation becomes more critical because it occurs with a considerable number of occupants who remain in the room. The influence on indoor air temperature and humidity did not undergo critical variations in the records, having a significant influence of voluntary natural ventilation, through the opening of windows and doors. The control of temperature variation inside buildings is crucial to the increase of energy efficiency and human thermal comfort [18, 19]. This test coincided with a written assessment exam of the 2nd year of the Computer Engineering course. The assessment test started at 10:00 am, which is translated in the graphs of CO_2 values, the number of occupants and the air purification actuator. Despite the considerable number of occupants inside the room, the system acted in such a way as to ensure a stability of the CO_2 concentration values. It has some peaks initially that have been stabilized over time.

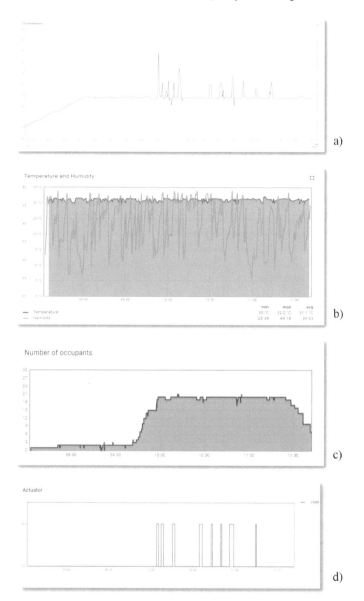

Fig. 3. Evolution of sensor data for the third test: a) CO_2 concentration, b) Temperature and Humidity, c) Number of occupants and d) Actuator

5 Conclusions and Future Work

The proposed AirVA system - Indoor Air Quality Index Monitoring and Control System, with alert to occupants, was successfully implemented, highlighting the principle of ubiquitous computing, without the need for direct human interaction.

Unlike traditional systems that only report the air quality value, the AirVA system not only reports the air quality, but also acts upon a situation of poor air quality (critical or bad) according to the information acquired by the sensors.

As future work, we intend the AirVA system to be able to count the number of people more accurately, resorting to image processing and analysis, using the OpenCV framework, to detect each occupant entering and exiting the space [20]. Another feature that we intend to develop in the future will be a direction detection system, which signals the direction of the occupant with different colors: the blue color signals the entrance, and the red color signals the exit of the occupant of the space [20].

Regarding intelligence features, we intend to provide the AirVA system with the ability to infer the number of occupants based on the values collected by the sensors.

Another interesting feature would be to add a motor in each window of the space, to open and close the windows depending on the IQA and the number of occupants, making the system as autonomous as possible.

References

1. Adochiei, F., et al.: Electronic system for real-time indoor air quality monitoring. In: IEEE International Conference on e-Health and Bioengineering (EHB), pp. 1–4, October 2020
2. Moharana, B., Anand, P., Kumar, S., Kodali, P.: Development of an IoT-based real-time air quality monitoring device. In: IEEE International Conference on Communication and Signal Processing (ICCSP), pp. 191–194, July 2020
3. Nigam, H., Saini, A., Banerjee, S., Kumar, A.: Indoor environment air quality monitoring and its notification to building occupants. In: IEEE Region 10 Conference (TENCON), pp. 2444–2448 (2019)
4. Jacobson, I., Ng, P.W., McMahon, P.E., Goedicke, M.: The Essentials of Modern Software Engineering: Free the Practices from the Method Prisons! Morgan & Claypool (2019)
5. Fernandes, S., Igrejas, G., Feliciano, M.: Monitorização e controlo de qualidade do ar interior em gabinetes de edifícios escolares. Revista de Ciências Agrárias **40**, 274–281 (2017)
6. Bales, E., Nikzad, N., Quick, N., Ziftci, C., Patrick, K., Griswold, W.G.: Personal pollution monitoring: mobile real-time air quality in daily life. Pers. Ubiquit. Comput. **23**(2), 309–328 (2019)
7. Chanthakit, S., Rattanapoka, C.: MQTT based air quality monitoring system using node MCU and node-red. In: IEEE Seventh ICT International Student Project Conference (ICT-ISPC), pp. 1–5 (2018)
8. Felix-Constantin, A.: Electronic system for real-time indoor air quality monitoring (2020)
9. Nasir, A., Gharib, N., Jaafar, H.: Automatic passenger counting system using image processing based on skin colour detection approach. In: IEEE International Conference on Computational Approach in Smart Systems Design and Applications (ICASSDA), pp. 1–8 (2018)
10. Li, T., Fong, S., Yang, L.: Counting passengers in public buses by sensing carbon dioxide concentration: data collection and machine learning. In: Proceedings of the 2018 2nd International Conference on Big Data and Internet of Things, pp. 43–48 (2018)
11. Myrvoll, T., Hakegard, J., Matsui, T., Septier, F.: Counting public transport passenger using WiFi signatures of mobile devices. In: IEEE 20th International Conference on Intelligent Transportation Systems (ITSC), pp. 1–6 (2017)
12. Kalikova, J., Krcal, J.: People counting in smart buildings. In: IEEE 3rd International Conference on Intelligent Green Building and Smart Grid (IGBSG), pp. 1–3 (2018)

13. Barbosa, M., Braga, P., Coelho, P.: Automatic system of monitoring the movement of passengers of collective transportation by bus. In: XX ANPET - Congress of Research and Education in Transportation, pp. 1–8 (2005)
14. ThingsBoard – Open-Source IoT Platform. https://thingsboard.io. Acedido em 10 05 2022
15. Hanninen, O.O.: WHO guidelines for indoor air quality: dampness and mold. In: Fundamentals of Mold Growth in Indoor Environments and Strategies for Healthy Living, pp. 277–302. Wageningen Academic Publishers, Wageningen (2011)
16. Satish, U., et al.: Is CO2 an indoor pollutant? Direct effects of low-to-moderate CO2 concentrations on human decision-making performance. Environ. Health Perspect. **120**(12), 1671–1677 (2012)
17. Fisk, W.J.: Is CO2 an indoor pollutant? Higher levels of CO2 may diminish decision making performance (2013)
18. Mandayo, G., et al.: System to control indoor air quality in energy efficient buildings. Urban Clim. **14**, 475–485 (2015)
19. Ahmed, K., Kurnitski, J., Sormunen, P.: Demand controlled ventilation indoor climate and energy performance in a high performance building with air flow rate controlled chilled beams. Energy Build. **109**, 115–126 (2015)
20. Din, M.M., Nordin, N.N., Siraj, M.M., Kadir, R.: IOT real-time people counting using raspberry PI (IOT-RepCO). IOP Conf. Ser. Mater. Sci. Eng. **864**(1), 012093 (2020)

A Federated Algorithm for the Lightweight Generation of High-Entropy Keys in Distributed Computing Systems

Borja Bordel[(✉)], Ramón Alcarria, and Tomás Robles

Universidad Politécnica de Madrid, Madrid, Spain
{borja.bordel,ramon.alcarria,tomas.robles}@upm.es

Abstract. To build robust secure channels for information exchange, distributed computing systems must generate and handle high-entropy secret keys. However, solutions to generate those high-entropy keys such as Physical Unclonable Functions or sensing devices are very dependent on the environment and the hardware performance. Thus, keys may not achieve the expected entropy or show uncontrolled behaviors that may prevent communicating remote nodes to synchronize with a shared key. Therefore, new solutions are needed to enable distributed computing nodes to generate high-entropy keys in a lightweight, consistent, and robust manner. In this paper we propose a federated algorithm to address this challenge. Remote nodes are provided with different physical devices to initialize with a random configuration a Fibonacci random number generator. The parameter set describing the configuration of the key generator is locally encoded using a gradient function and sent to an edge computing manager where different encoded configurations coming from different remote nodes are collected. The edge computing manager combines all these configurations considering different weights and an optimization target function based on the definition of mutual information. An experimental validation is also provided. Simulation tools are employed, and results show the long-term average entropy increases up to 23% when using the proposed solution.

Keywords: federated algorithms · key generation · distributed computing · entropy · mutual information · random number generators

1 Introduction

Encryption is a basic enabling technology in distributed computing systems [1]. In some applications, private personal information (such as pictures, sounds, or video recordings) is captured and processed, and must be protected against unauthorized accesses [2]. In some other applications, attackers can try to manipulate the behavior of the system (especially in critical scenarios) and legitimate data must be authenticated and encrypted to ensure their integrity and trustworthy origin [3]. But in all distributed computing systems, high-quality secrecy solutions are necessary, i.e., schemes producing encrypted messages from raw information, so the entropy of the encrypted message is maximum and no information from the clear original message is accessible [4].

© The Author(s), under exclusive license to Springer Nature Switzerland AG 2024
A. Rocha et al. (Eds.): WorldCIST 2023, LNNS 801, pp. 82–93, 2024.
https://doi.org/10.1007/978-3-031-45648-0_9

Different techniques can be applied to build those encryption schemes, but the limited computational power in most common distributed nodes prevents the implementation of complex solutions such as elliptic curves [5] or asymmetric encryption [6]. For example, the ROM memory required to implement such algorithms (at least 32 KB) is higher than the memory available in most microcontrollers [33]. Then, symmetric stream or block ciphers [7] are typically employed to protect communications in distributed computing systems. But the behavior of those ciphers is very dependent on the selected symmetric keys. Thus, to build robust secure channels for information exchange, remote nodes must generate and handle high-entropy secret keys [8]. Those keys may be obtained using only numerical methods, but these algorithms are usually computationally costly [9]. Besides, keys cannot be permanently stored into the nodes' memory, as they must be refreshed periodically and could be easily captured by any attacker with physical access to the distributed devices [10]. As a solution, remote nodes are usually provided with key generation algorithms based on physical phenomena and lightweight true random number generators [11]. Therefore, they can produce high-entropy symmetric secret keys to feed the stream ciphers in the communication subsystem, taking advantage of the random behavior of physical processes.

However, solutions to generate high-entropy keys such as Physical Unclonable Functions [12] or sensing devices [13] are very dependent on the environment and the hardware performance. For example, in geographical regions with very stable environmental conditions, keys may not achieve the expected entropy, as measurements are predictable and not enough random. Additionally, as hardware ages it shows uncontrolled behaviors that may cause the generation of keys with a low entropy and/or problems for remote devices to synchronize with a shared common key. Therefore, new solutions to enable distributed computing nodes to generate high-entropy keys in a lightweight, consistent, and robust manner are needed.

In this paper we propose a federated algorithm to address this challenge. The key generation process is divided into two phases. First, remote devices employ different hardware instruments to define the configuration of a Fibonacci random number generator, so they can generate a high-entropy number stream. This process is local and the secret configuration is never transmitted. Every distributed node gets its own and independent configuration. Later, in the second phase, distributed nodes behave as a federation and delegate the final calculation of the configuration of the key generators to an edge computing manager. To do that, every independent distributed node locally encodes the parameter set that describes the local configuration of the key generator using a gradient function. The edge computing manager collects different encoded configurations from different remote nodes, which individually are affected by local environmental conditions or local hardware performance. To compensate those effects and ensure all key generators are initialized with a configuration being able to produce high-entropy streams, the edge computing manager combines all the local configurations through an optimization target function based on the mutual information definition. In addition, local configurations are weighted to improve those with the highest entropy. The resulting optimum configuration is sent back to the remote nodes, so they can finally initialize the key generation algorithms.

The rest of the paper is organized as follows: Sect. 2 describes the state of the art on key generation in distributed computing systems; Sect. 3 describes the proposed solution, including the new federated algorithm and the encryption and optimization schemes; Sect. 4 presents an experimental validation; and Sect. 5 concludes the paper.

2   State of the Art on Key Generators

Key generators for distributed computing systems may be classified into two different groups. On the one hand, some generators produce keys with a limited length each time they are triggered [17]. Those keys are typically employed in block symmetric ciphers. On the other hand, we find generators producing continuous unlimited key flows [4], usually employed in stream ciphers, and based on random or pseudo-random sequences of numbers.

Typically, key generators of limited length employ complex cryptographic algorithms to transform some input information into a valid key. Solutions based on information from user profiles and identity [14, 15] have been reported. In addition, technologies supported by biometric information [16] may also be found and promise keys for high-security applications. The main disadvantage of all these schemes is their excessive computational cost, which typically cannot be managed by distributed computing nodes.

As a solution, hardware-supported technologies were proposed. Cryptographic algorithms integrated into reconfigurable hardware (FPGA) [18], or dedicated circuits [19] were reported. But all of them show a poor performance regarding resistance against most usual cyberattacks and, mainly, new cyber-physical attacks [20]. Then, in the last five years, Physical Unclonable Functions (PUF) [12] are the most promising approach. PUF are pieces of hardware with an unpredictable response when they are excited with a given challenge. They cannot be replicated and, if they are manipulated, their response change. Three different types of PUF have been reported [21]. The arbiter PUF employs the uncontrollable delays within silicon devices to produce binary random responses (usually thank to multiplexers) [22]. Memory-based PUF employs random phenomena in static RAM memories to create binary vectors [23]. Finally, optical PUF have been described but their complexity prevents their use is standard distributed computing nodes. But PUF cannot guarantee that responses are reproducible and that response entropy is not the maximum [25]. Numerical algorithms to mitigate these problems based on fuzzy logic [24] or pattern management [25] have been reported. But cannot fully solve the challenge.

In this paper we propose a different cryptographic scheme, so PUF are only employed as seed generators to initialize and configure more sophisticated (but lightweight) stream key generators being able to ensure a maximum entropy and reproducibility.

On the other hand, stream key generators are pseudo-random number generators (PRNG), designed to approach as much as possible to a truly random phenomenon. Quantum solutions have been reported [26], as well as technologies supported by low-level silicon circuits [32]. But they are complex and costly, so techniques based on logic gates are usually preferred [31]. Lagged Fibonacci generators are the most used [4], as they are very lightweight and, increasing the complexity of the interconnection among the gates, high-entropy sequences can be obtained. The main disadvantage is that these generators strongly depend on the initial configuration [4], and there is no clear strategy to find the one enabling sequences with expected entropy. Usually, instruments with erratic behaviors such as fractals [27] or discrete [28] or continuous chaos [29] are employed to improve the performance of PRNG, but, still, there is no guarantee the final keys have a high entropy. Besides, software models describing those erratic behaviors show a long-term deterministic behavior (so keys at long-term reduce their entropy), and hardware implementations cannot guarantee replicable results [30], as they are very dependent on the uncontrollable electric noise.

In this paper we address this problem using a federated key generation algorithm. Different configurations will be combined to optimize a target function based on the mutual information definition, so we can find the one ensuring the highest entropy and fully encryption of messages.

3   A Federated Algorithm for Key Generation

The proposed federated solution for key generation includes two different phases. Figure 1 shows a holistic representation for the proposed technology.

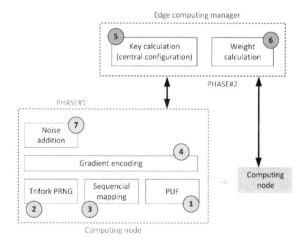

Fig. 1. Global architecture

As can be seen, the federated solution includes two subsystems. The local subsystem is integrated into each distributed computing node. It includes a PUF device ① and Lagged Fibonnaci PRNG (Trifork) ② [4] known for its potential good behavior (if a proper configuration is employed). PUF is employed to initialize the Trifok PRNG with a random (high entropy) configuration through a sequential mapping algorithm ③. But this configuration is not guaranteed to allow the PRNG to generate high-entropy keys. Encryption is performed using a Vernam cipher.

To solve this problem, all nodes encode the obtained PRNG's configuration using an irreversible gradient function ④. The results are sent to the second subsystem: an edge computing manager (ECM) which performs the second phase of the algorithm. This ECM will be typically deployed to cover a wide geographical area, but with a lower delay than the cloud server. This ECM receives all the encoded configurations and calculates a central configuration ⑤ by combing them. The central configuration is the one that makes minimum the average distance to all local encoded configurations. This average is weighted, so those configurations that enable keys with a higher entropy are strengthened. These weights are calculated ⑥ using the mean mutual information over a set of standard messages, so we guarantee that encrypted messages do not reveal any raw information.

Finally, the central configuration is sent back to the distributed computing nodes, which can use it in its original form, or include some additive noise ⑦ to avoid duplicated keys within the computing system.

Next subsections describe both subsystems with details.

3.1 Local Subsystem: Lightweight Cryptosystem and Gradient Encoding

The studied distributed computing system is composed of N different nodes. Each node n_i is provided with a cryptosystem composed of three basic elements: a hardware device for random number generation, a Lagged Fibonacci PRNG, and a Vernam cipher. Hereinafter we are considering raw messages msg in our system are composed of binary words with L bits.

The hardware device, usually a Physical Unclonable Function, has an unpredictable and random behavior. This behavior is represented by function $bv_i(\cdot)$. This function takes as input a set of p_i physical signals $s_j(t)$ and, as a result, generates a word W_i with a length of M_i bits (1).

$$W_i = bv_i\big(s_1(t), \ldots, s_{p_i}(t)\big) \tag{1}$$

As Lagged Fibonacci PRNG we propose the Trifork architecture [4]. This PRNG (see Fig. 2) is composed of three interconnected elemental Lagged Fibonacci PRNG (2). This PRNG is presented by function $Tf(\cdot)$. . This function takes as input eleven different values (3). d is a positive integer free parameter, as well as r_1, r_2, r_3, q_1, q_2 and q_3 parameters. Besides, m represents the discrete time instant. Moreover, three integer vectors x_{init}, y_{init} and z_{init} (4) must be also considered. These vectors are known as seeds or initialization vectors and are composed by numbers in the range $\big[0, 2^L - 1\big]$. As a result, a pseudo-random number flow $e[m]$ is obtained.

$$w[m] = LFG(j, k) = w\big[m - j\big] + w[m - k] \ mod \ L \tag{2}$$

$$e[m] = Tf\,(m, d, r_1, r_2, r_3, q_1, q_2, q_3, x_{init}, y_{init}, z_{init}). \tag{3}$$

$$
\begin{aligned}
x_{init} &= \left(x_{ini}^1, \ldots, x_{init}^{M_x}\right) \ M_x = max\{r_1, q_1\} \\
y_{init} &= \left(y_{ini}^1, \ldots, y_{init}^{M_y}\right) \ M_y = max\{r_2, q_2\} \\
z_{init} &= \left(z_{ini}^1, \ldots, z_{init}^{M_z}\right) \ M_z = max\{r_3, q_3\}
\end{aligned} \tag{4}
$$

The analytical description of function $Tf(\cdot)$ may be easily provided using L-modular arithmetic, the vector shift-right operation represented by symbol $\gg$, and the exclusive OR (or XOR) operation $\oplus$ (5). This notation does not allow an explicit representation of seed vectors, but they are necessary to fix the initial state of the PRNG.

$$x[m] = \left((x[m-r_1] + x[m-q_1]) \bmod L\right) \oplus z^{\gg}[m]$$
$$y[m] = \left((y[m-r_2] + y[m-q_2]) \bmod L\right) \oplus x^{\gg}[m]$$
$$z[m] = \left((z[m-r_3] + z[m-q_3]) \bmod L\right) \oplus y^{\gg}[m]$$

$$x^{\gg}[m] = \left((x[m-r_1] + x[m-q_1]) \bmod L\right) \gg d \qquad (5)$$
$$y^{\gg}[m] = \left((y[m-r_2] + y[m-q_2]) \bmod L\right) \gg d$$
$$z^{\gg}[m] = \left((z[m-r_3] + z[m-q_3]) \bmod L\right) \gg d$$

$$e[m] = x[m] \oplus z[m]$$

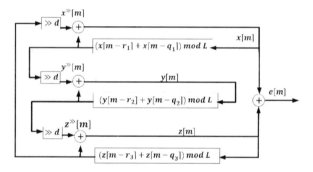

Fig. 2. Trifork internal structure

In the proposed cryptosystem, messages msg are encrypted using the Vernam cipher (also known as XOR cipher). In this approach, the encrypted messages $emsg$ are obtained through an XOR operation (6) where the pseudo-random number flow $e[n]$ coming from Trifork is used as secret key. The Vernam cipher is very strong, as it ensures a perfect encryption (i.e., the encrypted message and the raw message have a null mutual information $\mathfrak{I}$) if the key is with high entropy (ideally the maximum, i.e., $\mathfrak{E} = \frac{1}{2}$ so it is fully random), being $\mathfrak{E}$ the entropy operator as defined in the Shannon's theory and $P(\cdot)$ the probability operator (7).

$$emsg[m] = msg[m] \oplus e[m] \qquad (6)$$

$$\mathfrak{E}\{e[m]\} = \tfrac{1}{2} \Rightarrow \mathfrak{I}\{msg; emsg\} = 0$$
Being $\qquad\qquad\qquad\qquad\qquad\qquad\qquad\qquad\qquad (7)$
$$\mathfrak{E}\{e[m]\} = -\sum_m P(e[m]) \cdot log(P(e[m]))$$

But the characteristics of $e[n]$ flow depend on the input parameters in $Tf(\cdot)$ function. In general, previous works [4] have proved as these input parameters get a higher entropy, the resulting secret flow also increases its entropy. To ensure a good random configuration parameter set, we define a sequential mapping algorithm transforming the unpredictable random binary words W_i coming from the PUF into a valid PRNG random configuration. To do that, we serialize the PRNG's configuration parameter set, but m parameter is not included as we assume it'll start from the unit with each new configuration. The serialized configuration vector Q has a length of L_Q bits (8), which are obtained from M_W binary words coming from PUF (9). Those words are juxtaposed in the same order they are generated to create the configuration vector Q. If required, adjustments in the length of the final word may be done. Algorithm 1 describes the propose sequential mapping approach.

$$L_Q = (7 + M_x, M_y, M_z) \cdot L \tag{8}$$

$$M_W = \left| \frac{L_Q}{M_i} \right| \tag{9}$$

Algorithm 1. Sequential mapping algorithm node n_i

Input: Null
Output: Configuration vector Q
Create vector Q with L_Q binary positions
Create integer $L_{current}$ and initialize with value zero
while $L_{current} < L_Q$ **then**
   $W_i = bv_i(s_1(t_0), \dots, s_{p_i}(t_0))$
   if $L_{current} - L_Q \geq W_i$ **then**
      $Q \leftarrow \{Q, W_i\}$
   else
      $W_{cut} \leftarrow W_i(1 : L_{current} - L_Q)$
      $Q \leftarrow \{Q, W_{cut}\}$
   end if
   Update t_0 to current time instant
end while

However, the obtained configuration vector Q is not guaranteed that it is random or with high entropy. That will depend on the hardware and environment conditions we cannot control. To mitigate this problem a more complex analysis is needed, but it cannot be done locally. To enable the ECM a deeper processing, it must know the locally calculated configuration vectors. But this information is critical and secret, as it could be employed by an attacker to replicate the encryption system. To solve this challenge, the configuration vector Q is encoded using a gradient function $grad(\cdot)$. The resulting encoded vector θ_i is then sent to the ECM (10).

$$\theta_i = grag(Q) = \nabla Q \tag{10}$$

The gradient function, in this technology, is numerically obtained as progressive finite differences over the original configuration vector Q (11).

$$\theta_i(k) = \nabla Q(k) = Q(k+1) - Q(k) \tag{11}$$

3.2 Edge Computing Manager: Federated Key Calculation

The ECM receives a collection of N different encoded configuration vectors θ_i. Ideally, every configuration will have a similar entropy, but usually because of the hardware and environment conditions significant variations may be observed. To mitigate and correct this problem, the ECM combines all local configurations into a central configuration Q^* with the desired entropy.

This central configuration Q^* is the solution of an optimization problem (12), where the configuration Φ with the lowest weighted average distance to all encoded local configurations θ_i must be calculated. In this context the standard Euclidian distance is employed (13).

$$Q^* = \arg\min_{\Phi} \left(\sum_{i=1}^{N} \beta_i \cdot \|\nabla\Phi - \theta_i\| \right) \tag{12}$$

$$\|\nabla\Phi - \theta_i\| = \sqrt{(\nabla\Phi - \theta_i)^2} \tag{13}$$

Weights β_i must be obtained so configurations with the highest entropy are strengthened. But as vectors θ_i are encoded, the standard Shannon's definition for entropy in the information theory cannot be applied. Then we define a set $\mathcal{M}$ with M_{msg} messages msg_k (14). This dataset of raw messages must be created to represent in the best way the messages to be handled by the distributed computing system. Then, encrypted messages $emsg_k^{\Phi}$ obtained from the Vernam cipher and the Trifork PRNG initialized with configuration Φ can be calculated (15). Then, the mean mutual information $\mathfrak{T}$ between both datasets $\mathcal{M}$ and $\mathcal{E}$ (16), as defined in the Shannon's information theory can be obtained. Being ρ_m and ρ_e symbols in the raw and encrypted messages.

$$\mathcal{M} = \{msg_k \quad k = 1, \ldots, M_{msg}\} \tag{14}$$

$$emsg_k^{\Phi} = msg_k \oplus Tf(m, \Phi) \tag{15}$$

$$\mathfrak{T}(\mathcal{M}; \mathcal{E}) = \frac{1}{M_{msg}} \sum_{\forall msg_k \in \mathcal{M}} \sum_{\forall \rho_m \in msg_k} \sum_{\forall \rho_e \in emsg_k^{\Phi}} P(\rho_m, \rho_e) \cdot log\left(\frac{P(\rho_m|\rho_e)}{P(\rho_e)}\right) \tag{16}$$

Ideally, this mutual information should be null. Thus, configurations Φ so the mean mutual information $\mathfrak{T}$ is close to zero must be strengthened, as the keys associated to that configurations have a higher entropy. And then, weights β_i and mutual information $\mathfrak{T}$ are inversely proportional. But weights are associated to encoded configurations θ_i, not to the unknown optimization variable Φ. To solve this problem, weights β_i are defined as functions (17) so their final value depend on how close the encoded configurations θ_i are to the current value of the unknown optimization variable Φ.

In these functions, parameters τ_1 and τ_2 are real positive control parameters. As they get bigger, changes in the mean mutual information have a less significant impact on the weights β_i. And parameters h_1 and h_2 are real positive thresholds that can be freely chosen within the interval $[0, 1]$ according to the security requirements of our application scenario. As they get smaller, keys with a higher entropy are required. Being θ_{max} the maximum distance between two L_Q-bit vectors, organized in word with length L bits.

$$\beta_i = \begin{cases} \frac{1}{N}e^{\frac{\tau_1}{\mathfrak{I}(\mathcal{M};\mathcal{E})}} & if \quad \frac{\|\nabla\Phi-\theta_i\|}{\theta_{max}} < h_1 \\ \frac{1}{N}e^{-\frac{\mathfrak{I}(\mathcal{M};\mathcal{E})}{\tau_2}} & if \quad \frac{\|\nabla\Phi-\theta_i\|}{\theta_{max}} > h_2 \\ \frac{1}{N} & otherwise \end{cases} \tag{17}$$

$$\theta_{max} = \sqrt{L_Q} \cdot \left(2^L - 1\right) \tag{18}$$

The resulting central configuration Q^* is sent back to the distributed computing nodes. In this case, as the ECM has a great computational power, complex encryption mechanisms such as elliptic curves may be employed to protect this information against unauthorized accesses.

The distributed nodes, then, may employ the central configuration Q^* as received from the ECM. Then, all remote nodes have the same key. This is necessary in scenarios with mesh networks where all nodes must share a common key. But, if the system is structured with a start topology and it is preferred every node to have a different key, that may be achieved by adding to the central configuration Q^* a Gaussian additive noise (19). The probability distribution $\mathcal{N}(r)$ for this noise has a null mean and a very narrow standard deviation σ (at least one magnitude order below the parameter it is going to be added) and it is generated using software tools to ensure it is controllable with precision.

$$\mathcal{N}(r) = \frac{1}{\sigma\sqrt{2\pi}}exp\left(-\frac{r^2}{2\sigma^2}\right) \tag{19}$$

4   Experimental Validation: Simulation and Results

In order to analyze the performance of the proposed federated solution, an experimental validation was designed and conducted. Experiments were based on simulation tools, representing a typical distributed computing system. Simulation tools will store all the keys generated by remote nodes, so an offline scientific processing is enabled.

The proposed simulation scenario was supported by MATLAB 2017a software. All simulations were performed using a Linux architecture (Linux 16.04 LTS) with the following hardware characteristics: Dell R540 Rack 2U, 96 GB RAM, two processors Intel Xeon Silver 4114 2.2G, HD 2TB SATA 7,2K rpm. The simulation scenario represented a metropolitan distributed computing system deployed around an area of six hundred square kilometers, where a variable number of computing nodes are deployed. The number of nodes within the system was the independent variable for our experiment. Parameters h_1 and h_2 were fixed for all experiments to values 0.1 and 0.9 respectively. Parameters τ_1 and τ_2 had the same value, which was variable. PUF in this scenario were simulated using the proposed models for arbiter PUF [22]. The key generation process was triggered at any random time instant. The keys generated using the proposed federated algorithm and the original local keys were collected and, later, the mean entropy for both sets is obtained and compared. The relative difference between both entropies is finally calculated as the global experiment's result. Every simulation scenario represented twenty-four hours of network operations. Besides, in order to remove any exogenous effect, all simulations were repeated twelve times and final results were obtained as the average value.

Figure 3 shows the results of the proposed experiment. As can be seen, as more nodes participate in the federated algorithm, the improvement is more significant. The evolution, besides, is exponential. Actually, as more nodes participate in the key calculation process, it is more feasible to mitigate and compensate for inefficiencies due to hardware or environmental conditions. Additionally, as values for τ_1 and τ_2 parameters higher, improvement is slower, but more stable. While for small values of τ_1 and τ_2 parameters the evolution is much faster, but random variations appear because the proposed algorithm is much more sensitive. Anyway, an improvement of up to 23% can be clearly seen. This improvement reduces up to 15% in scenarios for high values in the τ_1 and τ_2 parameters. Both results are significant.

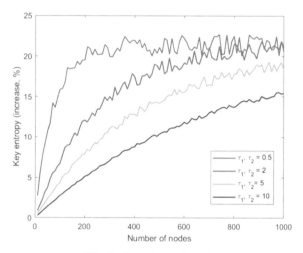

Fig. 3. Experimental results

5 Conclusions and Future Works

In this paper we propose a federated algorithm to enable distributed computing systems to generate high-entropy keys. Remote nodes are provided with different physical devices to initialize with a random configuration a Fibonacci random number generator. The parameter set describing the configuration of the key generator is locally encoded using a gradient function and sent to an edge computing manager where different encoded configurations coming from different remote nodes are collected. The edge computing manager combines all these configurations considering different weights and an optimization target function based on the definition of mutual information. The resulting configuration is sent back to the remote nodes. Experimental results show that the long-term average entropy increases by up to 23% when using the proposed solution.

Future works will consider a new experimental validation based on real hardware devices.

Acknowledgments. The research leading to these results has received funding from the Ministry of Science, Innovation and Universities through the COGNOS project. (PID2019-105484RB-I00).

References

1. Bordel, B., Alcarria, R., Morales, A., Castillo, I.: A framework for enhancing mobile workflow execution through injection of flexible security controls. Analog Integr. Circ. Sig. Process **96**(2), 303–316 (2018)
2. Robles, T., Bordel, B., Alcarria, R., Sánchez-de-Rivera, D.: Enabling trustworthy personal data protection in eHealth and well-being services through privacy-by-design. Int. J. Distrib. Sens. Netw. **16**(5), 1550147720912110 (2020)
3. Bordel, B., Alcarria, R., Robles, T., Sánchez-Picot, Á.: Stochastic and information theory techniques to reduce large datasets and detect cyberattacks in ambient intelligence environments. IEEE Access **6**, 34896–34910 (2018)
4. Bordel, B., Orúe, A.B., Alcarria, R., Sánchez-De-Rivera, D.: An intra-slice security solution for emerging 5G networks based on pseudo-random number generators. IEEE Access **6**, 16149–16164 (2018)
5. Lara-Nino, C.A., Diaz-Perez, A., Morales-Sandoval, M.: Elliptic curve lightweight cryptography: a survey. IEEE Access **6**, 72514–72550 (2018)
6. Meng, Z., Wang, Y.: Asymmetric encryption algorithms: primitives and applications. In: 2022 IEEE 2nd International Conference on Electronic Technology, Communication and Information (ICETCI), pp. 876–881. IEEE, May 2022
7. Hatzivasilis, G., Fysarakis, K., Papaefstathiou, I., Manifavas, C.: A review of lightweight block ciphers. J. Cryptogr. Eng. **8**(2), 141–184 (2018)
8. Bordel, B., Alcarria, R., Robles, T.: Lightweight encryption for short-range wireless biometric authentication systems in Industry 4.0. Integr. Comput.-Aided Eng., 1–21 (2022, Preprint)
9. Es-Sabry, M., El Akkad, N., Merras, M., Saaidi, A., Satori, K.: A new image encryption algorithm using random numbers generation of two matrices and bit-shift operators. Soft. Comput. **24**(5), 3829–3848 (2020)
10. Bordel, B., Alcarria, R., De Andrés, D.M., You, I.: Securing Internet-of-Things systems through implicit and explicit reputation models. IEEE Access **6**, 47472–47488 (2018)
11. Yu, F., Li, L., Tang, Q., Cai, S., Song, Y., Xu, Q.: A survey on true random number generators based on chaos. Discrete Dyn. Nat. Soc. (2019)
12. Pérez-Jiménez, M., Sánchez, B.B., Migliorini, A., Alcarria, R.: Protecting private communications in cyber-physical systems through physical unclonable functions. Electronics **8**(4), 390 (2019)
13. Rontani, D., Choi, D., Chang, C.Y., Locquet, A., Citrin, D.S.: Compressive sensing with optical chaos. Sci. Rep. **6**(1), 1–7 (2016)
14. Kate, A., Goldberg, I.: Distributed private-key generators for identity-based cryptography. In: Garay, J.A., De Prisco, R. (eds.) Security and Cryptography for Networks. LNCS, vol. 6280, pp. 436–453. Springer, Heidelberg (2010). https://doi.org/10.1007/978-3-642-15317-4_27
15. Wariki, K., et al.: Malicious Private Key Generators in Identity-Based Authenticated Key Exchange (2022)
16. Wu, Y., Qiu, B.: Transforming a pattern identifier into biometric key generators. In: 2010 IEEE International Conference on Multimedia and Expo, pp. 78–82. IEEE, July 2010

17. Mohammed, N.Q., Hussein, Q.M., Sana, A.M., Khalil, L.A.: A hybrid approach to design key generator of cryptosystem. J. Comput. Theor. Nanosci. **16**(3), 971–977 (2019)
18. Kumar, T.M., Reddy, K.S., Rinaldi, S., Parameshachari, B.D., Arunachalam, K.: A low area high speed FPGA implementation of AES architecture for cryptography application. Electronics **10**(16), 2023 (2021)
19. Goswami, J., Paul, M.: Symmetric key cryptography using digital circuit based on one right shift. Rev. Comput. Eng. Stud. **4**(2), 57–61 (2017)
20. Bordel, B., Alcarria, R., Sánchez-de-Rivera, D., Robles, T.: Protecting industry 4.0 systems against the malicious effects of cyber-physical attacks. In: Ochoa, S.F., Singh, P., Bravo, J. (eds.) Ubiquitous computing and ambient intelligence. LNCS, vol. 10586, pp. 161–171. Springer, Cham (2017). https://doi.org/10.1007/978-3-319-67585-5_17
21. Gao, Y., Al-Sarawi, S.F., Abbott, D.: Physical unclonable functions. Nat. Electron. **3**(2), 81–91 (2020)
22. Zerrouki, F., Ouchani, S., Bouarfa, H.: A low-cost authentication protocol using Arbiter-PUF. In: Attiogbé, C., Ben Yahia, S. (eds.) Model and Data Engineering. LNCS, vol. 12732, pp. 101–116. Springer, Cham (2021). https://doi.org/10.1007/978-3-030-78428-7_9
23. Chen, B., Ignatenko, T., Willems, F.M., Maes, R., van der Sluis, E., Selimis, G.: A robust SRAM-PUF key generation scheme based on polar codes. In: GLOBECOM 2017–2017 IEEE Global Communications Conference, pp. 1–6. IEEE, December 2017
24. Gao, Y., Su, Y., Yang, W., Chen, S., Nepal, S., Ranasinghe, D.C.: Building secure SRAM PUF key generators on resource constrained devices. In: 2019 IEEE International Conference on Pervasive Computing and Communications Workshops (PerCom Workshops), pp. 912–917. IEEE, March 2019
25. Delvaux, J., Verbauwhede, I.: Attacking PUF-based pattern matching key generators via helper data manipulation. In: Benaloh, J. (ed.) Topics in Cryptology – CT-RSA 2014. LNCS, vol. 8366, pp. 106–131. Springer, Cham (2014). https://doi.org/10.1007/978-3-319-04852-9_6
26. Abd EL-Latif, A.A., Abd-El-Atty, B., Venegas-Andraca, S.E.: Controlled alternate quantum walk-based pseudo-random number generator and its application to quantum color image encryption. Phys. A Stat. Mech. Appl. **547**, 123869 (2020)
27. Ayubi, P., Setayeshi, S., Rahmani, A.M.: Deterministic chaos game: a new fractal based pseudo-random number generator and its cryptographic application. J. Inf. Secur. Appl. **52**, 102472 (2020)
28. Sahari, M.L., Boukemara, I.: A pseudo-random numbers generator based on a novel 3D chaotic map with an application to color image encryption. Nonlinear Dyn. **94**(1), 723–744 (2018)
29. Cang, S., Kang, Z., Wang, Z.: Pseudo-random number generator based on a generalized conservative Sprott-A system. Nonlinear Dyn. **104**(1), 827–844 (2021)
30. Mareca, P., Bordel, B.: Robust hardware-supported chaotic cryptosystems for streaming commutations among reduced computing power nodes. Analog Integr. Circ. Sig. Process **98**(1), 11–26 (2019)
31. Karunamurthi, S., Natarajan, V.K.: VLSI implementation of reversible logic gates cryptography with LFSR key. Microprocess. Microsyst. **69**, 68–78 (2019)
32. Raffaelli, F., Sibson, P., Kennard, J.E., Mahler, D.H., Thompson, M.G., Matthews, J.C.: Generation of random numbers by measuring phase fluctuations from a laser diode with a silicon-on-insulator chip. Opt. Express **26**(16), 19730–19741 (2018)
33. Szczechowiak, P., Oliveira, L.B., Scott, M., Collier, M., Dahab, R.: NanoECC: testing the limits of elliptic curve cryptography in sensor networks. In: Verdone, R. (ed.) Wireless Sensor Networks. LNCS, vol. 4913, pp. 305–320. Springer, Heidelberg (2008). https://doi.org/10.1007/978-3-540-77690-1_19

Food Supply Chain Cyber Threats: A Scoping Review

Janne Alatalo[✉][iD], Tuomo Sipola[iD], and Tero Kokkonen[iD]

Institute of Information Technology, Jamk University of Applied Sciences,
Jyväskylä, Finland
{janne.alatalo,tuomo.sipola,tero.kokkonen}@jamk.fi

Abstract. Cyber attacks against the food supply chain could have serious effects on our society. As more networked systems control all aspects of the food supply chain, understanding these threats has become more critical. This research aims to gain a better understanding of the threat landscape by reviewing the existing literature about the topic. Previous research concerning food supply chain cyber threat was surveyed using the scoping review method. In total, 43 research articles focusing on different parts of the food supply chain were reviewed and summarized in this study. The most prominent identified cybersecurity topics include smart farming, cyber-physical systems, threats against industrial control systems and old unmaintained software.

Keywords: Food Supply Chain · Critical Infrastructure · Cybersecurity

1 Introduction

As an industry that affects the everyday life of everyone worldwide, the food supply chain is one of the most critical functions of a society. A.H. Lewis wrote already in 1896: *"the only barrier between us and anarchy is the last nine meals we've had"* [32]. Actors in the food supply vary from individual farms to logistic companies, food production companies and retail chains. Farm to Fork Strategy for a fair, healthy and environmentally-friendly food system, released by European Commission, states that *"The COVID-19 pandemic has underlined the importance of a robust and resilient food system that functions in all circumstances, and is capable of ensuring access to a sufficient supply of affordable food for citizens"* [19]. Systems and actors of the food supply can be valuable targets for a cyber attack because literally every human is dependent on food. Indeed, EU's Cybersecurity Strategy for the Digital Decade [20] and EU's directive on the resilience of critical entities [21] acknowledge the importance of farming, food production, processing and distribution. Similarly, a private industry notification from the FBI states that since 2021 ransomware attacks have impacted agricultural cooperatives and warns about probable ransomware attacks against agricultural cooperatives during the upcoming plant and harvest season [23].

A. Rocha et al. (Eds.): WorldCIST 2023, LNNS 801, pp. 94–104, 2024.
https://doi.org/10.1007/978-3-031-45648-0_10

There are several networked software and hardware components included in this system of systems [33], for example, automated robotic logistic systems, food production systems, refrigeration machines and milking robots. A cyber attack against one point of this crucial chain may cause cascade effects and deny the usage of the whole food supply chain [52]. Factors such as industrial espionage, criminal intent and hostile state activities could present motivations for various attacks. News outlets have already described attacks against advanced smart farming machinery [36], and, e.g., a ransomware attack against AGCO corporation [2]. Furthermore, Bowcut lists some notable cyber attacks against food and agriculture actors in 2020 and 2021 and introduces a case study of ransomware against the food company JBS in 2021 [11].

Related work in this research topic includes Latino and Menegoli, who have conducted a systematic literature review in the topic of "cybersecurity in the food and beverage industry" [31]. They used thematic analysis to analyze the results and built a reference framework for future research and for identifying the future research directions. They used Scopus knowledge database for finding the relevant articles with a final sample of 17 studies that were included in the final analysis. The small number of analyzed articles is probably the result of strict exclusion criteria that are dictated by the restrictions in the research methodology chosen by the authors. Although the topic of our paper is very similar to this paper, our scoping study research methodology is more flexible giving us the freedom to include many more studies into the final analysis. That way we bring more complete overview of the existing research about the topic and contribute new knowledge to the research area.

2   Methods

Our research question in this study is: "What literature exists about the food supply chain cybersecurity?" We wanted to get a high level overview of existing research about cybersecurity in food supply chain. For this purpose, the scoping review methodology was chosen [40]. We used the methodological framework defined by Arksey and O'Malley in [6] as the basis for our research. The framework is flexible enough to cast a wide enough net over the wide topic of food supply chain cybersecurity, and it allows us to use an iterative method to converge on the right keywords to identify the relevant studies.

The review started by searching all types of literature using general search terms encompassing the whole wide topic. The literature found this way was used to identify more keywords that are often used in specific parts of the food supply chain literature. The new keywords were added to the list of the searched keywords, and the process was iterated until the topic was sufficiently covered. In the end, the search phrases were constructed from the keywords by combining terms referring to the food supply chain: "food industry", "food production", "food production chain", "food logistics", "smart farming", "agribusiness", "food retail" and "catering industry", with a term that specifies the cybersecurity focus: "cybersecurity" and "cyber attacks".

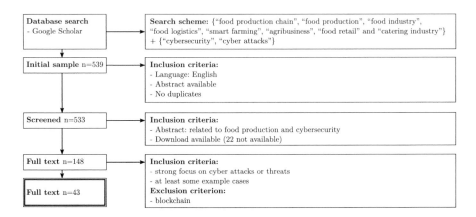

Fig. 1. Review protocol.

The Google Scholar service was used as the primary search engine. The search queries were constructed by joining the cybersecurity keyword to the food keyword using a space e.g. "smart farming cyber attacks". Special query characters such as quotes or logical operators like OR and AND were not used when constructing the search queries. The searches were carried out using the normal Google Scholar search user interface. From each query, the first 50 search results were taken under review. A program was developed for automating the search result extraction using the Playwright[1] browser automation framework. This was done to prevent human errors in the repetitive task of copying the search results from the browser to the Excel worksheet where the review of the search results was coordinated. All the search results were in English. No restriction was placed on the literature type: grey literature was also considered during the review process. Because of space restrictions, all papers about blockchains were deemed out of scope for this study. Figure 1 illustrates the different stages of the review protocol used during the scoping review.

3   Results

Smart farming cybersecurity is one of the main topics that was studied in multiple papers. Smart farming uses modern technologies such as the Internet of Things (IoT), artificial intelligence and robotics to increase crop yields, maximize production and streamline farming processes [51].

Chi et al. have defined a security framework for innovations that support smart farming in [12]. They identify that the data generated by smart farming sensors is a very valuable resource for all kinds of other purposes such as research of plants biology and genetics, forecasting the market and economics, and new farm equipment design. This makes the data a high value target for corporates, activists or even nation level adversaries to steal, sabotage or inject

---

[1] https://playwright.dev/.

misinformation. To mitigate these threats the paper defines the cybersecurity framework for smart farming to include three components: abnormal measurement detection, access control, and encryption.

Bogaardt et al. focused on dairy industry in their research report [9]. The study claims that 90% of new installations in dairy farms are robotic milking equipment, and by 2025 half of the cows in north-western Europe are milked by robots. The milking robots collect a large amount of data from the cows and the milk, and the normal everyday processes are starting to be dependent on this data. For that reason, it is important to protect the data against cyber threats. Additionally, the business management systems that some farmers use for food safety tracking reasons can be a tempting target for data theft. The report identifies unmaintained software and human errors as the main threats to these systems.

The attack types against the smart farming systems are well studied. Studies by Gupta et al. [27], Zanella et al. [42], Demestichas et al. [15], Koduru and Koduru [30], Yazdinejad et al. [54], Farooq et al. [22], Boghossian et al. [10], Okupa [39], Barreto and Amaral [7], Rosline et al. [43], Angyalos et al. [5], Akshatha and Poornima [53] and Racovita [41] try to identify the major cybersecurity threats in smart farming. Data security is one major security issue that is raised in some form by all the studies. Data theft is always possible in systems where data is collected and stored. Another threat is data forgery. Smart farms use the data collected by the sensors to make decisions. By injecting forged data to the system, a hacker can easily disrupt farm operations. The hacker can physically take control of a sensor and modify the hardware or software to transmit malicious data to the system, or in some cases the hacker can exploit the problems in authentication and authorization to inject the data remotely. In addition of these threats, the threat of autonomous vehicles and robots was identified in some of the studies. A hijacked autonomous tractor could cause real physical damage to the infrastructure and be even a life-threatening danger. More common attack types such as denial of service, phishing, RF jamming and malware attacks were also identified as threats to smart farms.

Some papers have a narrower scope. Cho et al. limited their study to the cyber threats in smart greenhouses [13]. Rouzbahani et al. studied the potential cyber attacks of smart farming communication technologies [44]. Linsner et al. tested the cybersecurity of wireless sensor networks using simulated attacks [35]. Alsinglawi et al. studied the cybersecurity threats and attack types of microservice based meat production smart farm. Alahmadi et al. researched side channel attacks in smart farming systems [3]. Nikander et al. summarized a case study that studied the cybersecurity of 6 dairy farms in Finland [38]. Dorairaju conducted a case study on the cybersecurity of an IoT enabled pest trap system that was targeted for agricultural use [16].

Studies of the existing publications about the topic have also been carried out. Nakhodchi et al. conducted a bibliometric analysis on the publications about the privacy and security in smart farming [37]. Rudrakar and Rughani have completed a systematic literature review about IoT based agriculture cybersecurity and forensics challenges [45].

Cyber-physical system cybersecurity in food production is another topic that is quite well researched. This topic includes studies about the cybersecurity of industrial control systems (ICS) that are used, for example, in food packaging plants.

Beluli has studied the possibility of cyber attack in the beer production industry [8]. The industry uses computer automation in the beer production process with high temperature and pressure tanks. Cyber attack against the process control systems could cause an explosion that damages the equipment, or even cause danger to human life.

Alim et al. studied the cybersecurity threats of a modeled canal SCADA system [4]. Water management systems are critical infrastructure for agriculture industry. They are used in crop irrigation and water processing. Attacks against these kinds of systems can cause major financial losses to the farmers. The authors tested multiple attack types against the model with successful results. One of the attack types was a message injection attack that caused flooding in the modeled farmland.

Freyhof studied the cybersecurity of agricultural machinery [24]. The focus of the study was to estimate the financial losses of a cyber attack against an electrically controlled variable rate nitrogen side-dressing equipment. The suggested attacks keep the cumulative quantity of the applied nitrogen the same but use different strategies to distribute the total nitrogen quantity so that actual application rates are different from the prescribed rates. This way the cyber attack can reduce the crop yields and thus cause financial losses.

Streng studied the cybersecurity of the ICSs used in food processing and manufacturing [50]. The study identifies common cybersecurity problems in these systems, such as old operating system versions, insecure protocols and old unmaintained software. The study concludes that cyber attacks targeting these cyber-physical systems are possible in food industry, and they can be even life threatening if they target equipment such as co-robots that work alongside people in production lines. Additionally, a cyber attack against the food production system can make the produced food somehow unsafe for consumption, which can cause danger to the consumers. This study can be seen as a continuum to a report written by the same author about an industry summit meeting in Washington USA [49]. In the meeting, the cyber-physical systems, such as the ICSs used in food packing plants, were identified a major security threat to the food industry.

Chundhoo et al. conducted a case study about the cybersecurity of a meat processing plant [14]. They identified serious threats in the meat processing system where the meat temperature must be closely monitored. IoT sensors monitor the meat temperature, and the thawing, chilling, freezing, cooking, and smoking rooms are controlled by the readings coming from these sensors. Spoofed sensor readings can change the actual temperature of the meat to a range that can make the product unsafe for consumption or even cause equipment damage further in the processing line, and that way contaminate the food with, for example, broken blades.

Survey and interview studies have also been conducted about the topic. All these studies targeted the farmers. One of the studies was written by Geil, who surveyed the cybersecurity awareness of the people working in the agriculture business ($N = 138$) [25]. The survey responders were farmers, producers and other workers from the industry in three different counties in Illinois USA. The study concludes that there are gaps in cybersecurity knowledge among the people working in the agriculture industry, and for that reason there is a need for more cybersecurity training for the industry workers. Also, see Geil et al. [26].

A similar survey was conducted by Spaulding and Wolf [48] ($N = 222$). They surveyed the farmers in Illinois USA, but targeted the survey towards beginning farmers. They conclude that even though the beginning farmers use computers more than experienced farmers, they still lack the skills to identify the cyber threats against their farming business accompanied with the computer usage.

Russell studied the cybersecurity risks at smart farms by interviewing farmers in Ontario Canada with supplemental interviews with five cybersecurity experts [46]. The paper includes a very detailed analysis of the interviews with some examples of phishing cyber attacks that the farmers have already experienced, and examples of attacks that the farmers see realistic and that could take place against their farm.

Linden et al. conducted a case study in the cybersecurity of dairy farming in Israel [34]. The authors interviewed a farmer about the usage of smart farming and threats that it poses. The authors discovered that in Israel it is common to share data between the farming community and the researchers, and for that reason the farmer did not identify data theft as a major threat. Additionally, losing the data was not seen as a high threat as the relevant data could be obtained from colleagues. The highest threat that the interviewed farmer identified was the injection of fraudulent data, or cases where data was otherwise inaccurate, because that directly impacts the productivity and welfare of the animals. The authors claim that the openness of sharing the data is not as common elsewhere, like in the UK, where leaking of the farm data is identified as a major threat. For that reason, the authors claim that the socio-cultural context matters when the cybersecurity of smart farming is considered.

The general state of the cybersecurity in food industry was also studied in many papers. Ajith et al. studied cyberespionage and cyberterrorism in the food industry [1]. They discuss the motives and review some of the existing research about the topic.

Hoffmann et al. studied cyber attacks against agribusiness industry [28]. They searched English news articles from the internet and found 31 reports of attacks against different parts of the food production infrastructure.

Jahn et al. conducted a high-level overview of cyber risks in north American food industry [29]. They identified some of the same cyber threats discussed in the previous sections caused by increased automation in farming and food processing processes. They also discuss how the food industry works by using the just-in-time principle. Farmers rely on the delivery of the fertilizer, fuel, seeds etc. in time when needed, and the stores cannot keep large stocks of perishable food products, hence they also rely on the timely deliveries of the products. This can make the whole food delivery chain especially vulnerable to cyber attacks.

Russell and Chow discussed food cybersecurity in general in their paper [47]. They present examples of possible attacks against home smart refrigerators and processing plant irradiation machines used to sterilize food products. Both cases are potential health hazards, as they can lead to food poisoning.

Duncan et al. also presented a high-level view of cybersecurity in food and agriculture industry [17]. The paper includes some concrete examples of possible attacks, such as attacks against genetic databanks that the breeders use to develop more productive dairy cows and other food animals. Some of the same authors contributed to the second paper where they considered these attacks more closely and suggested possible mitigations, such as cryptographic signatures for the data [18].

4   Conclusion

There is a large number of studies about the food supply chain cybersecurity. The threat of cyber attacks against food industry has been well identified by researchers. Especially the new cyber threats that come with the increasing popularity of smart farming are well studied. Many of the reviewed papers identified industrial control systems as a major security threat. Furthermore, old unmaintained software is vulnerable to attacks and is expensive to update. This is not unique to the food industry; in every industry where automation is used, there are also legacy systems that are vulnerable to attacks. Food industry is unique in the sense that these vulnerabilities can easily threaten human health and life. This was a well identified threat, but not many concrete examples exist in the literature. This could be an area where more research is needed. In addition, future research could study the topics excluded from this work, such as the use of blockchain technologies in food industry and the cybersecurity threats that they cause.

Acknowledgements. This research is funded by the Regional Council of Central Finland/Council of Tampere Region with fund of Leverage from the EU, European Regional Development Fund (ERDF), Recovery Assistance for Cohesion and the Territories of Europe (REACT-EU). Research is implemented as part of the Food Chain Cyber Resilience project of Jamk University of Applied Sciences Institute of Information Technology.

The authors would like to thank Ms. Elina Suni for identifying some relevant sources and Ms. Tuula Kotikoski for proofreading the manuscript.

References

1. Adetunji, C.O., et al.: Cyberespionage: socioeconomic implications on sustainable food security. In: Abraham, A., Dash, S., Rodrigues, J.J., Acharya, B., Pani, S.K. (eds.) AI, Edge and IoT-based Smart Agriculture, Intelligent Data-Centric Systems, pp. 477–486. Academic Press (2022). https://doi.org/10.1016/B978-0-12-823694-9.00011-6

2. AGCO Corporation: AGCO announces ransomware attack (2022). https://news.agcocorp.com/news/agco-announces-ransomware-attack
3. Alahmadi, A.N., Rehman, S.U., Alhazmi, H.S., Glynn, D.G., Shoaib, H., Solé, P.: Cyber-security threats and side-channel attacks for digital agriculture. Sensors **22**(9), 3520 (2022). https://doi.org/10.3390/s22093520
4. Alim, M.E., Wright, S.R., Morris, T.H.: A laboratory-scale canal scada system testbed for cybersecurity research. In: 2021 Third IEEE International Conference on Trust, Privacy and Security in Intelligent Systems and Applications (TPS-ISA), pp. 348–354 (2021). https://doi.org/10.1109/TPSISA52974.2021.00038
5. Angyalos, Z., Botos, S., Szilagyi, R.: The importance of cybersecurity in modern agriculture. J. Agric. Inform. **12**(2), 1–8 (2022). https://doi.org/10.17700/jai.2021.12.2.604
6. Arksey, H., O'Malley, L.: Scoping studies: towards a methodological framework. Int. J. Soc. Res. Methodol. **8**(1), 19–32 (2005). https://doi.org/10.1080/1364557032000119616
7. Barreto, L., Amaral, A.: Smart farming: cyber security challenges. In: 2018 International Conference on Intelligent Systems (IS), pp. 870–876 (2018). https://doi.org/10.1109/IS.2018.8710531
8. Beluli, V.M.: Smart beer production as a possibility for cyber-attack within the industrial process in automatic control. Procedia Comput. Sci. **158**, 206–213 (2019). https://doi.org/10.1016/j.procs.2019.09.043
9. Bogaardt, M.J., Poppe, K.J., Viool, V., Zuidam, E.V.: Cybersecurity in the agri-food sector. Technical report (2016). https://edepot.wur.nl/378724
10. Boghossian, A., et al.: Threats to precision agriculture (2018 public-private analytic exchange program report). Technical report, U.S. Department of Homeland Security (2018)
11. Bowcut, S.: Cybersecurity in the food and agriculture industry (2021). https://cybersecurityguide.org/industries/food-and-agriculture/
12. Chi, H., Welch, S., Vasserman, E., Kalaimannan, E.: A framework of cybersecurity approaches in precision agriculture (2017)
13. Cho, S.H., et al.: A study on threat modeling in smart greenhouses. J. Inf. Secur. Cybercrimes Res. **3**(1), 1–12 (2020). https://doi.org/10.26735/KKJN1042
14. Chundhoo, V., Chattopadhyay, G., Karmakar, G., Appuhamillage, G.K.: Cybersecurity risks in meat processing plant and impacts on total productive maintenance. In: 2021 International Conference on Maintenance and Intelligent Asset Management (ICMIAM), pp. 1–5 (2021). https://doi.org/10.1109/ICMIAM54662.2021.9715193
15. Demestichas, K., Peppes, N., Alexakis, T.: Survey on security threats in agricultural IoT and smart farming. Sensors **20**(22), 6458 (2020). https://doi.org/10.3390/s20226458
16. Dorairaju, G.: Cyber security in modern agriculture. case study: IoT-based insect pest trap system. Master's thesis, JAMK University of Applied Sciences (2021). https://urn.fi/URN:NBN:fi:amk-202105128397
17. Duncan, S.E., et al.: Cyberbiosecurity: a new perspective on protecting U.S. food and agricultural system. Front. Bioeng. Biotechnol. **7**, 63 (2019). https://doi.org/10.3389/fbioe.2019.00063
18. Duncan, S.E., et al.: Securing data in life sciences-a plant food (edamame) systems case study. Front. Sustain. **1**, 10 (2020). https://doi.org/10.3389/frsus.2020.600394

19. European Commission: Communication From the Commission to the European Parliament, the Council, the European Economic and Social Committee and the Committee of the Regions. A Farm to Fork Strategy for a fair, healthy and environmentally-friendly food system (2020). https://eur-lex.europa.eu/legal-content/EN/TXT/?uri=CELEX:52020DC0381
20. European Commission: Joint Communication to the European Parliament and the Council. The EU's Cybersecurity Strategy for the Digital Decade (2020). https://eur-lex.europa.eu/legal-content/ga/TXT/?uri=CELEX:52020JC0018
21. European Commission: Proposal for a Directive of the European Parliament and of the Council on the resilience of critical entities (2020). https://eur-lex.europa.eu/legal-content/EN/TXT/?uri=CELEX%3A52020PC0829
22. Farooq, M.S., Riaz, S., Abid, A., Abid, K., Naeem, M.A.: A survey on the role of IoT in agriculture for the implementation of smart farming. IEEE Access **7**, 156237–156271 (2019). https://doi.org/10.1109/ACCESS.2019.2949703
23. FBI: Ransomware attacks on agricultural cooperatives potentially timed to critical seasons (2022). https://www.ic3.gov/Media/News/2022/220420-2.pdf
24. Freyhof, M.T.: Cybersecurity of agricultural machinery: exploring cybersecurity risks and solutions for secure agricultural machines. Master's thesis, Department of Biological Systems Engineering, University of Nebraska-Lincoln (2022)
25. Geil, A.: Cyber security on the farm: an assessment of cyber security practices in the agriculture industry. Master's thesis, Illinois State University, School of Information Technology (2014). https://doi.org/10.30707/ETD2014.Geil.A
26. Geil, A., Sagers, G., Spaulding, A., Wolf, J.: Cyber security on the farm: an assessment of cyber security practices in the united states agriculture industry. Int. Food Agribus. Manag. Rev. **21**, 1–18 (2018). https://doi.org/10.22434/IFAMR2017.0045
27. Gupta, M., Abdelsalam, M., Khorsandroo, S., Mittal, S.: Security and privacy in smart farming: challenges and opportunities. IEEE Access **8**, 34564–34584 (2020). https://doi.org/10.1109/ACCESS.2020.2975142
28. Hoffmann, C., Haas, R., Bhimrajka, N., Penjarla, N.S.: Cyberattacks in agribusiness. In: Gandorfer, M., Hoffmann, C., El Benni, N., Cockburn, M., Anken, T., Floto, H. (eds.) 42. GIL-Jahrestagung, Künstliche Intelligenz in der Agrar- und Ernährungswirtschaft, pp. 117–122. Gesellschaft für Informatik e.V., Bonn (2022)
29. Jahn, M., et al.: Appendix: cyber risks in North American agriculture and food systems. Global Assessment Report on Disaster Risk Reduction (2019). Appendix of the article: Cybersecurity and its cascading effect on societal systems
30. Koduru, T., Koduru, N.P.: An overview of vulnerabilities in smart farming systems. J. Stud. Res. **11**(1) (2022). https://doi.org/10.47611/jsrhs.v11i1.2303
31. Latino, M.E., Menegoli, M.: Cybersecurity in the food and beverage industry: a reference framework. Comput. Ind. **141**, 103702 (2022). https://doi.org/10.1016/j.compind.2022.103702
32. Lewis, A.H.: Further facts: in the case of the labor record of Mark Hanna, the republican party's manager. The Owensboro Messenger, p. 2 (1896)
33. Lezoche, M., Hernandez, J.E., Díaz, M.D.M.E.A., Panetto, H., Kacprzyk, J.: Agri-food 4.0: a survey of the supply chains and technologies for the future agriculture. Comput. Ind. **117**, 103187 (2020)
34. van der Linden, D., Michalec, O.A., Zamansky, A.: Cybersecurity for smart farming: socio-cultural context matters. IEEE Technol. Soc. Mag. **39**(4), 28–35 (2020). https://doi.org/10.1109/MTS.2020.3031844
35. Linsner, S., Varma, R., Reuter, C.: Vulnerability assessment in the smart farming infrastructure through cyberattacks, pp. 119–124. Wien (2019). http://tubiblio.ulb.tu-darmstadt.de/116032/

36. Marshall, C., Prior, M.: Cyber security: global food supply chain at risk from malicious hackers. BBC (2022). https://www.bbc.com/news/science-environment-61336659

37. Nakhodchi, S., Dehghantanha, A., Karimipour, H.: Privacy and security in smart and precision farming: a bibliometric analysis. In: Choo, K.K., Dehghantanha, A. (eds.) Handbook of Big Data Privacy, pp. 305–318. Springer, Cham (2020). https://doi.org/10.1007/978-3-030-38557-6_14

38. Nikander, J., Manninen, O., Laajalahti, M.: Requirements for cybersecurity in agricultural communication networks. Comput. Electron. Agric. **179**, 105776 (2020). https://doi.org/10.1016/j.compag.2020.105776

39. Okupa, H.: Cybersecurity and the future of agri-food industries. Master's thesis, Kansas State University, Department of Agricultural Economics (2020). https://hdl.handle.net/2097/40529

40. Peters, M.D., Godfrey, C.M., Khalil, H., McInerney, P., Parker, D., Soares, C.B.: Guidance for conducting systematic scoping reviews. JBI Evid. Implement. **13**(3), 141–146 (2015). https://journals.lww.com/ijebh/Fulltext/2015/09000/Guidance

41. Racovita, M.: Cybersecurity for the internet of things and artificial intelligence in the agritech sector. Industry briefing, PETRAS National Centre of Excellence for IoT Systems Cybersecurity, London, UK (2021)

42. Rettore de Araujo Zanella, A., da Silva, E., Pessoa Albini, L.C.: Security challenges to smart agriculture: current state, key issues, and future directions. Array **8**, 100048 (2020). https://doi.org/10.1016/j.array.2020.100048

43. Rosline, G.J., Rani, P., Gnana Rajesh, D.: Comprehensive analysis on security threats prevalent in IoT-based smart farming systems. In: Karuppusamy, P., Perikos, I., García Márquez, F.P. (eds.) Ubiquitous Intelligent Systems, pp. 185–194. Springer, Singapore (2022). https://doi.org/10.1007/978-981-16-3675-2_13

44. Rouzbahani, H.M., et al.: Communication layer security in smart farming: a survey on wireless technologies (2022). https://doi.org/10.48550/ARXIV.2203.06013

45. Rudrakar, S., Rughani, P.: IoT based agriculture (IoTA): architecture, cyber attack, cyber crime and digital forensics challenges (2022). https://doi.org/10.21203/rs.3.rs-2042812/v1

46. Russell, C.: Cyber security in digital agriculture: investigating farmer perceptions, preferences, & expert knowledge. Master's thesis, The University of Guelph, Department of Geography, Environment and Geomatics (2022). https://hdl.handle.net/10214/27219

47. Russell, N., Chow, M.: Cybersecurity and our food systems. Technical report (2017)

48. Spaulding, A.D., Wolf, J.R.: Cyber-security knowledge and training needs of beginning farmers in Illinois. 2018 Annual Meeting, August 5–7, Washington, D.C. 273781, Agricultural and Applied Economics Association (2018). https://EconPapers.repec.org/RePEc:ags:aaea18:273781

49. Streng, S.: Food industry cybersecurity summit meeting report (2016). Retrieved from the University of Minnesota Digital Conservancy. https://hdl.handle.net/11299/217704

50. Streng, S.: Adulterating more than food: The cyber risk to food processing and manufacturing (2019). Retrieved from the University of Minnesota Digital Conservancy. https://hdl.handle.net/11299/217703

51. Sundmaeker, H., Verdouw, C., Wolfert, S., Pérez Freire, L.: Digitising the Industry - Internet of Things Connecting the Physical, Digital and Virtual Worlds, chap. Internet of food and farm, pp. 129–152. River Publishers Series in Communications and Networking. River Publishers (2016). https://doi.org/10.13052/rp-9788793379824

52. Urciuoli, L., Männistö, T., Hintsa, J., Khan, T.: Supply chain cyber security-potential threats. Inf. Secur. Int. J. **29**(1) (2013)

53. Akshatha, Y., Poornima, A.S.: IoT enabled smart farming: a review. In: 2022 6th International Conference on Intelligent Computing and Control Systems (ICICCS), pp. 431–436 (2022). https://doi.org/10.1109/ICICCS53718.2022.9788149

54. Yazdinejad, A., et al.: A review on security of smart farming and precision agriculture: security aspects, attacks, threats and countermeasures. Appl. Sci. **11**(16), 7518 (2021). https://doi.org/10.3390/app11167518

Review of Platforms and Frameworks for Building Virtual Assistants

Rodrigo Pereira[1](✉) ⓘ, Claudio Lima[1] ⓘ, Arsénio Reis[1,2] ⓘ, Tiago Pinto[1,2] ⓘ, and João Barroso[1,2] ⓘ

[1] Universidade de Trás-os-Montes e Alto Douro, Vila Real, Portugal
al68798@alunos.utad.pt, {claudiolima,ars,tiagopinto,
jbarroso}@utad.pt
[2] NESC-TEC, Vila Real, Portugal

Abstract. Virtual assistants offer a new type of solution to handle interaction between human and machine and can be applied in various business contexts such as Industry or Education. When designing and building a virtual assistant the developers must ensure a set of parameters to achieve a good solution. Various platforms and frameworks emerged to allow developers to create virtual assistant solutions easier and faster. This paper provides a review of available platforms and frameworks used by authors to create their own solutions in different areas. Big tech companies like Google with Dialogflow, IBM with Watson Assistant and Microsoft with Bot Framework, present mature solutions to build virtual assistants that provide to the developer all components of the basic architecture to build a fast and solid solution. Open-Source solutions focus on providing to the developer the main components to build a virtual assistant, namely language understanding and response generation.

Keywords: Conversational Agents · Development Frameworks · Review · Virtual Assistants

1 Introduction

In recent years there has been an increase in popularity of intelligent virtual assistants controlled by voice and/or text like Siri, Alexa, or Google Assistant. These virtual assistants are available to use in several devices such as smartphones, smart watches, smart home appliances and even cars with the aim of providing the user with all kinds of information that can be customized based on user's context (e.g., location) [1].

A virtual assistant is a conversational agent that can execute actions based on instructions or queries, which can organize natural conversations in a set of intents that can be triggered by a human or by a system and each of these intents realize a particular task [2]. Virtual assistants are evolving very rapidly to provide more capabilities through evolution of speech recognition and natural language processing algorithms that allows a viable alternative to the traditional methods of interaction (e.g., keyboard, mouse, touch display) [3]. The virtual assistants are generally used by humans in home context, but this technology can be applied to other business areas such as education [4], marketing [5], business-customer relationship [6] or industry [7, 8].

To build a virtual assistant several platforms and frameworks have emerged, these tools generally provide the user with the basic functionalities organized in building blocks. These building blocks are simple packages of functionalities designed to meet the business needs that can be used independently or can be integrated by connectors to work together with other blocks [9].

When developing a virtual assistant, it is necessary to have in account some key features to build a solution that enables to deliver to the user a human like conversation. The architecture for development can be divided into three components: user interface (UI), the application core and the external data sources and services (Fig. 1).

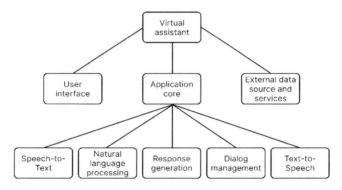

Fig. 1. Virtual Assistant architecture based on [10] and [11]

The UI is the main means of contact between the virtual assistant and the user, so it is important for the UI to suit the user target domain. If a platform can provide different UI modules, this should be done with little or no impact to the rest of the application. This makes it easy to interact to the application through various interfaces (e.g., Website, Messaging platforms, Mobile Applications).

The application core should contain all the logic of the assistant and it generally is implemented with a combination of various services. These services aim to complement the basic dialog, improve the user experience, and allow the conversation between user and assistant as human as possible. The application core should contain the following components [11]:

- **Speech-to-Text**: convert user speech to text.
- **Natural Language Processing**: responsible to transform unstructured text in structured representation of the text (Extract Entities and Intents).
- **Response Generation**: central component of a virtual assistant architecture which receive as input the structured representation of the spoken text, and as output the response generator generates a response to deliver to the user.
- **Dialog Management**: is responsible to maintain the conversation flow between a person and an assistant. Is also responsible to handle unexpected or unrelated cases.
- **Text-to-Speech**: convert text into speech.

Every virtual assistant application must access specific data sources or services. These sources can provide information that are not available at design time, or to provide the user with dynamic content, and to improve user engagement. Generally, the application core can connect to external sources via REST web services calls [10].

Plenty of platforms and frameworks have emerged to build virtual assistants, from Natural Language Processing services helping the decoding of user's conversations to easy-to-use low-code development platforms that cover all steps to build a virtual assistant [12]. In this paper we present a review and analysis of platforms and frameworks to build custom virtual assistants to help anyone to build their virtual assistant solution applied to any user or business context.

2 Methodology

To pursuit our goal, a query was constructed to do the search: ("virtual assistant") OR ("Artificial Intelligent Virtual Assistants") AND ("development") AND ("platform") OR ("framework"). The peer reviewed papers have been consulted in the Google Scholar database to find platforms and frameworks that were used by the authors to create their virtual assistants' solutions. Additionally, a Google search using the same search query has been performed, in order to include another platforms or frameworks not found in the previous search. Exclusion criteria were set for discarding all platforms or frameworks that do not have enough documentation to do a proper analysis or there are not available anymore, all the tools that are not dedicated to build new and customizable solutions with low expandability to new contexts, and tools that can publish virtual assistants only in company's ecosystem.

In order to perform a suitable examination, the gathered platforms and frameworks are analyzed under different parameters, which have been chosen based on the fundamental characteristics to build a complete virtual assistant. The analysis parameters are: type, which refers to how the developer interacts to build their solution, dialog management that refers how the built assistants can handle conversation in various user contexts making it possible to have a human like interaction, developer Integration/Expandability, 3rd Party UI integration, support the triggering of events from other services, the Natural Language processing (NLP) support, the supported interaction languages that user can interact, the Speech-to-Text(S-T) and Text-to-Speech(T-S) support, the type of software license and the cost.

3 Results

The search has resulted in a total of 28 platforms and frameworks. From these, 11 platforms and frameworks have been referenced and used in papers found in google scholar and 17 referenced in website articles. 4 tools that do not provide enough documentation to enable a proper analysis have been excluded, as well as 1 tool that is not available anymore, 6 tools that are not dedicated to build new and customizable solutions and already offer to the client ready to use solution with low expandability to new contexts, and 10 tools that are not dedicated to build new and customizable solutions. After applying the exclusion criteria, 7 tools have remained as the ones that suit the needs for this research. These are analyzed under different parameters based on the available online documentation (See table 1). The studied solutions are described as follows.

Amazon Lex

Amazon LEX is the platform presented by AWS (Amazon Web Services) to design, build, test and deploy virtual assistants using voice and text. Amazon Lex provide pre-made templates and customizable conversational interfaces that can execute different tasks through functions (AWS lambda functions). These functions can be written in different programming languages (C#, Go, Java, Node.js, Python, Ruby) and are executed when an event is triggered. It uses Deep Learning models for services like Speech recognition and natural language processing (NLP) to aid the assistant to understand the intention of the user. To learn, the assistant needs sample utterances, that is, a collection of possible expression that a user might say. The created models support various languages such as English, Spanish, French, German and Italian. The models created can be applied in chat platforms, Mobile Devices or IoT devices as well as integrated in websites, and messaging platforms such as Facebook or Twilio [13, 14].

Botpress

Botpress is an open developer tool focused on natural language understanding, that is process the incoming message and transform them in structured data ready to be consumed by the other(external) modules and components. Botpress exposes a simple API that can do automatic spell checking, intent extraction, language identification and entity extraction. This tool provides a Low code platform to build conversations where is possible to design conversations, write custom logic, and improve the conversations. Are available two versions: Botpress Open-Source, the free version that contain the main functionalities of this platform, has support only for English and can be deployed in any server; Botpress Enterprise, the paid version that have all the free version and add more functionalities such as support for 12 built-in languages [15].

Dialogflow

Dialogflow is the platform present by Google that has the focus in make easy to design and integrate a conversational user interface into mobile apps, web apps, and devices. It is possible to create conversation flows, that can be trained by machine learning models, capable to analyze multiple types of inputs and can also respond in different ways. This platform is available in 2 versions: Dialogflow ES and Dialogflow CX. This platform provides pre-built agents, and custom agents that can execute various tasks and interact with different version through custom functions (cloud functions) that can be written in different languages and make possible execute the business logic. The created models can be integrated automatically direct from the platform to many 3rd party platforms such as Facebook Messenger, Slack or Google Assistant and it's possible to trigger events based on these 3rd party platforms directly to the assistant [16, 17].

IBM Watson Assistant

IBM Watson Assistant is the platform present that combines Machine Learning and natural language understanding that provides customers with fast and easy to create solutions. This solution support universal language models that allows interact through any language even if is not supported by the platform. The model will learn the new languages that is not supported through some training examples or intent user examples to teach the model about the unique syntactic and grammatical rules of the language. There are several SDKs available that support the various AI services. Also is available an API that provides run-time methods to send user input to an assistant and receive a response. There are also ready-to-use integrations to some CRM's such as Salesforce or Zendesk and to several messaging platforms such as WhatsApp, Facebook Messenger, or Slack [18].

Microsoft Bot Framework (SDK + Compose + LUIS.AI + Azure Cognitive)
Microsoft Bot Framework is a framework released by Microsoft for building intelligent bots. Is possible to create an assistant with a lot of customization using the various services/components provided by Microsoft. Is also available a simple and easy-to-use platform (Microsoft Compose) for create and test the solutions via a drag-and-drop interface. An Open-Source, free SDK is available to a wide variety of programming languages (C#, JS, Python, Java). This framework has support to various languages, through the Translator Service. To understand what the user say is used the LUIS (Language Understanding Intelligent Service) that use language understanding models specific to their tasks. Is possible to train this model with custom data to understand all kind of languages, but also can be used pre-trained models. This framework also uses Azure cognitive services that is responsible to empower to assistant with the ability enhance the human-to-machine conversation [19, 20].

RASA
Rasa is a tool to build custom chatbots that have various components Rasa Open Source, Rasa X and Rasa enterprise. Rasa Open Source is free and Open Source, made up of two components that work together to deliver virtual assistants, Rasa NLU and Rasa Core. Rasa NLU directly deal with humans and converts the user input in structural data. Rasa Core is responsible for deciding what action must be taken once the intent is found and entities are extracted from user input. Rasa X a free but proprietary tool set that layer on top of Rasa Open Source. Is possible to create custom actions using the SDK for Python or provide an endpoint which accepts HTTP POST request from the Rasa Open Source. Is possible to connect Rasa to various messaging and voice channels through prebuild communication modules. Is possible to train Rasa NLU to accept any language and can use pre-trained language models [21, 22].

Wit.AI
Wit.AI is an online platform, very similar to Botpress that can create Natural language interface for apps, capable of turning sentences intro structured data. The assistant is trained by sample utterances, that don't need to describe all the possibilities to the user ask something because it uses natural language processing (NLP) capabilities to understand the user intention. This platform is free or personal and commercial use and with the support of more than 30 languages can be applied in wide variety of use cases. It is available a free platform to create and test natural language interfaces from pre-built templates and an SDK available in different programming languages (Node.js, Python,

Table 1. Framework and platforms analysis

	Amazon Lex	Botpress	Dialogflow	IBM Watson	Microsoft Bot Framework	RASA	Wit.AI
Type	Online low code platform	Online low code platform	Online low code platform	Online low code platform	Framework with Desktop App	Framework	Framework with online platform
S-T/T-S	S-T	No	Both	Both	Both	No	No
NLP	Yes	Yes	Yes	Yes	Yes	Yes	Yes
Languages	5	10 +	6	Any	20 +	Any	20 +
Dialog Management	Yes	Yes	Yes	Yes	Yes	Yes	Yes
Integration/Expandability	AWS functions	SDK and API	SDK, Rest API and cloud functions	SDK and Rest API	SDK and Rest API	SDK and HTTP endpoint	SDK
Trigger events	No	Yes	Yes	Yes	Yes	No	NO
3rd UI Integration	Web UI	Messenger, Slack, Teams	Web UI, Messenger, Google Assistant, Slack…	Salesforce, Zendesk, WhatsApp, Messenger, Slack	Web UI, Skype, Messenger, Teams, Cortana, Slack…	-	-
License	Proprietary	Open Source and proprietary	Proprietary	Proprietary	Proprietary	Open Source	Proprietary
Cost	Paid	Free and paid	Free and paid	Paid	Free and paid	Free	Free

Ruby, Go) where is possible to create customized interactions between various services or create new interactions and functionalities [23].

4   Discussion

The concept of virtual assistant can be easily confused with the concept of chatbot because these are software solutions powered by AI with the ability of interact with humans. Virtual assistants are developed to execute activities based on user inputs that do not need to be configured previously and can adapt their response to the user context. On the other hand, chatbots are developed to execute only limited activities that are previous configured. Despite having some technique functionalities, chatbots and virtual assistants differ on their functionalities.

As previously mentioned, when developing a virtual assistant it is necessary to have some considerations and integrate various functionalities to build a good virtual assistant. The most relevant ones refer to how the virtual assistant communicates with the user, how the assistant understands what the user says and their intentions and how to behave according to the user expectations and needs.

The interaction with a virtual assistant is done mainly by voice, however, it is also common to use text. So, it is necessary to translate what the user says into data that machines can understand (Speech-to-text) and how to give the answer back to the user (Text-to-Speech). To do these "translation" Machine Learning algorithms are commonly used, because these models can adapt to various languages as well as various dialects. Only three pf the studied platforms (Dialogflow, Microsoft bot Framework and IBM Watson Assistant) provide these Speech-to-Text and Text-to-Speech services based on Machine Learning algorithms. Amazon Lex only provides the user with Speech-to-text recognition. The remaining platforms do not provide these services to the developer but it is possible to integrate an external specific service to do these jobs; however, it is harder to develop and may cause compatibility problems.

When humans are talking something to someone it can be performed in different ways, so a good assistant must have the ability to handle the various ways how a user talks to the assistant, for example when asking to do a specific task. This feature is the core of a virtual assistant. Natural Language Processing (NLP) and Natural Language Understanding (NLU) algorithms, based mainly in Deep Learning, are applied with the aim of extracting and structuring the utterances that the user says.

Besides the Speech-to-Text services, these NLP and NLU algorithms make it possible the support of different interaction languages. In general, the platforms and frameworks focus on developing these modules and use pre-trained NLU and NLP models, but RASA and IBM Watson Assistant support the use of models built by the developer. This allows the developer to e.g. devise a solution in a language that is not yet supported by the platform or framework. It is important to note that all the platforms support some main languages (English, French, German).

Other aspect that is fundamental in human-to-machine conversation is machine's ability to contextualize the user conversation. The dialog management provides the assistant with characteristics to understand and contextualize what the user says or asks. For example, when someone asks for the weather, the assistant must contextualize this request to the user location automatically. If the assistant does know the location of the user, the assistant must have the ability to ask for it. The studied platforms and frameworks use machine learning to identify and handle the conversation contexts but in a very simple way, i.e., if we want more contextual information from and to the user this needs to be implemented by the developers. This feature is the main difference between chatbots and virtual assistants, because chatbots are developed to give answers to the user based in pre-made responses and generally cannot handle conversations in different contexts.

For a virtual assistant to be considered an assistant it must have the ability to communicate to other services to execute different tasks. The assistant needs to communicate to the corresponding services generically through a web API to be able to turn off the light. To do this type of integration, some platforms like Amazon Lex and Dialogflow make available to the developer an inline editor available in different languages such as C#, Java or Go. The code written is deployed in the platform and it is executed when the assistant detects the intent to do that task. However, this type of integration has associated costs, so all the tools analyzed, except the Amazon Lex, provide an SDK to be able to extend the assistant features and tasks in an independent platform. This SDKs also allow the developer to integrate their own business logic in their own private environment. Microsoft bot framework, IBM Watson Assistant, Botpress and Dialogflow have available an API REST where is possible to develop other type of interaction such as touch or gestures and to receive, from other services, information that can trigger an event to the assistant. This leads to being able to provide information to the user through the assistant without the involvement of the user.

When we make available our solution to the final user it is important to have a good and easy to use interface, so Amazon Lex, Dialogflow and Microsoft bot have a ready to use Web UI that can be deployed directly to any website. Generally, in the Web UI the most used interaction is text, which makes the assistant look like an advanced chatbot. However, the interaction can also be done by voice. Amazon Lex, Botpress, Dialogflow, IBM Watson and Microsoft bot framework have automatic deployments to mainstream messaging services such as Facebook Messenger, Slack or WhatsApp, however RASA and Wit.AI do not provide any automatic integration with 3rd party services. But with the available SDKs and API it is possible to extend and create new ways of interaction between the user and the virtual assistant.

The cost and type of license that these tools specify are also two important factors to have in consideration when choosing a tool to build a solution. Of all the tools analyzed only RASA and Botpress have Open-Source version of their software. Most of the proprietary solutions, except IBM Watson Assistant and Amazon Lex, do provide free versions to build virtual assistants with some limited functions. However, all of them provide to the developer only some days to try the free of charge tool version.

5   Conclusions

This review embraces the current tools to build custom virtual assistants that can be applied in any business context. Based on review of the articles that examined the tools utilized by authors, it seems clear that virtual assistants have exceptional characteristics that can make them highly relevant in the market. With the growth of this industry more companies are entering this business field and more building tools have emerged. The most widely used tools are the Dialogflow platform, the IBM Watson Assistant, and the Microsoft Bot framework. By analyzing these tools, we conclude that these are the most mature tools that provide the user with all the basic services, most of them with low code interfaces, to build a virtual assistant as well as to use the latest machine learning models to be able to create human-like conversations with all the users. The open-source platforms, with emphasis on RASA, provide the core services to be able to create a solution but lack the modules to interact with the user and a low code interface to build the solutions. Although it is possible to use services from other building solutions to fulfil the missing services, these represent an additional complexity of development.

We can conclude that with the advances of artificial intelligence applied in human-machine relationship, the popularity of virtual assistants brings great and easy to use platforms to build them, which can be applied in different areas such as industry, education, custom relationship management or marketing. However, virtual assistant development solutions still need work to improve the interaction between humans and machines.

Acknowledgment. The study was developed under the project A-MoVeR – "Mobilizing Agenda for the Development of Products & Systems towards an Intelligent and Green Mobility", operation n.º 02/C05-i01.01/2022.PC646908627-00000069, approved under the terms of the call n.º

02/C05-i01/2022 – Mobilizing Agendas for Business Innovation, financed by European funds provided to Portugal by the Recovery and Resilience Plan (RRP), in the scope of the European Recovery and Resilience Facility (RRF), framed in the Next Generation UE, for the period from 2021 -2026.

References

1. Perez Garcia, D.M., Saffon Lopez, S., Donis, H.: Everybody is talking about virtual assistants, but how are people really using them? In: Proceedings of the 32nd International BCS Human Computer Interaction Conference 32, pp. 1–5 (2018)
2. Bernard, D.: Cognitive interaction: Towards "cognitivity" requirements for the design of virtual assistants. In: 2017 IEEE International Conference on Systems, Man, and Cybernetics (SMC), pp. 210–215 (2017)
3. Ballati, F., Corno, F., De Russis, L.: "Hey Siri, do you understand me?": virtual assistants and dysarthria. In: Intelligent Environments 2018, pp. 557–566. IOS Press (2018). https://ebooks.iospress.nl/doi/10.3233/978-1-61499-874-7-557
4. Cóndor-Herrera, O., Jadán-Guerrero, J., Ramos-Galarza, C.: Virtual assistants and its implementation in the teaching-learning process. In: Karwowski, W., Ahram, T., Etinger, D., Tanković, N., Taiar, R. (eds.) IHSED 2020. AISC, vol. 1269, pp. 203–208. Springer, Cham (2021). https://doi.org/10.1007/978-3-030-58282-1_33
5. Jones, V.K.: Voice-activated change: Marketing in the age of artificial intelligence and virtual assistants. J. Brand Strategy 7, 233–245 (2018)
6. Kuligowska, K., Lasek, M.: Virtual assistants support customer relations and business processes. In: The 10th International Conference on Information Management, Gdańsk (2011)
7. Schmidt, B., et al.: Industrial virtual assistants: challenges and opportunities. In: Proceedings of the 2018 ACM International Joint Conference and 2018 International Symposium on Pervasive and Ubiquitous Computing and Wearable Computers, pp. 794–801. Association for Computing Machinery, New York (2018). ISBN: 978-1-4503-5966-5. https://doi.org/10.1145/3267305.3274131
8. Reis, A., Barroso, J., Santos, A., Rodrigues, P., Pereira, R.: Virtual assistance in the context of the industry 4.0: a case study at continental advanced antenna. In: World Conference on Information Systems and Technologies, pp. 651–662 (2022)
9. Radhakrishnan, R., Radhakrishnan, R.: IT Infrastructure Architecture Building Blocks. Sun Professional Services May (2004)
10. Cahn, J.: CHATBOT: Architecture, design, & development. University of Pennsylvania School of Engineering and Applied Science Department of Computer and Information Science (2017)
11. Di Prospero, A., Norouzi, N., Fokaefs, M., Litoiu, M.: Chatbots as assistants: an architectural framework. In: Proceedings of the 27th Annual International Conference on Computer Science and Software Engineering, pp. 76–86. IBM Corp., USA (2017)
12. Schmidt, B., et al.: Industrial virtual assistants: challenges and opportunities. In: Proceedings of the 2018 ACM International Joint Conference and 2018 International Symposium on Pervasive and Ubiquitous Computing and Wearable Computers, pp. 794–801 (2018)
13. Conversational AI and Chatbots - Amazon Lex - Amazon Web Services (2022). https://aws.amazon.com/lex/. Accessed 16 Nov 2022
14. Samuel, I., Ogunkeye, F.A., Olajube, A., Awelewa, A.: Development of a voice chatbot for payment using amazon lex service with eyowo as the payment platform. In: 2020 International Conference on Decision AidSciences and Application (DASA), pp. 104–108 (2020)

15. What is Botpress? | Botpress Documentation (2022). https://botpress.com/docs
16. Dialogflow Documentation | Google Cloud (2022). https://cloud.google.com/dialogflow/docs/. Accessed 16 Nov 2022
17. Sousa, D.N., Brito, M.A., Argainha, C.: Virtual customer service: building your chatbot. In: Proceedings of the 3rd International Conference on Business and Information Management, pp. 174–179 (2019)
18. Introduction to Watson Assistant – IBM Developer (2022). https://developer.ibm.com/articles/introduction-watson-assistant/. Accessed 16 Nov 2022
19. Azure Bot Service documentation - Bot Service | Microsoft Docs (2022). https://docs.microsoft.com/en-gb/azure/bot-service/
20. Williams, J.D., et al.: Fast and easy language understanding for dialog systems with microsoft language understanding intelligent service (LUIS). In: Proceedings of the 16th Annual Meeting of the Special Interest Group on Discourse and Dialogue, pp. 159–161 (2015)
21. Dinesh, T., Anala, M.R., Newton, T.T., Smitha, G.R.: AI bot for academic schedules using rasa. In: 2021 International Conference on Innovative Computing, Intelligent Communication and Smart Electrical Systems (ICSES), pp. 1–6 (2021)
22. Introduction to Rasa Open Source (2022). https://rasa.com/docs/rasa/
23. Wit.ai (2022). https://wit.ai/docs. Accessed 16 Nov 2022

Learning About Recyclable Waste Management Through Serious Games

Marco Iza[1] ⓘ, Kevin Chuquimarca[1] ⓘ, Eleana Jerez[1,2] ⓘ,
and Graciela Guerrero[1(✉)] ⓘ

[1] Departamento de Ciencias de la Computación, Universidad de Las Fuerzas Armadas ESPE,
Sangolquí, Ecuador
{maiza4,kschuquimarca,eijerez,rgguerrero}@espe.edu.ec,
eleana.jerez@estudiante.uam.es
[2] Escuela Politécnica Superior, Universidad Autónoma de Madrid, Madrid, Spain

Abstract. Nowadays, the process of sorting waste, which consists of the selection and grouping of solid and liquid waste generated in different human activities, has acquired great ecological and economic importance for the benefit of environmental conservation. In recent years, the use of technology to create solutions that generate environmental awareness in people has grown significantly. There are applications such as PadovaGoGreen, Save the Planets and Ant Forest, which are games to motivate users to recycle and take care of the planet's natural resources in a fun and creative way. However, they are only available and functional in the countries where they were developed and released. Also, their players experience trouble winning and advancing to the next levels. Hence, we carried out the development of the beta version of a video game, accessible and easy to use called "Cool Ways To Manage Waste" to promote and educate on the most appropriate practices for the correct management of waste and recycling, the game is focused on the scheme of a serious game. The game was evaluated by young users and we used Student's t-test for paired samples, making hypotheses comply right tailed test in order to check the significance of the performed process.

Keywords: Serious Game · Environmental Awareness · Recycling

1 Introduction

Nowadays, the amount of garbage people generate exceeds any previous estimates [1], it has become a main challenge to be tackled, hoping to, in some way, address the delicate environmental situation the entire world faces. In the last Joint Meeting between the United Nations' Energy and Mining and the Environment and Tourism Commissions [2], this constant increase in waste generation in the Latin American and Caribbean region was highlighted, projecting garbage production to up to 670 thousand tons per day by 2050, a rather overwhelming and contrasting figure, compared to the average percentage of apparent recycling of this waste, which ranges from 1% to 20% depending on the area in the region. In the same way, the National Institute of Statistics and Censuses of

Ecuador, in its most recent report published in 2018 [3], mentions that the collection of daily tons of solid waste on average was 12,897, this statistic, compared to the number of Ecuadorian households that carry out some waste classification work, which was 61.43% in 2019 [4], is a fairly low percentage for the country. This figure, about the number of households that classify their waste, is an important indicator of the low participation by the population in the planet's care, particularly, in the correct management of waste. On the other hand, in the report made by The World Bank [5]; it is mentioned that this generation of waste is a natural product of urbanization, economic development and population growth. Precisely, this is the most important reason for searching for new tools and strategies that adjust to the current reality, in terms of teaching and promoting recycling. As mentioned in the article, "Serious game on recognizing categories of waste, to support a zero waste recycling program" [6], despite the fact that the topic of recycling is a tedious task; it is important to generate means through which it can be promoted and, even better, teach people about recycling and the correct handling of the waste we generate. This objective can be achieved using current trends, such as the use of technology, in order to promote recycling as a daily task or activity.

With this background, the development of the beta version of a video game to promote and educate on the most appropriate practices for the correct management of waste and recycling was proposed, the game will be focused on the scheme of a serious game[1]. After development and implementation, the game will be evaluated by young users; to evaluate both the motivation generated in the participants, and the amount of data drawn from the practice.

This paper is structured as follows. In Sect. 2, related works. Section 3, development and implementation of the application. Section 4, assessment and analysis of results. This is followed by a discussion and conclusions in Sect. 5. This paper ends with future work in Sect. 6.

2   Related Works

The study developed by Chia-Lin et al. [7] mentions that recycling guarantees that products and materials can remain in use, thus extending their life cycle. The study by Janakiraman et al. [8] states that a growing world population and limited natural resources make it imperative to learn to live together sustainably. Ozgen et al. [9] mentions that serious games not only address the purposes of entertainment, but also the changes in the behavior of their players. On the other hand, PadovaGoGreen [10] is a serious game for mobile devices, focused on teaching people how to identify garbage bins with the corresponding type of waste. The game, which, based on a photograph of a recyclable object, can provide feedback for its classification, it also has questionnaires and six levels, with a difficulty that increases considerably starting from the third. This work has generated encouraging results, it is focused on the city of Padua, Italy and its recycling regulations, so its use is limited to this country.

---

[1] 1 A serious game is characterized by two main points: (1) It combines the video game and one or more utility functions: spreading a message, providing training, facilitating data exchange. (2) It is aimed at a market other than that of unique entertainment: defense, training, education, health, commerce, communication, etc. [20].

Ozgen et al. [9] propose the multipurpose game "Save the Planets", focused on creating environmental awareness and changing the way solid waste is classified, this serious game focuses on better informing and motivating people to take long-term environmental action. The results of this study showed that the game positively affected its players in their pro-environmental awareness, motivating them to carry out follow-up actions to manage the needs of their surroundings and the challenges learned from the game.

Mi et al. [11] conducted a study on the Ant Forest platform, the largest online environmental protection application in launched in China by Alipay, which offers its users a virtual platform to participate in the caring for the environment online, this game won the Champions of the Earth Award granted by the United Nations (UN) on September 19, 2019, some countries have tried to imitate Ant Forest's idea, developing similar programs, one of them is its Philippine version called GCash Forest.

Menon et al. [6] conducted a study to support a waste recycling program in public places in India, the program consists of an augmented reality game and a motion detection device, its objective is to train people in sorting waste from different categories like hard and soft plastic, paper and cardboard.

The serious game proposals [6, 9–11] are meant to work in their country or city of origin, so they cannot be used by users from other parts of the world who want to learn about waste management and taking care of the planet, due to the language of the game or recycling regulations. The video game's difficulty is another factor to take into account, in the study developed by Gaggi et al. [10], results show that from the third level of the game on, players experience trouble winning and advancing to the next levels, another factor to consider is accessibility, since games like Ant Forest [11] and PadovaGoGreen [10] are only available and functional in the countries where they were developed and released.

In this section, several game proposals and developments have been analyzed, which seek to attract a large number of players, motivating them to recycle and take care of the planet's natural resources in fun and creative ways. Therefore, the present research work (to develop a serious game focused on learning recyclable waste management) is achievable, since the previously mentioned studies focus on learning using serious games of simulation or association of objects, destined for users of different ages.

3   Development and Implementation

This section introduces the proposal for: design, architecture and description of the development process of a video game for web platforms, which we have called "Cool Ways To Manage Waste".

3.1   Proposal

The proposal consists of developing a video game that allows users to draw information about the management of recyclable waste in a more friendly way through the gamification[2] of activities, with the purpose of promoting and contributing to the care of the environment.

---

[2] Gamification of education is a developing approach to increase learner motivation and engagement by incorporating game design elements in educational settings [19].

The methodology for the development of the proposed video game is based on the following process: It begins with the design phase, in which the learning objectives and target population form the basis for the learning and creative design of the game. The game is created using combinations of these designs and validated through initial evaluations using a sample of the target population and also validated with professional users. This process is repeated until the game is fully verified. Then, once the game is validated, the application deployment is made available.

3.2   Game Design

The video game is responsible for gamifying the activity of classifying recyclable waste that is in use or are present in people's daily lives more frequently, in this way, associating the game with caring for the environment is encouraged, as mentioned earlier.

The video game works under certain parameters both in its design and in its functionality. The mechanics of the game is that the user must perform a sequence of activities in the proposed video game, these are: 1) The user accesses the video game through their web browser; 2) The user has a menu with the options: "Play", "Instructions", "Scores"; 3) Once the user chooses the "Play" option, the game starts with a 45-s timer; 4) Once the timer reaches zero, information is presented as feedback on practices linked to the management of recyclable waste; 5) The video game finally saves the score within an overall ranking.

3.3   Architecture

For the development of the video game, Unity's 2D game engine was used, C# as programming language, while for the rendering and web browser application compatibility, the WebGL API [12] compatible with Unity was used. Finally, Firebase [13] was used as host to access the game through the web. Figure 2 shows how these mentioned elements communicate and are distributed to achieve the objective described in the previous section.

Unity, being a centralized software that has all the necessary tools for the development of 2D video games, allows fast development cycles. In addition, Unity supports the rendering of 2D graphics, its physics engine will allow us to simulate and work with the physics of objects.

WebGL allows interactive 2D and even 3D graphics to be rendered in supported web browsers, and its API allows users to access content through WebGL-enabled browsers without the need to download any additional plug-ins. In addition, for developers, it provides low-level access to hardware with the familiar OpenGL code structure.

When working with web applications, it is necessary to have a cloud platform that users can access to use content, in this case, the developed video game. Firebase allows the automatic sending of events to applications when data changes, thus facilitating the use of the game in web browsers (Fig. 1).

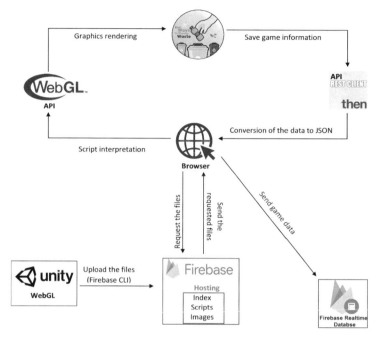

Fig. 1. Architecture.

3.4 Description of the Development Process

Assets for 2D games are used to implement user interface design in Unity. In this case, the implementation of original multimedia resources (images, text, animation) made with the free PIXLR tool [14] was chosen.

To generate the objects with which the user interacts (residues), Unity's 2D game engine was used, so that "Sprites" (reference to the game) can be used, which are provided with physical features that allow the user to interact with the game in a better way. Figure 2 shows such generated objects for the user to interact with.

Once the development process of the video game in Unity is finished, the WebGL platform is turned on and the video game is built for this API. The result of the previous process generates a folder with an index and the necessary scripts for the game to work in the browser. In the game folder, the necessary Firebase modules are implemented through the terminal, executing the command "firebase init".

Firebase allows to manage the video game through terminal commands, this allowed to perform a test of the game before uploading it to Firebase Hosting, the command for this was "firebase serve". Finally, the video game files generated by WebGL, were uploaded to the Firebase hosting service, the command "fire-base deploy" loads the files in the network and generates a link to access the game.

Fig. 2. Use of Unity 2D game engine.

4   Assessment and Analysis of Results

In the assessment process, it is important to control two fundamental research parameters. First, user satisfaction with the video game, taking into account that it is one of the factors that directly affects the proper and recurrent use of the tool. Besides the video game's scoring system, which seeks to assess functionality in terms of learning concepts related to waste recycling. In order to analyze the effect of these parameters on the fundamental objectives of the work, functional tests have been carried out to measure and relate the data obtained in the defined control group.

4.1   Definition of Scenarios

For testing the video game, two test scenarios were defined, both for the first and for the second scenario, a group of 20 students was assigned, who were evaluated before and after using the video game, in order to obtain significant samples for analysis.

The first scenario, referred to as Scenario 1 in the present work, assesses the group of students to whom traditional methods for associative learning are applied, such as slides and didactic material. The users of this scenario will only fill out the questionnaire aimed at evaluating the knowledge acquired about recycling, without playing the video game.

With the second scenario or Scenario 2, the aim is to apply the questionnaire on knowledge acquired, as well as the satisfaction generated after playing the video game, which is being used as an alternative tool for associative learning of concepts linked to waste recycling.

4.2   Evaluation Method

Firstly, to quantify the level of knowledge that users acquired in the application of the two proposed scenarios, a questionnaire of 10 questions focused on 3 axes was used: the identification of concepts about waste recycling, the association of elements from the video game with those of real life, and the environmental awareness produced in users.

Questions: 1, 2, 3, 4, 5 and 6; corresponding to the first two axes mentioned above, have a value of one point if the answer is correct and zero if it is incorrect. On the other hand, questions: 7,8,9 and 10; use the Likert scale [15] to measure the level of awareness created in users about recycling.

On the other hand, to quantify user satisfaction with the video game, the questionnaire devised by Mark James Parnell [16] was used. This questionnaire allows different aspects related to user experience to be measured. The main advantage of using this questionnaire is that it is already designed exclusively for video games, consequently, it is not necessary to adapt the questions for the present work. The questionnaire has 26 questions, based on the users to whom the questionnaire is directed to, 10 questions were taken into account and, as in the other questionnaire, they were assessed according to Likert's scale [15] so that the levels of satisfaction in quantifiable figures are obtained with a mark out of five points.

Both questionnaires were applied to the groups of 20 students ages 14 and 15 who at that time were in the tenth grade of an educational unit in Quito-Ecuador. The age of the users is based on the conclusions drawn from the study [17], which corresponds to the group of adolescents who interact more frequently with IT, in this way, we ensure that users have some notion on the use of online video games, as well as the ability to complete questionnaires in Google Forms.

The number of users was established thanks to the principles determined by Laura Faulkner, where it is mentioned that testing any type of software with 10 users provides with a minimum range of findings of 82% [18]. Under this premise, 20 users were selected to take the test, taking into consideration that, not only the video game's usability is to be measured, but also the participants' knowledge.

4.3 Results

Once the assessment of the proposed scenarios was finished, we continued with the analysis phase of the results obtained from the questionnaires and the video game's test. The method defined for this analysis focuses on checking the significance of the performed process, applying the Student's T-test for paired samples, making hypotheses comply with the right tailed test.

As shown in Fig. 3; At first glance, it is evident that there is indeed a significant difference between the results obtained from Scenario 1 and Scenario 2 in the recycling knowledge questionnaire. Applying the respective analysis, using the Microsoft Excel's Data Analysis tool, we assigned the null hypotheses H_0 and alternatives H_1 for each group of questions:

H_0: The mean of the differences will be less than or equal to zero, which would mean that there is no significant variance between Scenario 1 and Scenario 2.
H_1: The mean of the differences will be greater than zero, which would mean a significant variance between Scenario 1 and Scenario 2.

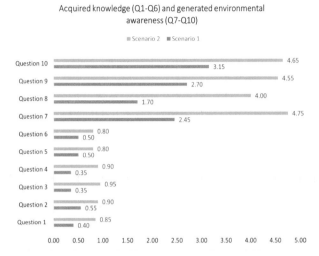

Fig. 3. Acquired knowledge (Q1–Q6) and generated environmental awareness (Q7–Q10).

Table 1. Results of the t-test applied to (Q1–Q6) and (Q7–Q10).

	T Stat	$P(T < = t)$	t Critical
Q 1–Q6	7.7679250344	0.0000001294	1.7291328115
Q7–Q10	17.6666666666667	0.000000000000150	1.7291328115

Since the T Stat value is located on the right side of the t Critical value in Table 1, we can say that: the statistical value of t is in the rejection region of H_0. Therefore, it indicates that the null hypotheses should not be accepted and the differences between the results of the first and second scenarios are significant.

Since the results obtained are significant to our work, it is important to relate these results to our assessment objectives, which correspond to the dependent variables that have been selected for analysis.

The first dependent variable corresponds to the knowledge acquired with the use of the video game; and the second variable corresponds to the environmental awareness created in the user. In the same way, the independent variables that affect dependent ones are; user satisfaction at playing the video game and the average score achieved.

In the case for the analysis of the relationship between the variables, we used a linear regression of data, as shown in Fig. 4. Using the Microsoft Excel Data Analysis tool once again, we arrived at the following results:

The influence of the score as described in the satisfaction questionnaire and the score achieved by the user after eight games, are directly proportional to a better performance at both, the knowledge test about recycling (1), and the level of environmental awareness raised by the user, demonstrated in the same questionnaire (2).

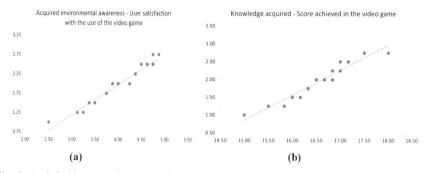

Fig. 4. (a. left) User satisfaction's influence with the video game on its raising of environmental aware-ness (b. right) influence of the score achieved in the video game on the knowledge acquired by users.

With the help of the previously referred tool, we can describe these results with the following mathematical expression, where the first coefficient corresponds to the intercept of the two independent variables, while the other two coefficients correspond to the coefficients of the independent variables:

$$C_R = -14.50 + 1.53\ S_V + 0.66\ P_V \tag{1}$$

$$C_A = -3.93 + 0.54\ S_V + 0.23\ P_V \tag{2}$$

C_R: Performance on the recycling knowledge quiz.
C_A: Performance on the environmental awareness quiz.
S_V: User satisfaction with the video game.
P_V: Average score achieved after eight games.

5   Discussion and Conclusions

The video game developed and implemented across web platforms constitutes a valid alternative for learning fundamental concepts about waste recycling, the classification of waste in different containers and a greater environmental awareness in users. As it is a video game implemented on web platforms, it's easily accessible to the general public; an example of this, is the way in which testing was carried out for this work, where as the video game's installation was not necessary for any the subjects.

The data collection and results of other serious games for learning to classify recy-clable waste have been successful. Spread out on a large number of platforms, such as web, mobile and virtual reality, thus increasing its reach for users in a given region, although it is mentioned that many of the applications focused on serious games do not have a global reach due to their implementation language.

The results when applying the video game in a group of students, to identify concepts and associate elements, were successful and significant for the present work, in part due to the user's satisfaction with the video game, added to the repetition of each activity during eight levels; which improved the performance of users in the questionnaire that assesses knowledge of activities related to recycling.

6   Future Work

As future work, the implementation of the video game into virtual reality technology can be considered, to better develop associative learning of recyclable waste, not only through the association of elements used within the video game, but through actions carried out in the video game that can be replicated in everyday life, improving the user's experience and interest in continuing to use the tool.

References

1. Información Ambiental en Hogares: Instituto nacional de estadística y censos (2016). https://www.ecuadorencifras.gob.ec/documentos/web-inec/Encuestas_Ambientales/Hogares/Hogares_2016/Documento%20tecnico.pdf. Accessed 19 Dec 2021
2. Perspectivas de la gestión de residuos en América Latina y el Caribe: Parlamento latinoamericano y caribeño (2021). https://parlatino.org/wp-content/uploads/2017/09/perspectiva-gestion-residuos.pdf. Accessed 19 Dec 2021
3. Instituto nacional de estadística y censos: Instituto nacional de estadística y censos (2018). https://www.ecuadorencifras.gob.ec/segun-la-ultima-estadistica-de-informacion-ambiental-cada-ecuatoriano-produce-058-kilogramos-de-residuos-solidos-al-dia/. Accessed 2 Dec 2021
4. Información Ambiental en Hogares ESPND 2019: Instituto nacional de estadística y censos (2020). Accessed 1 Dec 2021
5. Kaza, S., Yao, L., Bhada-Tata, P., Woerden, F.V.: What a Waste 2.0 A Global Snapshot of Solid Waste Management to 2050. International Bank for Reconstruction and Development/The World Bank, Washington (2018)
6. Menon, B.M., Unnikrishnan, R., Muir, A., Bhavani, R.R.: Serious game on recognizing categories of waste, to support a zero waste recycling program. In: 2017 IEEE 5th International Conference on Serious Games and Applications for Health (SeGAH), pp. 1–8 (2017)
7. Hsu, C.-L., Chen, M.-C.: Advocating recycling and encouraging environmentally friendly habits through gamification: an empirical investigation. Technol. Soc. **66**, 101621 (2021)
8. Janakiraman, S., Watson, S.L., Watson, W.R., Newby, T.: Effectiveness of digital games in producing environmentally friendly attitudes and behaviors: a mixed methods study. Comput. Educ. **160**, 104043 (2021)
9. Özgen, D.S., Afacan, Y., Sure, E.: Save the planets: a multipurpose serious game to raise environmental awareness and to initiate change. In: Proceedings of the 6th EAI International Conference on Smart Objects and Technologies for Social Good, pp. 132–137 (2020)
10. Gaggi, O., Meneghello, F., Palazzi, C.E., Pante, G.: Learning how to recycle waste using a game. In: Proceedings of the 6th EAI International Conference on Smart Objects and Technologies for Social Good, pp. 144–149 (2020)
11. Mi, L., et al.: Playing Ant Forest to promote online green behavior: a new perspective on uses and gratifications. J. Environ. Manag. **278**, 111544 (2021)
12. KHRONOS GROUP (2022). https://www.khronos.org/webgl/. Accessed 19 Jan 2022
13. Firebase: Firebase (2022). https://firebase.google.com/?hl=es-419. Accessed 19 Jan 2022
14. PIXLR: PIXLR (2021). https://pixlr.com/es/. Accessed 19 Jan 2022
15. Joshi, A., Kale, S., Chandel, S., Pal, D.K.: Likert scale: explored and explained. Br. J. Appl. Sci. Technol. **7**(4), 396 (2015)
16. Parnell, M.J., Berthouze, N., Brumby, D.: Playing with scales: creating a measurement scale to assess the experience of video games. University College London, pp. 1–190 (2009)

17. Servin, R.E., Monzón, V.E., Traverso, Y.A., Solís, A.A., Gómez, R.: Los adolescentes y la cultura de la informática: un fenómeno creciente. Rev Fac Med Unne **33**(2), 36–40 (2013)
18. Faulkner, L.: Beyond the five-user assumption: benefits of increased sample sizes in usability testing. Behav. Res. Methods Instrum. Comput. **35**(3), 379–383 (2003)
19. Dichev, C., Dicheva, D.: Gamifying education: what is known, what is believed and what remains uncertain: a critical review. Int. J. Educ. Technol. High. Educ. **14**(1), 1–36 (2017)
20. Alvarez, J., Djaouti, D.: An introduction to serious game definitions and concepts. Serious Games Simul. Risks Manag. **11**(1), 11–15 (2011)

Self-reporting Tool for Cardiovascular Patients

Hanna Vitaliyivna Denysyuk[1], João Amado[2], Norberto Jorge Gonçalves[2], Eftim Zdravevski[3], Nuno M. Garcia[4], and Ivan Miguel Pires[5(✉)]

[1] Instituto de Telecomunicações, Universidade da Beira Interior, Covilhã, Portugal
hanna.denysyuk@ubi.pt
[2] Escola de Ciências e Tecnologia, Universidade de Trás-os-Montes e Alto Douro, Vila Real, Portugal
joaoamado2001@gmail.com, njg@utad.pt
[3] Faculty of Computer Science and Engineering, University Ss Cyril and Methodius, Skopje, North Macedonia
eftim.zdravevski@finki.ukim.mk
[4] Faculdade de Ciências, Universidade de Lisboa, Lisbon, Portugal
nmgarcia@fc.ul.pt
[5] Instituto de Telecomunicações, Escola Superior de Tecnologia e Gestão de Águeda, Universidade de Aveiro, Águeda, Portugal
impires@ua.pt

Abstract. Cardiovascular diseases are one of the leading causes of death in the world. As there are more and more ways to prevent this kind of problem, people must start to take precautions and try, with the help of a doctor or even with some specific information about these problems. As with other diseases, it is important to discover and treat them as early as possible for a better quality of treatment. This paper aims to create an application for tracking and self-report cardiovascular diseases. The application was developed in Android Studio using Kotlin programming language.

Keywords: cardiovascular patients · mobile application · medicine · diseases

1 Introduction

Problems related to cardiovascular diseases are the leading cause of death around the globe, killing more lives than all types of cancer if put together. From 1990 to 2019, cardiovascular diseases increased from 271 million to 523 million, an increase of 93% [7]. In terms of deaths from 1990 to 2019, an increase of deaths from 12.1 to 18.6 million was verified, revealing an increase of 54%.

People must take precautions, especially if the person already has an established cardiovascular disease. According to the World Heart Federation, 35 million [3] people have an acute coronary or cerebrovascular event every year which 25% to 40% occur in people with established cardiovascular disease. If this problem were known and dealt with early on, it would be possible to do a better treatment to reduce death and disability provoked by cardiovascular problems.

Secondary prevention is needed to appear more and more. Secondary prevention refers to preventing this kind of problem through drugs or counseling

A. Rocha et al. (Eds.): WorldCIST 2023, LNNS 801, pp. 126–133, 2024.
https://doi.org/10.1007/978-3-031-45648-0_13

[4], specifically for individuals with a high risk of cardiovascular problems or previous events of cardiovascular diseases. The World Hearth Federation defines secondary prevention as any strategy to reduce the chance of recurrent cardiovascular pain in a patient with a known cardiovascular problem. Even though this can be very positive for people worldwide, it can be difficult for people to use self-management since there are a lot of complex treatments and medication [5]. In addition to that, to treat this problem, people need to change their lifestyle too, so it requires a lot o effort to change a thing that is so regular, like practicing physical exercise, stopping drinking, stopping smoking, or eating a healthy diet.

So how can we help in this cause? First of all, technology is a huge thing now. If we compare it to a few years ago, everything had a considerable upgrade. So we need to take advantage of these recent advancements, for example, in the field of mobile phones. A few years ago, these devices couldn't accomplish many things, but now, they are machines capable of doing a lot. This percentage is even higher if we only consider the more developed countries. It is a new era, and mHealth (mobile health) is next to us.

Even more mobile phones are being taken as a tool to use as secondary prevention [6]. This technology offers a non-expensive and personalized place to track and self-report and reminders for health behaviors. These devices promote a perfect approach when dealing with the prevention of cardiovascular diseases. There are over 350,000 mobile applications related to health [1], and there are a lot of studies where some of them have appeared in some studies related to this subject.

This paper will introduce a new application related to cardiovascular diseases. One of our goals in this application was to create a simple layout where everyone could understand and use the application without much effort. Furthermore, we tried to maximize the simplicity, even of how the blood pressure records are measured, by creating an intuitive way of using the smartwatch as a measurement device. So with this, we pretend the application is a tool for tracking and self-reporting cardiovascular diseases.

2   Related Work

In this paper, we are introducing a new application for patients with cardiovascular diseases (CVD). Several related works are connected with this one. First, the authors of [2] set a review about mobile applications that allow patients to monitor and report cardiovascular diseases. In this work, they selected the applications that fulfill their criteria, such as the fact that the applications needed to be free to use and register, available in the English language, updated within the last three years, and focused on health track and monitoring. With this, they analyzed all the applications that fulfilled this criterion and downloaded and tested them to see if they were what they were looking for. After this, a junction of characteristics was put together to find the pattern in this application.

Done the analysis, the authors pointed out some things the applications need to improve, such as how their "solutions" work. Finally, to conclude, they investigated how this application could help the user in their objective, which was to track and self-report cardiovascular diseases.

3 Methodology

3.1 Research Questions

For this paper, the three main questions were: (RQ1) what are the most important parts of developing this type of application? (RQ2) How can this mobile application help the user track and self-report? (RQ3)What is missing in other applications that can be fixed with this one?

3.2 Search Strategy

While making this paper, other applications were analyzed to create a better application for the user, including certain aspects that could miss in other applications or make a junction of all important contents needed to self-report and tracking of cardiovascular diseases.

3.3 Application Development

Developing this kind of mobile application is challenging because there are not many mobile applications we can inspire from. There are several other constraints that the developer needs to be aware of, such as:

- Capabilities of the mobile devices.
- Rapid evolution.
- Mobile specifications.

While developing the applications, we needed the physical characteristics, such as the display size and the data input mechanism, and the technical features, such as memory space and the operating system. This application was developed in the Android Studio using Kotlin programming language (Fig. 1).

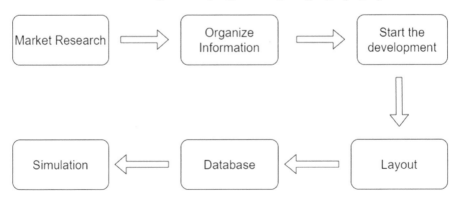

Fig. 1. Methodology Diagram

4   Discussion and Results

The application was successfully created by applying all the methodologies. Initially, the application has a login page (Fig. 2-a) and a register page (Fig. 2-b) where users can connect their accounts with the application.

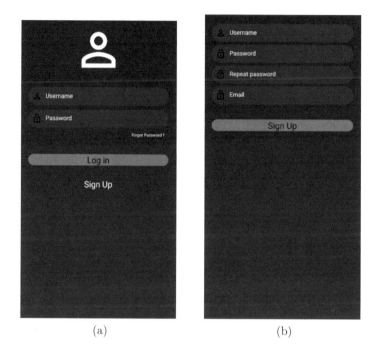

(a)   (b)

Fig. 2. Authentication Module. (a) Login. (b) Registration.

After the user connects their account to the application, the first thing he will see is the main dashboard menu. The dashboard menu (Fig. 3) is in the middle of the mobile application. From here, the user can connect with each part of the application.

Fig. 3. Dashboard Menu

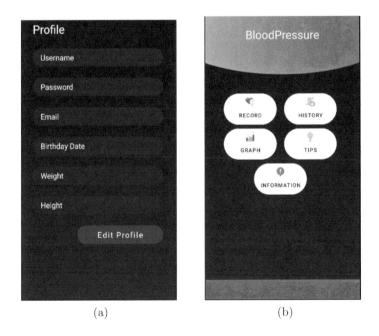

(a)   (b)

Fig. 4. Menu. (a) Profile. (b) Blood Pressure.

The dashboard allows the user to navigate the Profile Menu, the History Menu, the Settings Menu, and the Blood Pressure Menu. For example, in Fig. 4-a and Fig. 4-b, it's possible to see the Profile Menu and the Blood Pressure Menu.

Inside the application, specifically in the blood pressure menu (Fig. 4-b), we added a record menu where blood pressure measurements using the smartwatch would be. On this history menu, it will be all the data received by the application, a graph menu where will take the data saved in the history menu and build a graphic with the information over time.

Then there is information the user may want about cardiovascular problems: the tips and information menu. In this menu, the user can see information related to blood pressure (Fig. 5-a and Fig. 5-b).

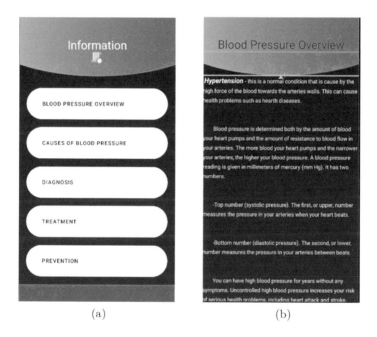

(a)   (b)

Fig. 5. Menu. (a) Information. (b) Blood Pressure Overview.

5   Conclusion

This paper led to the creation of an application to fulfill the needs of patients with cardiovascular disease problems. We answered the following questions:

– *(RQ1) What are the most critical parts in developing this type of application?* The essential part while developing this application is to have a notion of what cardiovascular diseases are. Knowing that the creator only needs to create software that connects the user data with the application and makes de application receive and analyzes the information giving feedback to the user.

- *(RQ2) How can this mobile application help the user track and self-report?* This mobile application retrieves data from the user. With this data, the application analyzes and sends a report to the user to let them know if their blood pressure values are reasonable and if there is a need to see a doctor. By doing this, the application helps the users track their health and self-report their problems with a professional's help.
- *(RQ3) What is missing in other applications that can be fixed?* Other mobile applications are very complex or lack some parts needed by people with this problem. This application connects a simple layout with what the user needs, a self-report and tracking tool that the user needs. Another critical thing other applications don't have is a medical feedback system. A few applications have that feature, but those who have are only from a few selected countries. So, the people of most countries don't have access to this kind of application.

In this area, there are still a lot of features that can be upgraded. So in the feature, there will always be space for improvement. In addition, this type of application needs to be more spread worldwide and needs to be more intuitive for the user.

Saying this, we can say that the main objective of this application was completed. With this application, we wanted to help fight this problem that is becoming bigger every day, using this application as secondary prevention to cardiovascular diseases. In the future, we want to add more features to the application to give the user a more specific result of their health, like adding a correlation between age and weight or steps performed by the user and cardiovascular diseases.

Acknowledgments. This work is funded by FCT/MEC through national funds and, when applicable, co-funded by the FEDER-PT2020 partnership agreement under the project **UIDB/50008/2020**. Hanna Vitaliyivna Denysyuk is funded by the Portuguese Foundation for Science and Technology under the scholarship number **2021.06685.BD**. This article is based upon work from COST Action CA19136 - NET4AGE-FRIENDLY - International Interdisciplinary Network on Smart Healthy Age-friendly Environments, supported by COST (European Cooperation in Science and Technology). COST is a funding agency for research and innovation networks. Our Actions help connect research initiatives across Europe and enable scientists to grow their ideas by sharing them with their peers. It boosts their research, career, and innovation. More information at www.cost.eu.

References

1. Byambasuren, O., Beller, E., Glasziou, P., et al.: Current knowledge and adoption of mobile health apps among Australian general practitioners: survey study. JMIR Mhealth Uhealth **7**(6), e13199 (2019)
2. Denysyuk, H.V., Amado, J., Gonçalves, N.J., Zdravevski, E., Garcia, N.M., Pires, I.M.: Monitoring of cardiovascular diseases: an analysis of the mobile applications available in the google play store. Electronics **11**(12), 1881 (2022). https://doi.org/10.3390/electronics11121881. https://www.mdpi.com/2079-9292/11/12/1881

3. Fuster, V., Voûte, J.: MDGs: chronic diseases are not on the agenda. The Lancet **366**(9496), 1512–1514 (2005)
4. Grundy, S.M., Hansen, B., Smith, S.C., Jr., Cleeman, J.I., Kahn, R.A., Participants, C.: Clinical management of metabolic syndrome: report of the American heart association/national heart, lung, and blood institute/American diabetes association conference on scientific issues related to management. Circulation **109**(4), 551–556 (2004)
5. Park, L.G., Beatty, A., Stafford, Z., Whooley, M.A.: Mobile phone interventions for the secondary prevention of cardiovascular disease. Prog. Cardiovasc. Dis. **58**(6), 639–650 (2016)
6. Unal, E., Giakoumidakis, K., Khan, E., Patelarou, E.: Mobile phone text messaging for improving secondary prevention in cardiovascular diseases: a systematic review. Heart Lung **47**(4), 351–359 (2018)
7. Ye, G., et al.: Integrated metabolomic and transcriptomic analysis identifies benzo [a] pyrene-induced characteristic metabolic reprogramming during accumulation of lipids and reactive oxygen species in macrophages. Sci. Total Environ. **829**, 154685 (2022)

Study of Detection Object and People with Radar Technology

Hugo Nogueira[1] , Dalila Duraes[1,2(✉)] , and Paulo Novais[1,2]

[1] Algorithm Centre, University of Minho, Braga, Portugal
a81898@alunos.uminho.pt, pjon@di.uminho.pt
[2] LASI - Intelligent Systems Associate Laboratory, Guimarães, Portugal
dalila.duraes@algoritmi.uminho.pt

Abstract. Street monitoring can be used as an excellent tool to decrease the number of incidents by, for example, giving information to street users (pedestrians, vehicles, cyclists, etc.) about the position of other street users and therefore helping prevent any possible harmful situation. Today, with the growing concern about privacy and data protection issues, the use of video and audio has become problematic in terms of street monitoring, so there is a need to find a solution to this problem. With that in mind, using radar is a possible solution since the data retrieved from it doesn't contain anything considered personal and could violate people's privacy. This paper presents a systematic review of pedestrian, vehicle and cyclist detection. The objective is to identify the main methods of radar target detection and the algorithms. With that in mind, a search in the SCOPUS repository identified thirteen papers as relevant to include in the review.

Keywords: Radar · Pedestrian Detection · Cyclist Detection · Vehicle Detection · Machine Learning · Deep Learning

1 Introduction

Technologies like Light detection and ranging (LIDAR), radio detection and ranging (Radar), visible spectrum cameras and ultrasonic sensors are the leading sensing technologies used to detect objects in several environments. In street monitoring, those technologies can help detect road users like pedestrians, cyclists and vehicles and give valuable information to improve the quality and safety of those users [6]. While cameras and LIDAR systems are commonly used in those detections, many manufacturers are increasingly adopting radar sensors to aid the perception of, for example, vehicle surroundings [7]. Radars have been employed in a variety of fields, including homeland security, local automotive radar, city monitoring, the military, and even healthcare [3]. In challenging conditions like nighttime, glaring sunlight, snow, rain or fog, the use of cameras and LIDAR sensors substantially affected [3,5,7,12,17]. There is also a growing concern in terms of privacy and data protection issues, which makes using technologies like video and audio problematic.

A. Rocha et al. (Eds.): WorldCIST 2023, LNNS 801, pp. 134–143, 2024.
https://doi.org/10.1007/978-3-031-45648-0_14

With this in mind, a systematic review of pedestrian, vehicle, and cyclist detection was conducted in this document. The research questions to be addressed are: RQ1) How is made the acquisition and data processing of radar sensor data? RQ2) What are the most important features for detecting cyclists, pedestrians and vehicles? RQ3) What AI models are used for object detection with radar?

The document is divided into three main sections. First, the methodology used for the systematic review is described, including the research strategy and exclusion criteria. Next, the results found was presented, followed by a discussion of the findings. Finally, all the preceding themes are summarised and the main conclusions are drawn, providing some directions for future research.

2   Methodology

The PRISMA statements and checklist[1] (Preferred Reporting Items for Systematic Reviews and Meta-Analyses) was the starting point for this literature review. The scientific community in engineering and computer science commonly accepts PRISMA. The following steps were taken into consideration in this methodology: defining the exclusion criteria to filter the articles and reduce the sample; analyzing the selected studies; and, finally, presenting and discussing the findings. The study's research questions and pertinent keywords were identified.

The literature search was conducted on 23 November 2022 in the popular databases for computer science: *SCOPUS* and *ACM digital Library*. The relevant literature documents were found via *SCOPUS*, and no literature was selected from the others databases.

The keywords *radar*, *radar detection*, *radar monitoring* were used to specify the scope of action of this work and the *pedestrian**, *object**, *vehicle**, *car**, *bicycle** and *bike** were added to focus the search on the targets that should be detected. Next, the keywords *Machine Learning*, *Machine Automated*, *Neural Network**, and *Deep Learning* were also added to define the technologies that this study hopes to find.

At last, each keyword related to a subject was joined by using disjunctions, and every subject was linked by the use of conjunctions. In the *SCOPUS* repository, the search started by applying the query below to the documents' title, keywords and abstract.

---

[1] http://www.prisma-statement.org.

```
1 TITLE-ABS-KEY (
2 ("radar" OR "radar detection" OR "radar monitoring")
3 AND
4 ("pedestrian*" OR ("object*" OR (("vehicle*" OR "car*") OR
  ↪ ("bicycle*" OR "bike*"))))
5 AND
6 ("Machine Learning" OR "Machine Automated" OR "Neural Network*" OR
  ↪ "Deep Learning")
7 )
```

In order to screen the articles found, some exclusion criteria were defined. Thus, the documents are excluded if they fall into one of these:

EC1 Don't come from the fields of computer science or engineering;
EC2 Haven't been made in the last three years (2020 to 2022);
EC3 Not open access;
EC4 Not written in English or Portuguese, the two languages that the authors understand;
EC5 Do not focus on the variables studied or are out of context (*i.e.* LIDAR instead of Radar).

3   Results

Using the query defined above, the search in the *SCOPUS* repository identified 5533 articles where exclusion and inclusion criteria were then applied. Firstly, the documents that met the first four exclusion criteria (EC1, EC2, EC3, EC4) were filtered out of the search, limiting the search to computer science and engineering papers that were written in the last three years (2020–2022) in either English or Portuguese and that was freely accessible. Then the 791 documents that remained, a lot of them were not really about radar detection but more about LIDAR. Therefore, it was necessary to discard all the papers related to LIDAR that focused more on the radar detection of the intended targets. In order to achieve this, another keyword was added to the set of keywords that specify the scope of action of this work. Still, instead of using that keyword to find papers, we use that keyword to exclude them (*i.e.* ("radar" OR "radar detection" OR "radar monitoring" AND NOT "LIDAR")). Finally, we were left with 21 documents for a full reading. In the end, 13 papers were included. Figure 1 shows a PRISMA flowchart that provides a general overview of the reported process.

Seungheon *et al.* [2] propose a new target classification method based on point cloud data using the spatial characteristics (i.e. length, height and width) of the target. Those spatial characteristics were calculated by using information retrieved from the FMCW Radar signal received. Then it's possible to estimate Distance, velocity and angles (azimuth and elevation), allowing a generation of 3D point cloud data that is then orthogonally projected onto the XY, YZ, and ZX planes, generating three images. For the model, the authors proposed a multi-view convolutional neural network (CNN)-based target classifier using those three images as inputs. CNN proposed has achieved the best results because although the model's accuracy was similar to well-known deep learning methods for image classification, the training time of the model was significantly shortened.

Buchman *et al.* [3] uses data from the MAFAT Radar Challenge, where the goal is a binary classification of either people or animals. Using a ground Doppler-pulse radar, the dataset contains real data collected from different geographical locations, with different times, sensors and quality. The data includes segments consisting of a matrix with I/Q values and metadata, where the x-axis represents the pulse transmission time, and the Y-axis represents the reception time of signals concerning pulse transmission time. Then the Fast Fourier Transform (FFT) was applied to the I/Q matrices generating spectrograms of the signals. As for the models, the authors propose an ensemble of two models: a main CNN model inspired by the ResNet and a secondary classification model that used a shallower network and improved the total accuracy and especially the accuracy of true negatives in the animal class. AUC of the ROC was used as an evaluation metric of the model's performance, and the proposed model achieved around 0.95 on the public test set and above 0.85 on the final (private) test set. Image augmentation techniques such as hyper-parameter tuning using K-fold validation were used to obtain such results.

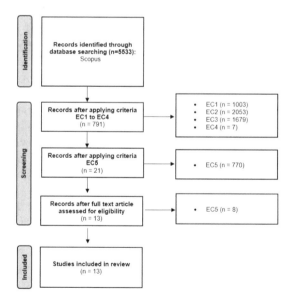

Fig. 1. *PRISMA flowchart* adapted to this study [1]

Jyoti Bhatia *et al.* [4] obtained data using an FMCW mmWave radar equipped with four transmitting and three receiving antennas. The received signal is converted from the time domain to the frequency domain by applying the

FFT algorithm, where the range FFT spectrum is obtained, which is then converted to an amplitude (dBFS) versus range (m) plot. The range FFT plot is then used to detect peaks in the spectrum, and those peaks in the range FFT spectrum represent targets in the mmWave radar's field of view. From the detected peaks in the FFT spectrum, features for each peak are extracted, such as the target's radial range, the peak's height, the peak width, the standard deviation, and the area under the peak. These features provide distinguishable information about the targets. Targets with a large cross-section reflect more power, resulting in more significant peaks in the range FFT plot. The models used are machine learning models such as Logistic Regression, Support Vector Machine, Light Gradient Boost methods, and Naive Bayes, which produced accuracies of 86.9%, 87%, 95.6% and 73.9%, respectively. When compared in terms of recall, precision and F1-score, the Light Gradient Boost methods still performed the best.

Jin-Cheol Kim *et al.* [5] proposed classifying the target type and estimating its moving direction using a millimetre-wave FMCW radar system. The proposed method is supposed to differentiate between the class targets, pedestrians, cyclists, and cars, by telling which way they are moving/facing (straight, left and right). With the signal received from the FMCW radar and by applying a Fourier transform, the distance to the object and the object's velocity can be estimated. Using multiple-input and multiple-output (MIMO) FMCW radar, i.e. using an FMCW radar that consists of several transmit and receiving antenna elements, allows for an estimate of the angle information of multiple targets. With the target's estimated distance, velocity, and angle information from the received radar signal, the target can be expressed in the 2D x-y distance plane as multiple points. Then the accumulated radar detection result is converted into an RGB image by saving the detection result in the format JPEG in order to be used as input data for CNN-based image classifiers. The model proposed by the authors is a Darknet-53 for feature extraction by using the converted image as input and a YOLOv2 that performs target identification based on the features extracted from the Darknet-53. The proposed model showed more than 85% recognition and classification accuracy for a new data set that was not used for training.

Qisong Wu *et al.* [6] proposed a hybrid support vector machine (SVM)-CNN classification method to distinguish vehicles from pedestrians. An LFMCW Radar with 1 Transmitter antenna and four receiver antennas (not MIMO) is used to collect radar data of the wanted targets. A two-dimension Fourier Transform (2D-FFT) is performed with the raw signal data received to acquire a Range-Doppler image. Data processing is performed by using constant false alarm rate (CFAR) and density-based spatial clustering of applications with noise (DB-SCAN) in the Range-Doppler images so that these physical features of underlying targets of interest are acquired for the subsequent classification. For the modified SVM, features like Range extension, Variance estimation in range, Radial velocity, Velocity extension and Variance estimation in the Radial velocity are extracted from the Range-Doppler images. In the case of the modified

SVM, it intends to predict that the target is a vehicle to alleviate the imbalance ratio of cars to pedestrians in the data. The rest of the unclassified targets have their Range-Doppler images used as input for the CNN. This hybrid method obtained an accuracy of 96%, an f1-score of 0.90 and an AUC of 0.99.

A. Moussa *et al.* [7] propose a compact set of convolutions to capture the primary features needed for pixel classification. The suggested method stacks measurements like occupancy, height, cross-section, power, and noise level into 2D grids representing the radar reflection's characteristics and computes the gradients of each pixel by using Haar-like kernels that represent average, first-order gradients in x and y directions. Then, a pixel classifier provides a vehicle classification probability/score to each pixel by using the outputs of the kernel convolutions.

Emanuele Tavanti *et al.* [8] use Doppler signatures and target reflectivity (RCS Signatures) to build a feature vector. These signatures are obtained by applying the signal infinite impulse response (IIR) filter to tackle clutter and leakage issues, calculating Range Doppler maps and applying both clusterings (DBSCAN) and tracking (Benedict-Bordner Smoothing) algorithms to then extract those signatures. For the classification, the k-nearest neighbour (k-NN) algorithm was used, showing 90% overall accuracy in recognition of single targets and an overall accuracy near 80% in recognition of multiple targets at once.

Michael Ulrich *et al.* [9] use a list of detected reflections extracted by calculating Range-Doppler spectrum and CFAR thresholding and direction-of-arrival (DOA) estimation. A global context layer (GCL) was proposed to integrate the architecture of the model (initiated by 1D-convolution) in order to learn abstract structural features, followed by another 1D-convolution to learn more about abstract features, which capture the structural relationships (e.g. length and width of the object), The reflection features used are x and y position of the reflection of the tracked object, radar cross-section (RCS) of the reflection, range of the reflection and radial velocity. The model proposed offered the best result with an accuracy of 93.45% with GCL and 92.20% without it.

A categorization system for human-driven vehicles is put forth by Eugin Hyun *et al.* [10] employing Doppler-spectrum properties. The authors introduced new three new features: scattering point count (SPC), scattering point difference (SPD) and magnitude difference rate (MDR), features based on the characteristics of the Doppler spectrum in two successive frames. SPC counts the Doppler reflection points of the detected target with power exceeding the reference threshold. SPD counts the difference between the number of SPC points of the current frame and the number of SPC points of the last frame in order to measure the time variance of the Doppler spectrum shape. MDR is the difference between the maximum power of the current frame (MPc) and the maximum power of the last frame, divided by MPc. From the two models proposed (SVM-support vector machine and BDT-binary decision tree), the binary decision tree obtained the best results by classifying humans and vehicles with an accuracy of 99.27% and 96.70%, respectively.

Ali Walid Daher *et al.* [11] use features that consist of the mean, variance and standard deviation of the range and Doppler profiles, along with their reflectivity (RCS signatures) and the estimated velocity of the target. These features are obtained by producing range-Doppler maps by performing Fast Fourier Transforms (FFT) on the raw data measured using FMCW radar, like in [15,16]. As for the model, a logic learning machine (LLM) was used where humans were classified with an accuracy of 100% and vehicles with an accuracy of 96.67%.

A 3D data matrix with axes corresponding to a range, azimuth, and velocity (also known as Doppler) and values that represent the measured radar reflectivity in that range/azimuth/Doppler bin is used by Andras Palffy*et al.* [12] in their study (known as radar cube). Therefore, a complete distribution of speed (i.e. Doppler vector) is given at multiple 2D range-azimuth locations, which can capture modulations of an object's main velocity caused by its moving parts and be valuable information for the classification. The proposed model is a specially designed CNN that obtained an average F1 score of 0.70 versus the 0.68 of the baseline used and also outperformed the same baseline in terms of accuracy.

Marcio Oliveira*et al.* [13] propose the use of Deep Convolutional Autoencoders (CAE) instead of Constant False Alarm Rate (CFAR) technique to reduce the noise in Range-Doppler maps (obtained by applying the 2D FFTs to the raw radar data). According to the authors, CAE outperforms CFAR, especially in highly noisy situations, because it can learn patterns and discern between noise and a moving object, thus being more robust. CAE system showed an improvement of 76.8% in correctly reconstructing objects and 188% in not reconstructing non-existing objects comparative the OS-CFAR method.

Ushemadzoro Chipengo *et al.* [14] propose a five layer Convolutional neural network (CNN) was used for classifying the different targets. Spectrograms were calculated by applying the Short-time Fourier transform (STFT) to the synthetic signal data generated by High-Frequency Structure Simulator (HFSS) Shooting and Bouncing Rays (HFSS SBRC) simulations. Those spectrograms contain valuable information about Doppler and micro-Doppler data of the detected targets. After just five epochs, the CNN achieved an accuracy of nearly 100% using 100% of the data generated. Using just 10% of the data generated and after 30 epochs, a classification accuracy of 90% was achieved.

4   Discussion

This section explores and discusses the review's findings against the research questions. The first subsection addresses the first research question and discusses how different computer science authors make the acquisition and data processing of radar sensor data. The second presents the most important features for detecting cyclists, pedestrians and vehicles. The last is related to the Artificial Intelligence strategies (methods and algorithms) used in Radar object detection, more precisely, pedestrians, vehicles and cyclists.

4.1   Acquisition and Data Processing of Radar Sensor Data

A Radar sends a signal through the transmitter antenna that, in case of an encounter with an object, will reflect in the body of that object and be received by the radar through the receiver antenna. Two types of radars exist to do this job. A pulse radar emits a signal from time to time, and a continuous wave radar, as the name says, emits a continuous signal. After the raw signal is received from the radar, the processing is needed in order to obtain meaningful information about the targets detected. Table 1 presents the types of acquisition and processing used by the different researchers.

Table 1. Acquisition and Data Processing

Author(s)/Article	Acquisition	Processing
Seungheon *et al.* [2]	FMCW Radar	FFT, Beamforming
Buchman *et al.* [3]	Ground Doppler-pulse radar	Fast Fourier Transform (FFT)
Jyoti Bhatia *et al.* [4]	FMCW mmWave radar	Fast Fourier Transform (FFT)
Jin-Cheol Kim *et al.* [5]	FMCW mmWave radar	Fast Fourier Transform (FFT), Beamforming
Qisong Wu *et al.* [6]	FMCW radar	2D-FFT, CFAR and DB-SCAN

4.2   Most Important Features for Radar Detection

Different approaches can be used to build a set of features to use in Radar Detection. The authors of the study's papers used different ways to express meaningful information to use in that detection. The most important features of some of the papers could be identified from the different datasets used, as presented in Table 2.

Table 2. Features for Radar Detection

Author(s)/Article	Features
Seungheon *et al.* [2]	3D point cloud data orthogonally projected onto the xy, yz and zx planes
Buchman *et al.* [3]	spectograms of the signals
Jyoti Bhatia *et al.* [4]	peaks characteristics in the FFT spectrum
Jin-Cheol Kim *et al.* [5]	estimated distance, velocity and angle information expressed in the 2D x-y distance plane as multiple points (RGB image)
Qisong Wu *et al.* [6]	Range-Doppler images and features extracted from the Range-Doppler images (Radial velocity, etc.)

4.3   AI Models in Radar Object Detection

In Table 3, we present some of the AI models implemented in other papers to Detect Radar objects.

Table 3. Models to Detect Radar Object

Author(s)/Article	Class Targets	Algorithms (Best)
Seungheon *et al.* [2]	pedestrians, cyclists, sedans and SUV's	multi-view CNN with an accuracy of 99.30%
Buchman *et al.* [3]	people or animals	ResNet and CNN with AUC of the ROC of 0.85 on the private test set
Jyoti Bhatia *et al.* [4]	Human, Drone and Car	Light Gradient Boost methods with an accuracy of 95.6%
Jin-Cheol Kim *et al.* [5]	pedestrian, cyclist, and car	YOLOv2 with an accuracy of 85%
Qisong Wu *et al.* [6]	vehicles and pedestrians	Hybrid SVM-CNN with an accuracy of 96%

5   Conclusion

Street monitoring aims to effectively identify road users (pedestrians, vehicles, cyclists). With the information provided by it, we can apply that information to various levels of our quotidian street life and try to improve it (e.g. the number of incidents can be decreased, etc.). So it becomes important to improve that monitoring.

In this paper, a systematic review was conducted to analyse the literature written by computer science and engineering authors on radar detection of Pedestrians, vehicles and cyclists. First, to guide the search on the SCOPUS database, research questions were defined, and the methodology followed in the search was explained. The PRISMA method was the inspiration for the review. A search query was constructed using keywords related to the scope of this study, and triage of the articles was done following the exclusion criteria created. In the end, 13 articles were chosen to answer the research questions. It was also analysed if data acquisition is made by using either Pulse Doppler radars or Frequency Modulated Continuous Wave Radar or FMCW. That data is processed by applying techniques like Fast Fourier transform (FFTs) and density-based spatial clustering of applications with noise (DBSCAN) for clustering or Beamforming for angle estimation. With those transformations, various features, like spectrograms, spectrum peaks or images with point cloud data projected, were used. Different Machine learning/Deep Learning algorithms for radar object detection are presented in this paper. Although another digital library was also searched, it is necessary to point out that only the SCOPUS repository was considered in this systematic review, so there is a chance that relevant studies are not included.

Future work includes data analysis, development and validation of AI models for radar detection, and presentation of a radar-based pedestrian monitoring architecture.

Acknowledgments. This work has been supported by FCT - Fundação para a Ciência e Tecnologia within the R&D Units Project Scope: UIDB/00319/2020.

References

1. Page, M.J., McKenzie, J.E., Bossuyt, P.M., et al.: The PRISMA 2020 statement: an updated guideline for reporting systematic reviews. Syst. Rev. **10**, 89 (2021). https://doi.org/10.1186/s13643-021-01626-4
2. Kwak, S., et al.: Multi-view convolutional neural network-based target classification in high-resolution automotive radar sensor. IET Radar Sonar Navig. 1–12 (2022). https://doi.org/10.1049/rsn2.12320
3. Buchman, D., Drozdov, M., Krilavičius, T., Maskeliūnas, R., Damaševičius, R.: Pedestrian and animal recognition using doppler radar signature and deep learning. Sensors **22**, 3456 (2022). https://doi.org/10.3390/s22093456
4. Bhatia, J., et al.: Classification of targets using statistical features from range FFT of mmwave FMCW radars. Electronics **10**(16), 1965 (2021)
5. Kim, J.-C., Jeong, H.-G., Lee, S.: Simultaneous target classification and moving direction estimation in millimeter-wave radar system. Sensors **21**(15), 5228 (2021)
6. Wu, Q., et al.: Hybrid SVM-CNN classification technique for human-vehicle targets in an automotive LFMCW radar. Sensors **20**(12), 3504 (2020)
7. Moussa, A., El-Sheimy, N.: Automotive radar based lean detection of vehicles. Int. Arch. Photogramm. Remote Sens. Spatial Inf. Sci. **43**, 257–262 (2022)
8. Tavanti, E., et al.: A short-range FMCW radar-based approach for multi-target human-vehicle detection. IEEE Trans. Geosci. Remote Sens. **60**, 1–16 (2021)
9. Ulrich, M., Gläser, C., Timm, F.: Deepreflecs: deep learning for automotive object classification with radar reflections. In: 2021 IEEE Radar Conference (Radar-Conf21). IEEE (2021)
10. Hyun, E., Jin, Y.S.: Doppler-spectrum feature-based human-vehicle classification scheme using machine learning for an FMCW radar sensor. Sensors **20**(7), 2001 (2020)
11. Daher, A.W., et al.: Pedestrian and multi-class vehicle classification in radar systems using rulex software on the raspberry PI. Appl. Sci. **10**(24), 9113 (2020)
12. Palffy, A., et al.: CNN based road user detection using the 3D radar cube. IEEE Robot. Autom. Lett. **5**(2), 1263–1270 (2020)
13. de Oliveira, M.L.L., Bekooij, M.J.G.: Deep convolutional autoencoder applied for noise reduction in range-Doppler maps of FMCW radars. In: 2020 IEEE International Radar Conference (RADAR). IEEE (2020)
14. Chipengo, U., et al.: High fidelity physics simulation-based convolutional neural network for automotive radar target classification using micro-doppler. IEEE Access **9**, 82597–82617 (2021)
15. Rizik, A., et al.: Feature extraction for human-vehicle classification in FMCW radar. In: 2019 26th IEEE International Conference on Electronics, Circuits and Systems (ICECS). IEEE (2019)
16. Wojtkiewicz, A., et al.: Two-dimensional signal processing in FMCW radars. In: Proceedings of XX KKTOiUE, pp. 475–480 (1997)
17. Gao, X., et al.: RAMP-CNN: a novel neural network for enhanced automotive radar object recognition. IEEE Sens. J. **21**(4), 5119–5132 (2020)

Voice Identification of Spanish-Speakers Using a Convolution Neural Network in the Audio Interface of a Computer Attack Analysis Tool

Andrey Vishnevsky📵 and Nadezda Abbas(✉) 📵

University of Bernardo O'Higgins, Avenida Viel 1497, Santiago, Chile
nadezda.abbas@ubo.cl

Abstract. This work is dedicated to the elaboration of a machine learning model for identifying users by voice. The voice identification module complements the previously developed prototype of a software solution for professional monitoring and response to information security incidents for visually impaired people. A brief characteristic of existing assistive technologies is presented. The article highlights the conditions and procedure of the experiment that was carried out in order to implement voice authorization in every day information security routine at one of the Universities. The main aim of the elaborated voice recognition module is, on the one hand simplify usage of information security software, on other is to implement the principal of employee inclusion in cybersecurity. Speech recordings of nine Spanish-speaking university employees were used as a training sample. Combinations of features were studied, according to which the convolution neural network has classified speakers with 90% accuracy.

Keywords: Inclusive technologies · Sonification · Assistive technologies · Voice recognition · Machine learning

1 Introduction

Modern technologies make it possible to manage complex software complexes by ear. Data sonification refers to the act of turning data into sound. There are various applications of sonification from seismology measurements to helicopter flight parameters audialization and assistive technologies.

In this work we have developed voice identification module for implementation in sonified user interface of computer attack detection system. The core of our assistive technology is based on sound synthesizer and specially crafted sound recordings connected with web-application which audializes text features obtained from reverse engineering tools and network activity monitoring tools. The sound interface is based on web browser abilities of text vocalization [1, 2]. We propose to supplement the previously developed set of sonified tools for analyzing computer attacks with a voice identification module to evaluate the effectiveness of several security operators – users of our information security system. In presented study, the voice identification is considered as a function that primarily increases the convenience of using the audio interface of information security tools. We study a scenario with a small data (nine users, five-six seconds of speech recordings) to check the applicability of standard speaker identification scenario in case close to small team of information security specialists. As a consequence, the 90% accuracy of speaker recognition was considered as enough. The elaborated system of software components will contribute to the solution of up-to date task of organizing the auditory work of information security specialists by taking into account the activity of identified users.

This article is formed up by five sections. The first one – the introduction. The second section provides an overview of the materials and methods of the subject area. The third part describes the methodology of the presented study. In the fourth part, the formulation of the experimental problem and the results of the voice recognition experiment are presented. The conclusions are presented in the final fifth section.

2 Materials and Methods

2.1 State of Art

Modern inclusive and assistive technologies could be divided into two groups, based on software and hardware solutions.

Among the hardware solutions, a special interest is a neurointerface, which reads electrical impulses of skin and transmits them as control signals to a computer [3]. Another hardware solution is a tactile Braille display, which allow reading from 40 to 80 characters. Although it allows people with certain vision problems to work with text information, it is not interactive enough to work with multi-level branched interfaces of computer programs and applications [4]. Another breakthrough solution is an invasive version of a hardware solution for restoring vision that is presented nowadays in the form of a bionic eye has not yet been sufficiently tested on humans for practical use [5].

Among the software solutions implemented in inclusive technologies and, in particular, among the assistive technologies for the blind and visually impaired people, auditory displays or audio user interfaces play a special role. Most of the audio user interfaces are implemented in the form of computer programs and are based on the methods of sonification and speech synthesis. Machine learning methods make it possible to name objects recognized by the smartphone camera in a voice, that helps visually impaired people navigate in space [6]. Screen readers pronounce text from a computer monitor and are the most popular and universal solution, but the speed of information exchange via the audio channel is significantly lower than the visual channel, which reduces the intensity of user interaction with the computer [7]. A number of works highlights various attempts that have been carried out in recent researches, aimed to improve the ergonomics of audio interfaces using music, in particular for encoding graphs, network activity and information security events [8–10].

Adaptive software solutions in most of the cases are based on machine learning libraries such as Tensorflow, Keras, DeepLearning4J, Theano, Pytorch, MXNet, Caffe, which allow users create and configure neural networks [11]. Each of these frameworks contains classification algorithms for recognition of various types of data patterns, including audio.

Recently various studies on voice identification were carried out in various scientific and research centres, their results show that the accuracy of speaker recognition depends not only on the machine learning model, but also on the sample that was used for its training, and varies from 60% to 90% [12–14].

2.2 Applied Methods

In the considered scenario of interaction with the system, presented in this publication, the user is identified by voice, recorded by a microphone. The size of chunk to divide large speech recording into the collection of training fragments is chosen to one second similarly as in [14]. One second if enough small size to be captured in authorization stage and also enough small to divide 5–6 min of recorded speech of each user into few hundreds of training samples.

The voice recording is divided into fragments with a duration of one second, and then transmitted to the Librosa library [15], which calculates the mel-frequency cepstral coefficients (MFCC) [16–18] using the formula 1:

$$\widehat{c_i} = \sum_{i=1}^{k} (log S_k) \, cos\left(\frac{\pi i\left(k - \frac{1}{2}\right)}{k}\right), \qquad (1)$$

where $\widehat{c_i}$ – final mel-frequency cepstral coefficients, S_k – filter set output, k – number of mel-frequency cepstral coefficients.

MFCC are the cepstral coefficients obtained on the basis of a distorted frequency scale based on human auditory perception. When calculating the low-frequency cepstral coefficients, the speech signal is divided into frames using windowing in Fourier transform. Then a fast Fourier transform is applied to find the power spectrum of each frame. The discrete cosine transform is applied to the speech signal after converting the power spectrum into a logarithmic domain to calculate the coefficients MFCC [17].

The obtained mel-frequency cepstral coefficients form a feature vector for the input of a convolutional neural network. Convolutional neural network (CNN), predicting the speaker's number by low-frequency cepstral coefficients, contains five dense layers and three dropout layers, as shown in Fig. 1. The convolutional neural network consisting of five layers of neurons predicts the speaker's identifier by mel-frequency kepstral coefficients. The first layer is convolutional, with a convolution core of 3 × 3 dimension and 128 filters. The second layer performs a pooling operation, divides the output of the first layer into 2 × 2 blocks. The third layer is convolutional, with 64 filters and a convolution core of 3 × 3 dimension. The fourth layer (Flatten layer) translates the output matrices of the third layer into one-dimensional vectors, which are fed to the input of the fifth fully connected layer. Activation functions of the first and third layers are ReLU. The last fifth layer uses the softmax activation function.

One of ours main objectives was to experimentally select the optimal values of the number of MFCC coefficients and the number of components of the Fourier transform (parameter **n_fft**) to achieve 90% accuracy in speakers recognition. In this case, the accuracy was calculated as the ratio of all correct predictions of the model to the total number of predictions according to formula 2.

$$Accuracy = \frac{TP + TN}{TP + TN + FP + FN}, \quad (2)$$

where TP – number of true positives, TN – number of true negatives, FP - false positives number, FN - false negatives number.

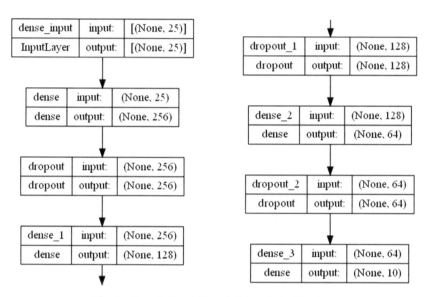

Fig. 1. Convolutional Neural Network Architecture

Number of losses in the model was approximated by Categorical Cross Entropy Loss function according to formula 3.

$$L_{CE} = -\frac{1}{N} \sum_{i=1}^{N} log p_{model}\left[y_i \in C_{y_i}\right], \quad (3)$$

where N – the number of samples in testing dataset, p_{model} – probability concluded by classification model that predicted label y_i belongs to real class C_{y_i}.

3   Experiment

We conducted an experiment on an isolated voice identification component written in the Python programming language. Convolutional neural network was created using Tensorflow machine learning library on a KVM virtual machine with two Xeon CPU 2.20 GHz cores.

The initial data for the analysis were nine speech recordings of Spanish-speaking speakers (university employees) lasting five to six minutes. All the participants of the experiment read the same text in Spanish. For these purposes we used Spanish version of Quentin Bells' book 'Bloomsbury Recall' ('El grupo de Bloomsbury' in Spanish). The marked-up sound recordings were saved in WAV format with a frequency of 16 kHz and a depth of 16 bits and broken into fragments with a duration of 1 s.

We sequentially iterated over the number of low-frequency cepstral coefficients **mfcc_n** and the number of components of the Fourier transform **n_fft** to minimize the time to achieve 90% speaker recognition accuracy.

The experiment was carried out twice. In the first case, noise was added to the voice recordings to increase the generalizing ability of the neural network. In the second case, no any noise was added.

3.1 Experiment Results

The duration of each calculation epoch in seconds was approximately equal to the number of mel-frequency cepstral coefficients **mfcc_n**. Empirically, it was found that between the 30th and 40th epochs, the accuracy decreases, and by the 50th epoch it is restored, as it's shown in Fig. 2 a, b. Therefore, we set the training duration to 50 epochs to stabilize the resulting speaker recognition accuracy.

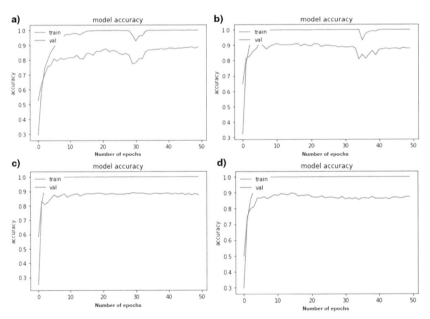

Fig. 2. The dependence of classification accuracy on the number of training epochs with a different number of low-frequency cepstral coefficients.

As it shown in Fig. 2 the following dependence of classification accuracy on the number of training epochs with a different number of low-frequency cepstral coefficients could be observed: a) **mfcc_n** is 10, resulting accuracy is 81%, b) **mfcc_n** is 20, resulting accuracy is 83%, c) **mfcc_n** is 40, resulting accuracy is 88%, d) **mfcc_n** is 80, resulting accuracy is 87%.

As shown in Table 1, when noise (sound of streaming water) is added to the recordings, 20 mel-frequency cepstral coefficients with 1024 Fourier transform components are sufficient to achieve 90% speaker recognition accuracy. The number of components of the Fourier transform less than 512 allowed to obtain the necessary accuracy in only one case with 50 MFCC coefficients, which increased the amount of calculations by three times compared to the optimal case.

Table 1. Dependence of the speaker classification accuracy on the number of mel-frequency cepstral coefficients when adding noise to the original signal

MFCC N	10	20	30	40	50	60	70	80
n_fft = 128	0.7812/8s	0.8555/20s	0.8086	0.8560	0.8745	0.7961	0.8275	0.8118
n_fft = 256	0.8863	0.8784	0.8549	0.8824	**0.9098/58s**	0.8677	0.8235	0.8745
n_fft = 512	0.8711	0.8555	0.8911	**0.9105/45s**	0.9333	0.9333	0.9294	0.9176
n_fft = 1024	0.8794	**0.9066/21s**	0.9023	0.9180	0.9219	0.9414	0.9490	0.9216

The results of experiments without adding noise are shown in Table 2.

Table 2. The dependence of the speaker classification accuracy on the number of mel-frequency cepstral coefficients without adding noise to the original signal

MFCC N	10	20	30	40	50	60	70	80
n_fft = 128	0.8175/8s	0.8333/20s	0.8214/32s	0.8651/44s	0.8770/57s	0.8492/69s	0.8333/84s	0.7381/87s
n_fft = 256	0.8571/7s	0.8849/17s	0.8968/29 s	0.8968/44s	0.8611/56s	**0.9087/68s**	0.8611/71s	0.8810/79s
n_fft = 512	**0.9008/7s**	0.9484/17s	0.9365/21s	0.9405/28s	**0.9802/49s**	0.9206/57s	0.9643/78s	0.9325/58s
n_fft = 1024	**0.9246/7s**	0.9127/17s	0.9524/20s	0.9524/44s	0.9762/49s	0.9683/58s	0.9683/80s	0.9643/55s

Without adding noise, the accuracy of speaker classification more than 90% was consistently obtained with the number of components of the Fourier transform n_fft from 512 and above. At the same time, the neural network achieved the required accuracy level with the number of mel-frequency capstral coefficients **mfcc_n** from 10 and above.

4   Conclusions

In this we present the results of estimation of the classification of speakers' accuracy by a convolutional neural network to implement voice identification in assistive technologies. Experiments have shown that with the least expenditure of computing resources, 90%

accuracy of speaker recognition by voice can be obtained with a small number of low-frequency cepstral coefficients from 10 to 20 and the number of components of the Fourier transform from 512 to 1024. The developed speaker recognition module shares the access to the sonified information security tools but is not accurate enough to be used as the only authorization mechanism. It opens the opportunity to evaluate the working process performed by each operator separately, but some additional logic is required (e.g. pronouncing a passphrase or user name for accurate authorization).

In the future, we plan to use the elaborated user recognition voice model to separate access to information security tools by configuring it for the voices of security system operators.

References

1. Vishnevsky, A., Abbas, N.: Sonification of information security incidents in an organization using a multistep cooperative game model. In: Rocha, A., Adeli, H., Dzemyda, G., Moreira, F. (eds.) WorldCIST 2022. LNNS, vol. 468, pp. 306–314. Springer, Cham (2022). https://doi.org/10.1007/978-3-031-04826-5_30

2. Vishnevsky, A., Ruff Escobar, C., Ruiz Toledo, M., Abbas, N.: Sonification of information security events in auditory display: text vocalization, navigation, and event flow representation. J. Access. Des. All **12**(1), 116–133 (2022). https://doi.org/10.17411/jacces.v12i1.359

3. Iskhakova, A.O., Volf, D.A., Iskhakov, A.Y.: Non-invasive neurocomputer interface for robot control. High-Perform. Comput. Syst. Technol. **5**(1), 166–171 (2021)

4. Bobko, R.A., Chepinskiy, S.A.: Multiline braille display construction model. Sci. Tech. J. Inf. Technol. Mech. Opt. **20**(5), 761–766 (2020). (in Russian). https://doi.org/10.17586/2226-1494-2020-20-5-761-766

5. Mirochenkov, M.V., Bozhinskaya, E.S.: Achievements in bionic eye implantation. In: Topical Issues of Modern Science and Education: Collection of Articles of the VIII International Scientific and Practical Conference, Penza, 20 February 2021. "Science and Education", Penza, pp. 232–235, EDN TINWOM, IP Gulyaev (2021)

6. Granquist, C., Sun, S.Y., Montezuma, S.R., Tran, T.M., Gage, R., Legge, G.E.: Evaluation and comparison of artificial intelligence vision aids: Orcam MyEye 1 and seeing AI. J. Vis. Impair. Blind. **115**(4), 277–285 (2021). https://doi.org/10.1177/0145482X211027492

7. Babikova, E.V.: Interfaces for communicating with local residents, blind and visually impaired users. Cult. Technol. Stud. **6**(4), 215–224 (2021). https://doi.org/10.17586/2587-800X-2021-6-4-215-224

8. Falk, C.: Sonification with music for cybersecurity situational awareness. In: The 25th International Conference on Auditory Display (ICAD 2019), pp. 50–55. Northumbria University, Newcastle upon Tyne, UK (2019). https://doi.org/10.21785/icad2019.014

9. Su, I., Hattwick, I., Southworth, C., et al.: Interactive exploration of a hierarchical spider web structure with sound. J. Multimodal User Interfaces **16**, 71–85 (2022). https://doi.org/10.1007/s12193-021-00375-x

10. Polaczyk, J., Croft, K., Cai, Y.: Compositional sonification of cybersecurity data in a baroque style. In: Ahram, T.Z., Karwowski, W., Kalra, J. (eds.) AHFE 2021. LNNS, vol. 271, pp. 304–312. Springer, Cham (2021). https://doi.org/10.1007/978-3-030-80624-8_38

11. Wafo, F., et al.: An evaluation of machine learning frameworks, pp. 1411–1416. ICIEA (2021). https://doi.org/10.1109/ICIEA51954.2021.9516253

12. Garain, A., Ray, B., Giampaolo, F., et al.: GRaNN: feature selection with golden ratio-aided neural network for emotion, gender and speaker identification from voice signals. Neural Comput. Appl. **34**, 14463–14486 (2022). https://doi.org/10.1007/s00521-022-07261-x
13. Shahin, I., Nassif, A.B., Nemmour, N., et al.: Novel hybrid DNN approaches for speaker verification in emotional and stressful talking environments. Neural Comput. Appl. **33**, 16033–16055 (2021). https://doi.org/10.1007/s00521-021-06226-w
14. Rituerto-González, E., Peláez-Moreno, C.: End-to-end recurrent denoising autoencoder embeddings for speaker identification. Neural Comput. Appl. **33**, 14429–14439 (2021). https://doi.org/10.1007/s00521-021-06083-7
15. McFee, B., et al.: librosa: audio and music signal analysis in python. In: Proceedings of the 14th Python in Science Conference, pp. 18–25 (2015)
16. Picone, J.W.: Signal modeling techniques in speech recognition. Proc. IEEE **81**(9), 1215–1247 (1993). https://doi.org/10.1109/5.237532
17. Gurtueva, I.A., Bzhikhatlov, K.: Analytical review and classification of methods for features extraction of acoustic signals in speech systems. News Kabardino-Balkarian Sci. Center RAS **1**(105), 41–58 (2022). https://doi.org/10.35330/1991-6639-2022-1-105-41-58
18. Zakovryashin, A.S., Malinin, P.V., Lependin, A.A.: Speaker recognition using mel-frequency cepstral coefficient distributions. Izvestiya Altai State Univ. **1**(84), 156–160 (2014). (in Russian). https://doi.org/10.17586/2226-1494-2020-20-5-761-766

Software and Systems Modeling

Task Scheduling in Cloud Computing Using Harris-Hawk Optimization

Iza A. A. Bahar[1], Azali Saudi[1], Abdul Kadir[2], Syed Nasirin[1(✉)], Tamrin Amboala[1], Esmadi A. A. Seman[1], Abdullah M. Tahir[1], and Suddin Lada[1]

[1] Universiti Malaysia Sabah, 87000 Labuan, Malaysia
{dssrg,snasirin}@ums.edu.my
[2] Universitas Sari Mulia, Banjarmasin City, Indonesia

Abstract. This paper presents a simulation of the Harris-Hawk Optimization (HHO) algorithm, which aims to minimize the makespan of a specified task set in a cloud computing environment. The algorithm is inspired by the team association of hawks in hunting and escaping prey. It has gained significant attention from researchers due to its effectiveness in solving real-world problems across different applications. As a result, the HHO algorithm has been widely applied to various optimization problems. In this study, the proposed HHO algorithm is simulated and compared with other well-known swarm intelligence algorithms, including Bat Algorithm (BA), Grey Wolf Optimization (GWO), and Particle Swarm Optimization (PSO). The simulation results demonstrate that the HHO algorithm surpasses the other three in producing better results. Furthermore, given the HHO algorithm's reliability in solving single-objective problems, this study further validates its effectiveness in addressing various optimization concerns.

Keywords: Task Scheduling · BA · GWO · PSO · HHO · Cloud Computing

1 Introduction

Since the rise of several cloud services, end-users' cloud-based data has rapidly grown. This is because different users require different types of cloud services. These demands are usually solved using task scheduling algorithms that may galvanize the cloud data centre resources. However, massive cloud data centre growth requires more refined task scheduling algorithms. Task scheduling is the process of organizing incoming requests (tasks) in such a way that the available cloud data centre resources are optimally utilized. If jobs are properly scheduled, resources will be free before deadlines, resulting in faster response times.

This process is significant as task scheduling may properly utilize limited resources (CPU, VMs, etc.). By considering some parameters, the task-scheduling algorithm may select available resources. These algorithms are rules that assign appropriate resources to get the most elevated performance possible. Various swarm algorithms, such as Bat Optimization (BA), Grey Wolf Optimization (GWO), and Particle Swarm Optimization (PSO), were formed to address task scheduling issues [7]. Recently, Harris-Hawks

A. Rocha et al. (Eds.): WorldCIST 2023, LNNS 801, pp. 155–166, 2024.
https://doi.org/10.1007/978-3-031-45648-0_16

Optimization (HHO) is becoming a prevalent swarm-based optimization as it could provide solutions to single-objective problems. The algorithm mimics the action of the hawk's team partnership in hunting and fleeing prey. As a result, the algorithm has received widespread attention among researchers regarding its performance in dealing with other systems in real-world problems. This interest has led to the emergence of HHO applications in diverse optimization issues.

Given the strength of this emerging algorithm in solving single-objective problems, this paper simulates the makespan performance of VMs' placement in a cloud data centre environment in which an HHO-based scheduler is proposed against three other well-known heuristic algorithms (BA, GWO and PSO).

2 Related Works

A group of researchers proposed a methodology for discovering a nearly optimal explanation for multi-objective task scheduling concerns in a cloud environment. The approach presented was a swarm-intelligence-based technique, the hybridized bat algorithm, designed for multi-objective task scheduling. The researchers compared the results of this approach to those of similar metaheuristic-based techniques evaluated under identical conditions. The simulation results demonstrated the significant potential of this approach for addressing task scheduling problems [1].

Several task-scheduling techniques have been proposed for cloud computing environments. One of these techniques is the Grey Wolf Optimizer (GWO), which has been shown to solve task scheduling potentially problems with varying tasks ranging from 100 to 600 [2]. Another technique is based on the particle swarm optimization (PSO) algorithm, where non-preemptive tasks are considered [3]. Significant improvements have been made to this method by integrating load-balancing techniques, resulting in a 22% increase in resource utilization and a 33% reduction in makespan compared to the basic PSO algorithm. Additionally, the proposed method displays faster convergence towards near-optimal solutions, demonstrating its superior performance.

Kumar, Nagaratna & Marrivada proposed a PSO-based method. He compared it with existing scheduling algorithms such as First Come, First Serve (FCFS) scheduling and Round Robin, Min-Min scheduling policy [4]. The proposed approach demonstrated superior performance in terms of both execution time and makespan compared to the other algorithms. Moreover, Alsaidy, Abbood & Sahib proposed improved initialization of the PSO algorithm using heuristic algorithms such as the longest job to fastest processor (LJFP) and minimum completion time (MCT) [5]. The LJFP-PSO and MCT-PSO algorithms were tested to see how well they could minimize the time it takes to complete a task, the total time it takes to do all the tasks, how evenly the tasks were distributed, and how much energy was used. The results showed that the new algorithms performed better than the old ones that were already being used.

Amer, Attiya Zedan & Nasr presented a modified Harris Hawks Optimizer (HHO) for multi-objective scheduling problems [6]. The modification, called elite opposition-based learning, enhances the exploration stage of the standard HHO algorithm. The algorithm was used to minimize schedule length and execution cost while maximizing resource utilization and was found to perform better than other algorithms. Other task scheduling techniques, such as Ant Colony Optimization, Genetic Algorithm, Clustering-based scheduling, and Fuzzy-based scheduling, are available [7]. In this study, we only examined the performance of standard HHO against the other three most common swarm-based algorithms, BA, GWO, and PSO, in solving a single-objective task scheduling problem.

3 The Task Scheduling Algorithms

3.1 Bat Algorithm

Bat Algorithm (BA) is a metaheuristic algorithm for optimization. It was inspired by the behaviour of microbats, with differing emission and loudness pulse rates. Xin-She Yang designed the BA in 2010. The algorithm is used in different optimization situations because of its excellent performance. Moreover, the BA bears many advantages, i.e., it can deliver fast intersections for services, much like classifications. The simulation also revealed that good function optimization is accomplished with the Bat Algorithm. Ulah, Navi and Khan further examine 17 papers on the Bat algorithm for enhancing resource allocation in cloud VMs. BA advances cloud computing to improve energy oversight and load balancing [8].

One of its significant advantages is its ability to handle energy-aware virtual machine placement tasks in cloud data centres. The algorithm uses microbats' loudness and pulse rates to search for optimal solutions efficiently. This approach can help optimize the allocation of virtual machines in the cloud data centre, thereby reducing energy consumption and improving load balancing. Furthermore, the BA has been successfully applied in optimizing resource allocation in cloud VMs, which is essential in cloud computing [15]. The algorithm can efficiently manage the energy consumption of virtual machines and balance the workload across servers. It can also adapt to changing workloads, making it a valuable tool for managing cloud resources.

3.2 Grey Wolf Optimization Algorithm

Grey Wolf Optimization (GWO) is a meta-heuristic optimization algorithm that imitates the social hunting behaviour of grey wolves. It is a relatively new algorithm, but it has shown promising results in solving various optimization problems. In the context of energy-aware virtual machine placement tasks in cloud data centres, the GWO algorithm can find an optimal placement of virtual machines that minimizes energy consumption while satisfying users' Quality of Service (QoS) requirements. The strength of GWO in this application lies in its ability to handle the complex and dynamic nature of cloud data centers [14].

The GWO algorithm can efficiently search the large solution space of virtual machine placement tasks to find the optimal placement that minimizes energy consumption. Furthermore, the algorithm can also adapt to changes in the workload and resource availability of the data centre, which is essential for ensuring efficient operation and minimizing energy consumption.

3.3 Particle Swarm Optimization (PSO) Algorithm

Particle Swarm Optimization (PSO) has proven to be a successful approach in optimizing a wide range of problems, including the placement of virtual machines in cloud data centers to improve energy efficiency. The strength of the PSO algorithm in this particular task lies in its ability to search for optimal solutions in a high-dimensional search space while considering the constraints and objectives related to energy consumption, resource utilization, and performance metrics [17].

The PSO algorithm is well-suited for such optimization problems. It can explore the search space efficiently using a swarm of particles that communicate and adapt their behaviour based on the best solution found so far. In the context of virtual machine placement, the PSO algorithm can be used to optimize the allocation of virtual machines to physical servers, taking into account the energy consumption of each server, the workload of each virtual machine, and the communication overhead between virtual machines. Furthermore, the PSO algorithm is easily parallelizable, which is particularly important for cloud data centres with many servers and virtual machines. This allows the algorithm to scale to larger problem sizes and exploit the computational resources in a cloud environment.

3.4 Harris Hawk Optimization (HHO) Algorithm

Harris Hawk Optimization (HHO) is a new metaheuristic algorithm designed to solve the global optimization problem [9]. The algorithm imitates the behaviours of the hawks in nature while searching and catching their prey. HHO executes the examination process during exploration and exploitation based on different strategies. The primary strength of HHO in this context is its ability to balance exploration and exploitation. During exploration, HHO can search the solution space effectively, while during exploitation, it can refine the solutions found to converge towards the optimal solution [19].

In the context of energy-aware virtual machine placement, HHO can optimize the placement of virtual machines on physical hosts to minimize energy consumption while meeting application performance requirements. This can be achieved by formulating the placement problem as an optimization problem and using HHO to find an optimal solution. HHO's ability to balance exploration and exploitation and handle multi-objective optimization problems makes it a strong candidate for solving energy-aware virtual machine placement tasks in cloud data centers.

4 Experimental Setup

The experiments were conducted on a laptop with Intel i5-1021U CPU @1.60GHz with 16.0GB RAM running Windows 10 and using CloudSim toolkit. All algorithms (PSO, BA, GWO & HHO) were implemented using Apache Ant. It is a Java library and command-line tool designed to perform operations described in build files as targets and extension points that depend on each other. Its primary use is for building Java applications, but it is versatile enough to handle other applications, such as C or C++. Ant is written in Java, offering great flexibility, and comes equipped with various built-in tasks for gathering, creating, testing, and executing Java applications. It can also manage any process described in terms of targets and tasks.

The performance evaluation and comparison of task scheduling algorithms were conducted using the CloudSim simulator, an open-source framework for modelling and simulating cloud computing environments [10]. The CloudSim toolkit [11] supports the modelling of various cloud system components, such as data centres, hosts, virtual machines (VMs), cloud service brokers, and resource provisioning strategies [11, 12].

The simulation environment was set up using two host machines in a data centre, hosting 25 virtual machines (VMs). The processing capacity of each VM was measured in Million Instructions Per Second (MIPS). To evaluate the performance of the distributed system, two commonly used parallel workloads, namely NASA Ames iPSC/860 and HPC2N, were employed. These workloads are recognized benchmarks for assessing the efficiency of distributed systems.

The VM and cloudlet information in the simulated workloads was extracted from two real log traces, namely NASA Ames iPSC/860 and HPC2N Seth log-trace [24], available in Feitelson's Parallel Workloads Archive (PWA) [25], to model High-Performance Computing (HPC) tasks. The jobs were assumed to be independent and non-preemptive. CloudSim Plus was employed to schedule CPU resources at the VM level using the BA, GWO, PSO, and HHO algorithms to minimize VMs makespan (optimum duration) time for a population of 100, 200, and 300 tasks. The experiments were performed ten times, and the average performance of task scheduling was recorded and compared. The specific parameter settings for each metaheuristic (MH) method are presented in Table 3.

Table 1. Table captions should be placed above the tables.

Cloud Entity	Parameters	Value
Data centre	No. of data centres	1
Host	No. of hosts	2
VM	Storage	1TB
	RAM	16GB
	Bandwidth	10Gb/s
	Policy type	Time shared
	No. of VMs	25
	MIPS	100 to 5000
	RAM	0.5GB
	Bandwidth	1Gb/s
	Size	10GB
	VMM	Xen
	No. of CPUs	1
	Policy type	Time shared

Table 2. Description of the real parallel workloads used in performance evaluations.

Log	Duration	CPUs	Jobs	Users	File
NASA iPSC	Oct 1993 – Dec 1993	128	18239	69	NASA-iPSC-1993–3.1-cln.swf
HPC2N	Jul 2002 – Jan 2006	240	202871	257	HPC2N-2002–2.2-cln.swf

Table 3. Parameter settings of each meta-heuristics method evaluated.

Algorithm	Parameter	Value
PSO	Swarm size	100, 200, 300
	Generation	50
	Inertia weight	0.9
	Cognitive coefficient c1	1.20
	Social coefficient c2	1.20
	Random cognitive	1.0
	Random social	1.0
BA	Population size	100, 200, 300
	Generation	50
GWO	Population size	100, 200, 300
	Generation	50
HHO	Population size	100, 200, 300
	Generation	50
	Lower bound	0
	Upper bound	7

Makespan: The makespan is a widely utilized metric for evaluating scheduling effectiveness in cloud computing. It refers to the time taken for the most recently completed job to finish. A smaller makespan implies that the cloud broker is proficiently assigning jobs to the appropriate virtual machines (VMs). The goal is to minimize the completion time of jobs, which is equivalent to minimizing the makespan. The objective is to identify the optimal allocation of virtual resources to jobs, reducing the makespan of job schedules [10]. The HHO algorithm demonstrated the most notable accomplishments in the evaluations conducted.

Pseudocode for PSO, BA, GWO & HHO is presented in Algorithm 1, 2, 3 & 4.
Algorithm 1: PSO Pseudocode [20].

```
1    Initialize population
2    for t = 1 : maximum generation
3        for i = 1 : population size
4            if  f(x_{i,d}(t)) < f(p_i(t))  then  p_i(t) = x_{i,d}(t)
5                f(p_g(t)) = min_i(f(p_i(t)))
6            end
7            for d = 1 : dimension
8                v_{i,d}(t+1) = wv_{i,d}(t) + c_1 r_1(p_i - x_{i,d}(t)) + c_2 r_2(p_g - x_{i,d}(t))
9                x_{i,d}(t+1) = x_{i,d}(t) + v_{i,d}(t+1)
10               if  v_{i,d}(t+1) > v_max  then  v_{i,d}(t+1) = v_max
11               else if  v_{i,d}(t+1) < v_min  then  v_{i,d}(t+1) = v_min
12               end
13               if  x_{i,d}(t+1) > x_max  then  x_{i,d}(t+1) = x_max
14               else if  x_{i,d}(t+1) < x_min  then  x_{i,d}(t+1) = x_min
15               end
16           end
17       end
18   end
```

Algorithm 2. BA Pseudocode [19, 21]

```
While (t < maximum number of iterations)
    For i = 1:N
        Generate a new bat (B_new) using (8), (9) and (10)
        If rand > r_new
            Select one among the best solutions and
            generate a local solution around this one, using (11)
        Else
            Select randomly a solution and generate a local
            solution around this one, using (11)
        End if
        Evaluate the bats
        If (rand < A_i) and (B_new < x_i)
            x_i = B_new
            Increase r_i and reduce A_i, using (12) and (13)
        End if
    End for
    Rank bats to find the best solutions in population
    Find the best bat
End while
```

Algorithm 3. GWO Pseudocode [22].

```
Initialize the grey wolf population Xₜ (i = 1, 2, ..., n)
Initialize a, A, and C
Calculate the fitness of each search agent
Xₐ=the best search agent
Xᵦ=the second best search agent
Xᵟ=the third best search agent
while (t < Max number of iterations)
    for each search agent
            Update the position of the current search agent by equation (3.7)
    end for
    Update a, A, and C
    Calculate the fitness of all search agents
    Update Xₐ, Xᵦ, and Xᵟ
    t=t+1
end while
return Xₐ
```

Algorithm 4. HHO Pseudocode [23].

Algorithm 1 Pseudo-code of HHO algorithm
***
Inputs: The population size N and maximum number of iterations T
Outputs: The location of rabbit and its fitness value
Initialize the random population $X_i(i = 1, 2, \ldots, N)$
while (stopping condition is not met) **do**
   Calculate the fitness values of hawks
   Set X_{rabbit} as the location of rabbit (best location)
   for (each hawk (X_i)) **do**
      Update the initial energy E_0 and jump strength J ▷
E_0=2rand()-1, J=2(1-rand())
      Update the E using Eq. (3)
      if $(|E| \geq 1)$ **then**   ▷ Exploration phase
         Update the location vector using Eq. (1)
      if $(|E| < 1)$ **then**   ▷ Exploitation phase
         if $(r \geq 0.5$ and $|E| \geq 0.5$) **then**   ▷ Soft besiege
            Update the location vector using Eq. (4)
         else if $(r \geq 0.5$ and $|E| < 0.5$) **then**   ▷ Hard besiege
            Update the location vector using Eq. (6)
         else if $(r < 0.5$ and $|E| \geq 0.5$) **then**   ▷ Soft besiege
with progressive rapid dives
            Update the location vector using Eq. (10)
         else if $(r < 0.5$ and $|E| < 0.5$) **then** ▷ Hard besiege
with progressive rapid dives
            Update the location vector using Eq. (11)
Return X_{rabbit}

5   Results and Discussions

All algorithms, BA, GWO, PSO and the proposed HHO were evaluated on 25VMs with 100, 200 and 300 tasks. Two public datasets were used, are obtained from NASA and HPC2N. All algorithms were examined based on ten runs, and the average results were taken to calculate the average value (Mean) and standard deviation (SD). Tables 4, 5, and 6 show the obtained Mean and SD of the four algorithms with 100, 200, and 300 tasks, respectively. As shown in Tables 4, 5, and 6, the HHO recorded the shortest makespan for both datasets. The other three algorithms also performed well by providing an acceptable length of makespan for different numbers of tasks on both datasets.

Table 4. Makespan performance comparison for all approaches with 100 tasks.

Algorithm	NASA dataset		HPC2N dataset	
	Mean	SD	Mean	SD
PSO	32.2176	10.9621	469.0628	91.8795
BA	32.1276	9.2246	464.3825	156.8146
GWO	30.0729	4.8688	399.1129	86.5015
HHO	26.6456	4.2822	382.1820	93.0792

Table 5. Makespan performance comparison for all approaches with 200 tasks.

Algorithm	NASA dataset		HPC2N dataset	
	Mean	SD	Mean	SD
PSO	30.9100	5.1291	444.5650	101.5169
BA	29.9823	10.2358	451.3150	134.8656
GWO	27.9023	6.0685	389.5701	69.4464
HHO	25.1203	4.2526	367.9577	78.1252

Table 6. Makespan performance comparison for all approaches with 300 tasks.

Algorithm	NASA dataset		HPC2N dataset	
	Mean	SD	Mean	SD
PSO	29.9865	5.9968	432.0106	52.4225
BA	29.7258	5.5160	458.8009	110.9215
GWO	26.9092	3.7989	385.914	51.0460
HHO	25.4294	4.3026	381.4154	118.0227

The results of HHO tasks 100, 200, and 300 indicated that when the population size is increased, the performance also improves. However, increasing the number of populations has little impact on the results.

Table 7. Average percentage reduction against PSO (using NASA dataset).

Population	BA	GWO	HHO
100	0.28%	6.66%	17.29%
200	3.00%	9.73%	18.73%
300	0.87%	10.26%	15.20%

Table 7 showed that HHO performs better than the other three algorithms for each population. This indicates that HHO is far better in task scheduling, followed by GWO and BA for NASA datasets.

Table 8. Average percentage reduction against PSO (using HPC2N dataset).

Population	BA	GWO	HHO
100	1.00%	14.91%	18.52%
200	−1.52%	12.37%	17.23%
300	−6.20%	10.67%	11.71%

Table 8 above shows that HHO performs better than the other three algorithms for each population. This indicates that HHO is far better in task scheduling, followed by GWO and BA for HPC2N datasets. The performance of different scheduling algorithms was compared based on their makespan, which refers to the time taken to complete all the tasks. Table 7 and Table 8 present the results, and they indicated that the HHO algorithm performs better than other algorithms like PSO, BA, and GWO. This suggests that the HHO algorithm is faster in completing tasks in the test cases. Additionally, the HHO algorithm performs significantly better in the NASA Ames iPSC/860 test case than in the HPC2N test case.

6   Discussions and Conclusions

A cloud data centre must manage resources efficiently for better user service and performance. Task scheduling plays an important role in achieving this. The HHO algorithm has been tested against other algorithms, such as BA, GWO, and PSO, to minimize the time a task takes to complete VM makespan. The results show that the HHO algorithm is better than others.

In conclusion, this study presents a simulation of the Harris-Hawk Optimization algorithm for minimizing the makespan of a specified task set in a cloud computing environment. The results show that the HHO algorithm surpasses known swarm intelligence algorithms, including the Bat Algorithm, Grey Wolf Optimization, and Particle Swarm Optimization. The findings of this study have significant implications for cloud data centres that require efficient resource management for performance. The HHO algorithm's ability to optimize task scheduling and minimize VM makespan can reduce costs. However, future work should test the HHO algorithm against other constraints, such as energy consumption, to assess its true capabilities comprehensively.

References

1. Bezdan, T., Zivkovic, M., Bacanin, N., Strumnberger, I., Tuba, E., Tuba, M.: Multi-objective task scheduling in a cloud computing environment hybridized bat algorithm. J. Intell. Fuzzy Syst. **42**, 411–423 (2022)

2. Bacanin, N., Bezdan, T., Tuba, E., Strumberger, I., Tuba, M., Zivkovic, M.: Task scheduling in a cloud computing environment by Grey Wolf Optimizer. In: 2019 27th Telecommunications Forum (TELFOR), pp. 1–4 (2019)

3. Ebadifard, F., Babamir, S.: A PSO-based task scheduling algorithm improved using a load-balancing technique for the cloud computing environment. Concurr. Comput. Practice Expertise. **30**(12), e4368 (2018)

4. Kumar, S., Nagaratna, M., Marrivada, I.: Task scheduling in cloud computing using PSO Algorithm. In: Smart Intelligent Computing and Applications, vol. 1, 541–550, Springer Nature Singapore (2022)

5. Alsaidy, S.A., Abbood, A.D., Sahib, M.A.: Heuristic initialization of PSO task scheduling algorithm in cloud computing. J. King Saud Univ. – Comput. Inf. Sci. (2020)

6. Amer, D., Attiya, G., Zeidan, I., Nasr, A.: Elite learning Harris hawk's optimizer for multi-objective task scheduling in cloud computing. J. Supercomput.Supercomput. **78**(2), 2793–2818 (2022)

7. Arunarani, A.R., Manjula, D., Sugumaran, V.: Task scheduling techniques in cloud computing: a literature survey. Future Gener. Comput. Syst. **91**, 407–415

8. Ullah, A., Nawi, N.M., Khan, M.H.: BAT algorithm used for load balancing purpose in cloud computing: an overview. Int. J. High-Perform. Comput. Network. **16**(1), 43–54 (2020)

9. Heidari, A.A., Mirjalili, S., Faris, H., Aljarah, I., Mafarja, M., Chen, H.: Harris hawks optimization: algorithm and applications. Futur. Gener. Comput. Syst.. Gener. Comput. Syst. **97**, 849–872 (2019)

10. Huang, X., Li, C., Chen, H., An, D.: Task scheduling in cloud computing using particle swarm optimization with time-varying inertia weight strategies. Clust. Comput.. Comput. **23**(2), 1137–1147 (2020)

11. Calheiros, R.N., Ranjan, R., Beloglazov, A., De Rose, C.A., Buyya, R.: CloudSim: a toolkit for modelling and simulating cloud computing environments and evaluating the evaluation of resource provisioning algorithms. Softw. Practice Exp. **41**(1), 23–50 (2011)

12. Attiya, I., Abd Elaziz, M., Xiong, S.: Job scheduling in cloud computing using a modified Harris Hawks optimization and simulated annealing algorithm. Computational Intelligence and Neuroscience. (2020)

13. Pirozmand, P., Hosseinabadi, A.A.R., Farrokhzad, M., Sadeghilalimi, M., Mirkamali, S., Slowik, A.: Multi-objective hybrid genetic algorithm for task scheduling problem in cloud computing. Neural Comput. Appl.Comput. Appl. **33**, 13075–13088 (2021)

14. Calheiros, R.N., Ranjan, R., Beloglazov, A., De Rose, C.A., Buyya, R.: CloudSim: a toolkit for modeling and simulation of cloud computing environments and evaluation of resource provisioning algorithms. Softw. Pract. Exp. **41**(1), 23–50 (2011)

15. Zbakh, M., Bakhouya, M., Essaaidi, M., Manneback, P.: Cloud computing and big data: Technologies and applications. Concurr. Comput. Practice Exp. **30**(12), e4517 (2018)

16. Quang-Hung, N., Thoai, N.: Eminret: heuristic for energy-aware vm placement with fixed intervals and non-preemption. In: 2015 International Conference on Advanced Computing and Applications (ACOMP), pp. 98–105. IEEE (2015)

17. Alsaidy, S.A., Abbood, A.D., Sahib, M.A.: Heuristic initialization of PSO task scheduling algorithm in cloud computing. J. King Saud Univ.-Comput. Inf. Sci. **34**(6), 2370–2382 (2022)

18. Mansoor, M., Mirza, A.F., Ling, Q.: Harris hawk optimization-based MPPT control for PV systems under partial shading conditions. J. Clean. Prod. **274**, 122857 (2020)

19. Yang, X.S.: A new metaheuristic bat-inspired algorithm. Nature inspired cooperative strategies for optimization (NICSO 2010), 65–74 (2010)

20. Dai, H.P., Chen, D.D., Zheng, Z.S.: Effects of random values for particle swarm optimization algorithm. Algorithms **11**(2), 23 (2018)

21. Severino, A.G., Linhares, L.L., de Araújo, F.M.: Optimal design of digital low pass finite impulse response filter using particle swarm optimization and bat algorithm. In: 2015 12th International Conference on Informatics in Control, Automation and Robotics (ICINCO), vol. 1, pp. 207–214. IEEE (2015)

22. Mirjalili, S., Mirjalili, S.M., Lewis, A.: Grey wolf optimizer. Adv. Eng. Softw. **69**, 46–61 (2014)

23. Heidari, A.A., Mirjalili, S., Faris, H., Aljarah, I., Mafarja, M., Chen, H.: Harris Hawk optimization: algorithm and applications. Future Gener. Comput. Syst. **97**, 849–872 (2019)

24. Yamany, W., Emary, E., Hassanien, A.E.: New rough set attribute reduction algorithm based on grey wolf optimization. In: The 1st International Conference on Advanced Intelligent Systems and Informatics (AISI2015), November 28–30, 2015, Beni Suef, Egypt, pp. 241–251. Springer, Cham (2016)

25. Shazly, K., Eid, M., Salem, H.: An efficient hybrid approach for Twitter sentiment analysis based on bidirectional recurrent neural networks. Int. J. Comput. Appl. **175**(17), 32–36 (2020)

Identifying Valid User Stories Using BERT Pre-trained Natural Language Models

Sandor Borges Scoggin and Humberto Torres Marques-Neto[✉]

Department of Computer Science, Pontifical Catholic University of Minas Gerais
(PUC Minas), Belo Horizonte, Brazil
`sandor.borges@sga.pucminas.br`, `humberto@pucminas.br`

Abstract. Currently, in the software industry, agile methodologies are vastly adopted. These methods use natural language for Requirement Engineering (RE) and are mostly done through the usage of User Stories (UST). This translates into a high cost to elicit and map the software requirements. Pre-trained language models have been shown to be able to verify the validity of requirements in UST format. Aiming to further understand their effectiveness, this paper presents a comparison between different BERT models to validate UST and discusses which model is the most efficient for RE. The main contributions contemplate an enhancement of the available datasets and validation models with up to 98% accuracy. We understand that further analyses should consider which aspect is more important in real-world applications: recall or precision.

Keywords: User Stories · BERT · NLP

1 Introduction

In the current landscape of the software industry, some of the most applied techniques are the ones that follow agile development methodologies such as SCRUM[1] and Kanban[2]. These methods can already reduce the cost of software development, especially regarding the cost of change throughout the development schedule. One of the reasons for this derives from how software requirements are defined and tracked.

Usually, Requirement Engineering (RE) starts a software process involving all stakeholders to define how the software would be established. Generally, agile methodologies use the User Story (UST) template to register a user's expectations. One of the great advantages of USTs is their clarity, which makes it easier for stakeholders to learn how to use them and understand what will be done.

In a general sense, the larger a project is the greater the number of requirements. This opens the possibility for requirement gathering with more expansive groups (e.g. student bodies, employees of companies, and citizens of a given

---

[1] https://www.scrum.org/.
[2] https://kanbanize.com/.

© The Author(s), under exclusive license to Springer Nature Switzerland AG 2024
A. Rocha et al. (Eds.): WorldCIST 2023, LNNS 801, pp. 167–177, 2024.
https://doi.org/10.1007/978-3-031-45648-0_17

area). With many advancements done in the field of Natural Language Processing (NLP), USTs could be pre-processed. In other words, both cost and time to deliver could be greatly reduced by reducing the need for many conversations between the dispersed parties involved. This paper intends to sort through large groups of UST and support practitioners, based upon state-of-the-art NLP techniques, such as attention mechanisms [10] and transformers [20].

Given this paper's intent, three methods to classify the USTs are presented. All of them are based on the Bidirectional Encoder Representations from Transformers (BERT) [5] pre-trained language model: (i) the BERT base, (ii) the DistilBERT [16] which is a distilled version of the original BERT model, and (iii) the Robustly Optimized BERT Pretraining Approach (RoBERTa) [8] an improved version of BERT with a larger amount of data used in pre-training and slight modifications in the training process. The work yields three final models with a high capacity to distinguish between the USTs, with accuracies of up to 98%.

The remainder of the paper is organized as follows: Sect. 2 presents a background of methods and technologies used here, Sect. 3 gives an overview of the related work, Sect. 4 describes the methodology applied, with a description of the data and implementation of each model, Sect. 5 describes and discusses the results, and finally Sect. 6 draws conclusions and possible future works.

2   Background

2.1   User Stories

These are concise text descriptions of functionality that should be present in the software product. This method clearly states who benefits, the benefit itself, and how it generates value using a simple and concise format. There is no fixed template and the format may vary among scholars and practitioners, but around 70% [9] use Cohn's template [2]: *"As a <role>, I want <benefit> [, so that <reason>]"*, the third part explaining why it generates value is optional and not always used.

2.2   Natural Language Processing

This is a field in Artificial Intelligence (AI) that performs the extraction of meaning or generation of texts. Many techniques can be applied such as ruled-based extraction or the application of Machine Learning (ML) for larger and more complex tasks.

One of the most common ways to treat NLP classification problems is through the use of Neural Networks (NN), where the natural language text is represented in a numerical form with the use of embeddings. Each representation extracts features from the text and creates a numerical form. One representation process is through self-attention for each token, normally applied in transformer architectures. Self-attention is an attention mechanism [10] that relates part of a sequence to the remaining parts of the sequence to generate its representation.

In [5], BERT is presented. The project expands upon transformers and their self-attention layer, more specifically, the tokens can attend to other tokens present on both sides of the text data, which incorporates greater context to the model. Another difference is the two steps implementation, using *pre-training* and *fine-tuning*. The former emphasizes two unsupervised tasks: masking of random tokens to allow bidirectional representation and next-sentence prediction to define and strengthen sentence relationships. The latter relates to the downstream task and its specific patterns. The results show improvements on several benchmarks that include the General Language Understanding Evaluation (GLUE [21]) and the Stanford Question Answering Dataset (SQuAD [14]).

2.3 Random Search

This is an approach that can be used to find the best hyperparameters, which are configuration aspects of the NN to generate the best model. To find the best possible values in a viable time frame many approaches have been developed, among them grid search and random search. As shown in [1], grid search presents reliable results in low dimensional spaces, with one or two parameters, but as more variables are added the results start to diminish. The second approach, using random values for each parameter, has shown to be more effective at identifying what are the best values for each parameter and therefore which set of values for all dimensions will yield the best-resulting model. This approach shows itself more effective the higher the dimensional space, as grid search has difficulty to find certain intervals that would generate the best outcomes.

3 Related Work

In past work, researchers have given great importance to the extraction of meaning and evaluation of a UST. The focus is mainly on the different AI methods that can be used for the analysis of different aspects of the USTs. The work of [19] presents a comparison between models that can identify valueless and untestable USTs. Using ML techniques (e.g. SVM, Decision Trees, and K-Nearest Neighbor) to analyze if a given UST has the needed characteristics to be valuable and testable. Their results show a tendency for decision trees to better classify USTs as they would mimic the sorting process of an actual developer based on the identification of keywords. Their conclusion emphasizes the importance, in their perspective, of not classifying bad USTs as valuable and testable since it would disturb the developers. In future work, they suggest the further classification of other aspects, such as size, independence, and estimability.

Another similar approach is presented in [7], where the authors discuss the importance of RE in the development process and how crowdsourcing could be implemented to help with this software process stage. The authors compare several techniques, such as SVM and Naïve Bayes classifiers, and keyword analyses. Their approach starts with the identification and preprocessing of keywords and labeling of user requests which are later fed to a classifier. Their results compare techniques, such as Unigrams and TF-IDF, to pass to the classifiers. Among them, the SVM had the best results along with the use of keywords. In conclusion, the keywords help in the classification process and increase metrics, such as accuracy, and it also shows that ML algorithms are more capable of dealing with this type of task.

In [13], a comparison between text classifiers is presented. Both models presented in the paper focus on classifying texts regarding if they are USTs or not. Their proposal stems from NLP resources, including a model that utilizes the ELMo architecture, focused on feature-based representations, and a BERT model that utilizes a masked linguistic model. The two models implemented consist of LSTM networks that have either an ELMo or a BERT embedding layer. The training data consists of both valid USTs and other text with a varying format or no requirement, these were labels with positive and negative labels so the models treat them as a binary classification problem. Results for both models showed high results in accuracy surpassing 96% and the BERT model outperforms the ELMo model, as it also provides analyses of semantic and syntactic roles. As suggestions for future work, the authors suggest improving the dataset used for training by adding more cases and adjusting the balance between both classes.

In this paper, a similar approach to [13] is presented by building upon it. The dataset is expanded and new classifiers are implemented. An improvement is made in the technologies used, with a deeper comparison among BERT models.

4   Methodology

The approach taken to achieve the goal of the paper follows certain steps. Firstly, the data used is structured and defined by applying several approaches, these include data gathering, data generation, and data augmentation. Secondly, the implementation of the models is described, with the architectures used and how the best hyperparameters for each model were searched and defined. Finally, the methods utilized to interpret and evaluate the results are described. The final results are generated by the best models trained in the implementation step, and the models are chosen based on the evaluation methods. In the following, we explain each step of our proposed methodology.

4.1   Data

The data used in this paper was structured based on the general process presented in [15]. The final dataset used to train the classifier is the aggregation of 3 different sources. Firstly, the final dataset presented in [13] which expands another dataset [4], secondly, part of the data extracted in [17] and, finally, newly generated data using openAI's[3] GPT-3 models.

The structure of the final data set counts with a split between training and testing sets, with 6,877 and 1,720 instances respectively, and a total of 8,896. The split is made with 80% of the data for training and 20% for testing.

---

[3] https://beta.openai.com/overview.

The dataset used as a base for this work is the one presented in [12,13]. As no dataset was available for the study the researchers created their own formatted dataset that positive and negative examples. Their data for valid USTs is extracted from two sources. Firstly, they extracted from [4] a dataset, which is an aggregation of USTs from actual software projects, and groups them by which project they belong. Secondly, they extract USTs from other projects, which also contributes to real samples. To generate the invalid stories an algorithm implementation to permute sub-sequences of the valid USTs was done. The permutation is done through a tokenizer and randomly selecting parts to form a new sequence of varying lengths.

As suggested in the future works of [13] new valid USTs were added. Considering how imbalanced the dataset was, many additions of valid USTs were necessary. For this, a part of the data extracted in [17], which compiles data from three open-source projects, hosted on Jira[4], is used. After filtering instances where USTs were not used or did not respect the established template, a total of 599 USTs were added to the dataset.

The other addition to the dataset is invalid USTs that still follow [2] structure, as the great majority of the samples with a negative label in the dataset are text with little grammatical sense and limited vocabulary. For this procedure, two methods were applied: generation of USTs with no meaning using the OpenAI API and rule-based augmentation [3].

Using OpenAI models, 250 invalid USTs were generated, these follow a cohesive idea such as "As a user, I want to be able to die and come back to life so that I can get away with anything" or "As an unemployed person, I want the ability to create money so that I don't have to work", but have no software requirement. For each call of the model, a prompt is passed with the instructions to "write a user story that makes no sense and is filled with nonsense", along with examples of invalid stories. The generation process focused on having the most diverse data possible, so frequency and presence penalties from the prompt were set to the highest possible values.

For ruled-based augmentation, a permutation of invalid USTs was made. The USTs were split between their 3 parts and mixed with others keeping the structure of the USTs. From the mixed invalid USTs, a random sample of 50 new invalid ones was picked.

In total, the data gathering and generation resulted in 599 instances of new valid USTs and 250 invalid cohesive USTs. The augmentation stage created 50 invalid USTs, resulting in 899 new instances.

4.2   Implementation

Given the data, different BERT pre-trained models were fine-tuned to differentiate between valid and invalid USTs. The BERT pre-trained models were chosen given their state-of-the-art performance in several tasks, and their great popularity [5].

---

[4] https://www.atlassian.com/software/jira.

Table 1. List of all hyperparameter groups.

	M1	M2	M3	M4	M5	M6	M7	M8	M9
Learning Rate	2e−5	5e−5	2e−5	2e−5	4e−5	5e−5	3e−5	5e−5	3e−5
Number of Epochs	4	3	3	3	4	3	2	3	3
Batch size	48	16	32	16	32	32	16	16	32
Weight decay	0.051	0.059	0.090	0.048	0.024	0.042	0.090	0.028	0.027

In more detail, a valid UST is a text that follows both the structure of a UST, presented in Sect. 2, and contains a valid requirement, like (i) *As a user, I want to revert to an older version of an uploaded file* and (ii) *As a Broker user, I want to create content mockups, so that I can submit my data.* Invalid USTs are all texts that fail to follow the structure or are unable to provide a valid requirement within the structure, like (i) *Add the Spectral LDA algorithm* or (ii) *As a researcher, I want to develop.*

To find the best model to classify the possible USTs, three different pre-trained models were fine-tuned and compared. The chosen pre-trained models were: BERT-base [5], DistilBERT-base [16], and RoBERTa-base [8]. The chosen models are among the most popular among the BERT family, shown by the number of uses in the Hugging Face platform[5], and give a better understanding of how different language model sizes affect the results. For each base model, an optimal hyperparameter search was made, with varying learning rates, a number of epochs, batch size, and weight decay.

The learning rate is the velocity at which the parameter vector is adjusted as the best value is searched, the optimal rate varies according to the dataset and other parameters. The number of epochs is the number of times a full pass is done over the entirety of the dataset. The batch size is the number of examples of data passed to the model in training at once. And weight decay is a type of regularization which adjusts the importance of parameters throughout the training process, this helps to avoid overfitting, which is when the model fits the data and cannot well identify new inputs [11].

The search for optimal hyperparameters was made through random search, since it has shown the best results in comparison to other methods [1]. The possible values for each training parameter are as follows: learning rate, batch size, and the number of epochs are the ones used in the BERT paper, and additionally, the search for the best weight decay is also done with a random generation of a value within the specified interval. The chosen values are from the default 0 up to 0.1, as in smaller datasets a higher weight decay can better suit the problem [18]. For comparison purposes, a group of 9 sets of hyperparameters was defined, named M1–M9, shown in Table 1. These sets were used to finetune models for BERT, DistilBERT, and RoBERTa.

The BERT base model is the original pre-trained model presented in its proposed work. The pre-training total parameters are similar to OpenAI GPT

---

[5] https://huggingface.co/.

with 110M for the base model. The utilized pre-training data includes two sets, BooksCorpus with the majority of words and English Wikipedia for complementary vocabulary. The BERT model has shown great performance for small datasets when fine-tuning for a specific task.

DistilBERT is a distilled version of the original BERT model. The major difference between them is the size of the model. Distillation is the process to reduce the size of models by compressing a larger model. The Distilled model trains to reproduce the original model but with fewer parameters, in this case, the DistillBERT model is 40% smaller than BERT and maintains 97% of the performance while being 60% faster. It also includes the same ability to fine-tune the model for a specific task.

RoBERTa is a proposed improvement upon the BERT model, as the authors of RoBERTa judged it to be significantly undertrained. The model addresses two major aspects, the data used in pre-training and the number of training steps, related to the size of each batch. The former is improved by expanding the data tenfold, the latter is modified by enlarging the number of steps five times. The resulting model sets new state-of-the-art results for the main NLP datasets (e.g. GLUE, RACE [6], and SQuAD).

5   Results

For all models, the training was made with the training split, which was split again in a training subset with an evaluation subset split in 90% for training and 10% for validation. After all models were trained with the different sets of hyperparameters (M1–M9), the best one among them was chosen to be evaluated against the test set. To choose which was the best model, the one with the highest accuracy was taken.

BERT Models had high scores across all metrics in their evaluation. Accuracy was on average around 98% which shows great performance overall, that is also the case for precision. On the other hand, recall is clearly the weakest metric across the board shown in Fig. 1. The best model achieved 97.73% accuracy, 98.15% precision, 95.94% recall, and 97.03% f1-score on the test set.

As shown in Fig. 1, BERT has mostly stable metrics, models M5, M6, M8, and M9 are on the higher end of the performance spectrum and are also the ones with lower weight decay values, up to 0.043. The other parameters varied among those four models but do not seem to affect the outcome as much, and is difficult to correlate the metrics directly to them.

DistilBERT Models achieve, on average, very close results to the BERT models, noticeable in Fig. 2. The average accuracy was slightly above 98%. In the case of precision, although it showed a similar average to BERT, it had a much broader range of values with a maximum of 99.21% and a minimum of 97.30%. The recall also varied, about 1.5%, but remained the lowest value. The best model achieved 97.61% accuracy, 97.27% precision, 96.54% recall, and 96.91% f1-score on the test set.

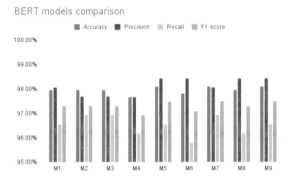

Fig. 1. Metrics of 9 BERT models

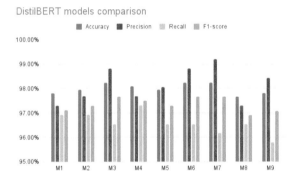

Fig. 2. Metrics of 9 DistilBERT models

Compared to BERT, DistilBERT got higher metrics in correlation with higher weight decay, models M3 and M7, shown in Fig. 2, have weight decays of 0.09. Also in contrast to BERT, DistilBERT has a greater variance which might come from it being a smaller language model. Furthermore, precision seems to work against the recall, which means the models neglect true data much easier with the high weight decay.

RoBERTa Models seem to, in general, outperform the BERT and DistillBERT models by a slight margin, achieving up to 98.4% in accuracy during training, visible in Fig. 3. However, in the averages of the trained RoBERTa models, they are not much more efficient than the other models and all metrics stay within a 0.2% difference. Nevertheless, the best RoBERTa model outperforms the final BERT and DistilBERT in the test set on all metrics, with 98.14% accuracy, 98.47% precision, 96.69% recall, and 97.57% f1-score.

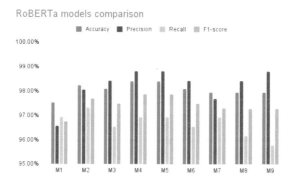

Fig. 3. Metrics of 9 RoBERTa models

RoBERTa diverges from both BERT and DistilBERT. It keeps some of the volatility of DistilBERT but has better performance, especially in the recall. Given how much larger the pre-trained model is RoBERTa was expected to have better performance. This might relate to its broader vocabulary and stronger semantic evaluation.

After testing a number of models in the training and validation set the one with the highest accuracy was taken. In this regard, the RoBERTa model achieved 98.14% and outperformed the BERT model which had 97.73% accuracy. The DistilBERT model had the weakest performance by a slight margin at 97.61% accuracy. While the difference between the models is not very big, a parallel can be drawn between the size of the language model used and the final performance. RoBERTa is designed to outperform BERT given the larger vocabulary which provides further context and facilitates the semantic examination. Considering USTs tend to have a broad set of words since they can specify a plurality of contexts, the RoBERTa model is more prepared to process them. The clearly superior performance of the RoBERTa models is shown in Fig. 4.

Although all models had great general performance, an important aspect is how much lower the recall score was for all models across the board. Combined with a high precision that also occurs, the highest metric in most cases, the data indicates a tendency for generating few false positives but also many false negatives. Given the objective of classifying USTs to eliminate poorly written ones, the models are quite effective. However, for real-world applications, it might make more sense to focus on improving the recall, as developers need the entirety of the requirements, and manually sorting out a few bad USTs that passed the model is a smaller workload than later adding requirements that were lost during elicitation.

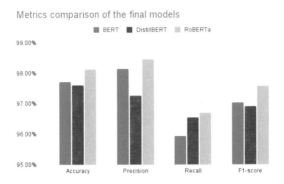

Fig. 4. Comparison of the final metrics for each model

6   Conclusion

In this paper, three different pre-trained language models were used to implement USTs classifiers to identify new ways to assist in RE. Based on the results for each model, it is possible to observe that BERT-based pre-trained models are highly effective in the classification of USTs, and among the ones presented, RoBERTa stands out as the most capable, with 98.14% accuracy, 98.47% precision, and 96.69% recall.

Regarding the effectiveness of the models in real-world applications, the importance given to precision and recall should be discussed. If missing a UST is too costly then there is much to improve in relation to the recall, on the other hand, if detecting bad USTs is more problematic, then the models present great performance but could be slightly improved.

For future works, there are mainly two fronts that could be improved. Firstly a great expansion of the dataset could be done. There is an imbalance in the type of data, and in general, the false examples could be expanded with texts with greater resemblance to USTs, such as the ones added in this paper with GPT-3 generation. But the number of valid USTs is of higher concern given the imbalance and the small current availability of datasets. Secondly, the models could be tested with larger BERT pre-trained models and a focus on improving the recall should be done. Finally, the validated USTs could be further analyzed, with the extraction of attributes, functions, and entities to generate structured software models in an automated procedure. This would also greatly reduce the cost of the software development process.

Acknowledgments. This work was supported by FAPEMIG (PPM- 00253-18) and by PUC Minas.

References

1. Bergstra, J., Bengio, Y.: Random search for hyper-parameter optimization. J. Mach. Learn. Res. **13**(2) (2012)
2. Cohn, M.: User Stories Applied: For Agile Software Development. Addison-Wesley Professional, Boston (2004)
3. Connor, S., Khoshgoftaar, T.M., Borko, F.: Text data augmentation for deep learning. J. Big Data **8**(1), 1–34 (2021)
4. Dalpiaz, F.: Requirements data sets (user stories). Mendeley Data, v. 1 (2018)
5. Devlin, J., Chang, M.-W., Lee, K., Toutanova, K.: BERT: pre-training of deep bidirectional transformers for language understanding. arXiv preprint arXiv:1810.04805 (2018)
6. Lai, G., Xie, Q., Liu, H., Yang, Y., Hovy, E.: Race: large-scale reading comprehension dataset from examinations. arXiv (2017)
7. Li, C., Huang, L., Ge, J., Luo, B., Ng, V.: Automatically classifying user requests in crowdsourcing requirements engineering. J. Syst. Softw. **138**, 108–123 (2018)
8. Liu, Y., et al.: RoBERTa: a robustly optimized BERT pretraining approach. arXiv preprint arXiv:1907.11692 (2019)
9. Lucassen, G., Robeer, M., Dalpiaz, F., Van Der Werf, J.M.E., Brinkkemper, S.: Extracting conceptual models from user stories with visual narrator. Requirements Eng. **22**(3), 339–358 (2017)
10. Niu, Z., Zhong, G., Yu, H.: A review on the attention mechanism of deep learning. Neurocomputing **452**, 48–62 (2021)
11. Patterson, J., Gibson, A.: Deep Learning: A Practitioner's Approach. O'Reilly Media Inc., Sebastopol (2017)
12. Peña Veitía, F.J.: Identifying User Stories in Issues Records. Mendeley (2020)
13. Peña Veitía, F.J., Roldán, L., Vegetti, M.: User stories identification in software's issues records using natural language processing. In: 2020 IEEE Congreso Bienal de Argentina (ARGENCON), pp. 1–7 (2020)
14. Rajpurkar, P., Zhang, J., Lopyrev, K., Liang, P.: Squad: 100,000+ questions for machine comprehension of text. arXiv (2016)
15. Roh, Y., Heo, G., Whang, S.E.: A survey on data collection for machine learning: a big data-AI integration perspective. IEEE Trans. Knowl. Data Eng. **33**(4), 1328–1347 (2019)
16. Sanh, V., Debut, L., Chaumond, J., Wolf, T.: DistilBERT, a distilled version of BERT: smaller, faster, cheaper and lighter. CoRR, abs/1910.01108 (2019)
17. Shahid, M.B.: Splitting user stories using supervised machine learning (2020)
18. Smith, L.N.: A disciplined approach to neural network hyper-parameters: Part 1-learning rate, batch size, momentum, and weight decay. arXiv preprint arXiv:1803.09820 (2018)
19. Subedi, I.M., Singh, M., Ramasamy, V., Walia, G.S.: Classification of testable and valuable user stories by using supervised machine learning classifiers. In: 2021 IEEE International Symposium on Software Reliability Engineering Workshops (ISSREW), pp. 409–414. IEEE (2021)
20. Vaswani, A., et al.: Attention is all you need (2017)
21. Wang, A., Singh, A., Michael, J., Hill, F., Levy, O., Bowman, S.R.: GLUE: a multi-task benchmark and analysis platform for natural language understanding. arXiv (2018)

Modelling Adaptive Systems with Maude Nets-within-Nets

Lorenzo Capra[1(✉)] and Michael Köhler-Bußmeier[2]

[1] Dipartimento di Informatica, Università degli Studi di Milano, Milan, Italy
capra@di.unimi.it
[2] University of Applied Science Hamburg, Hamburg, Germany
michael.koehler-bussmeier@haw-hamburg.de

Abstract. Systems able to dynamically adapt their behaviour gain growing attention to raising service quality by reducing development costs. On the other hand, adaptation is a major source of complexity and calls for suitable methodologies during the whole system life cycle. A challenging point is the system's structural reconfiguration in front of particular events like component failure/congestion. This solution is so common in modern distributed systems that it has led to defining ad-hoc extensions of known formal models (e.g., the pi-calculus) But even with syntactic sugar, these formalisms differ enough from daily programming languages. This work aims to bridge the gap between theory and practice by introducing an abstract machine for the "nets-within-nets" paradigm. Our encoding is in the well-known Maude language, whose rewriting logic semantics ensures the mathematical soundness needed for analysis and an intuitive operational perspective.

Keywords: Dynamically-reconfigurable systems · Maude ·
Nets-within-nets

1 Introduction

Adaptable or self-organizing cyber-physical systems are spreading to raise service quality and decrease development costs. But adaptation, as a major source of complexity, calls for suitable models both at design time and runtime. In this context, formal methods play an important role: [7,16] present a survey of self-adaptive systems' models. In this paper, we focus on *structural reconfiguration* of a system. Its relevance has led to extending formal models with ad-hoc features. Two examples are the π-calculus and Nets-within-Nets.

But even with syntactic sugar, these models are different enough from "daily" reflective programming languages. We aim to bridge the gap between theory and practice by presenting an abstract machine for Nets-within-Nets. The encoding in Maude enjoys the rewriting logic's soundness and intuitive operational perspective. The use of rewriting logic as a unifying framework for distributed systems dates back to [14]. Ours is the first efficient/extensible formalization of Nets-within-Nets. We focus on *Elementary Object Systems*, which achieve an acceptable trade-off between expressivity and analysis capability.

© The Author(s), under exclusive license to Springer Nature Switzerland AG 2024
A. Rocha et al. (Eds.): WorldCIST 2023, LNNS 801, pp. 178–189, 2024.
https://doi.org/10.1007/978-3-031-45648-0_18

2   Background

Petri Nets PN are a central model for distributed systems. Here we refer to *Turing-powerful* Place-Transitions nets. A PT net is a bipartite graph $N := (P, T, I, O, H)$, where: P and T are finite, disjoint sets holding *places* and *transitions*. Letting $Bag[P]$ denote the set of multisets on P, $\{I, O, H\} : T \to Bag[P]$ represent weighted input/output/inhibitor edges. The state of a PT net (*marking*) is $m \in Bag[P]$. PT nets have intuitive interleaving semantics: $t \in T$ is *enabled* in m if and only if: $I(t) \le m \wedge H(t) > m$ ('$>$' being restricted to the support of $H(t)$). If t is enabled in m it may *fire*, leading to $m' = m + O(t) - I(t)$. A PT *system* is a pair $\langle N, m_0 \rangle$.

Nets-within-Nets Adaptable/mobile systems have been deeply studied in the context of *Object-net* formalism [11], which follows Valk's *Nets-within-Nets* paradigm [15]. Object-nets are PN where tokens (graphically denoting a marking) are nets again, i.e., we have *nested* markings. *Elementary Object-net Systems* (EOS) [11] are the two-level specialization of object-nets [11]. Most problems for *safe* EOS (e.g., reachability and liveness) are PSPACE-complete [9]. Namely, safe EOS are no more complex than PT nets, as for these aspects. An EOS consists of a *system-net* whose places hold net-tokens of a certain type. In our encoding, the graph structure of both the system-net and net-tokens is a PT net. Net-tokens represent *PT systems*. In the example in Fig. 1 there are two different types of net-tokens (net_1, net_2); Both the system-net and the net-tokens consist of a single transition; Weight-one I/O edges are used. Events are accordingly nested and can be of three kinds:

1. System-autonomous: A system-net transition t fires in autonomy, by consistently moving net-tokens from the preset (places p_i, $i : 1 \ldots 3$) to the postset (p_j, $j : 4 \ldots 6$).
2. Object autonomous: A net-token, e.g., that of type net_1 in system-net place p_2, fires transition t_1 by "moving" a black (anonymous) token from a_1 to b_1.
3. Synchronisation (illustrated in Fig. 1): Whenever we add matching labels between a system-net transition (t) and nested transitions (t_1, t_2), then they fire synchronously: Net-tokens move from the preset to the postset of t; at the same time, in nested nets black tokens move from the preset to the postset of t_1, t_2.

There may be several *firing instances* for a system-net transition: If many net-tokens of a certain type are in the preset, their *cumulative* marking is distributed on output places of the same type.

Maude [6] is an expressive, purely declarative language with Rewriting Logic semantics [2]. Statements are *equations* and *rules*. Both sides of a rule/equation are terms of a given *kind*. A *functional* module specifies an *equational theory* $(\Sigma, E \cup A)$ in membership equational logic [1]: Σ declares (sub)sorts, *kinds* (implicit equivalence classes grouping related sorts) and operators; E is the set

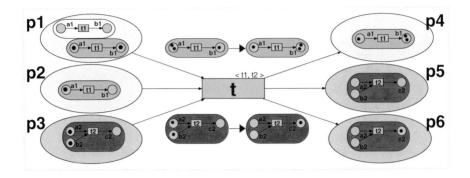

Fig. 1. Elementary Object Net System (Eos)

of equations and membership axioms, and A is the set of operator equational attributes (e.g., `assoc`). The model of $(\Sigma, E \cup A)$ is the *initial algebra*, i.e., the quotient of the ground term algebra $T_{\Sigma/E \cup A}$. Under some conditions, $T_{\Sigma/E \cup A}$ is isomorphic to the *canonical term algebra*, i.e., the denotational and operational semantics coincide.

A *system module* specifies a *rewrite theory* $\mathcal{R} = (\Sigma, E \cup A, \phi, R)$ [2]: $(\Sigma, E \cup A)$ is the algebraic structure of a system; R and ϕ define the rewrite rules (concurrent transitions). The model of $\mathcal{R}$ associates to each kind k a labeled transition system taking the form: $[t] \overset{[\alpha]}{\rightarrow} [t']$, with $[t], [t'] \in T_{\Sigma/E \cup A, k}$, and $[\alpha]$ is an equivalence class of rewrites.

3   Eos Implementation of in `MAUDE`

The Eos formalization relies on and extends that of *rewritable* PT systems [5]. In Eos, however, dynamic adaptation comes down to net-tokens manipulation. The full list of `Maude` source files is available at github.com/lgcapra/rewpt/tree/main/new/EOS.

We use three generic, functional modules: `BAG{X}`, `MAP+{X,Y}`, `SET+{X}`. With respect to other `Maude` formalizations [13,14], we do not represent mulisets merely as free commutative monoids. Specific operators guarantee more abstraction and efficiency: `_._`, `_+_`, `_[_] _-_`, `_<=_` , `_>'_` . We represent bags as associative weighted sums, e.g.: 3 . p(2,"net1") + 1 . p(1,"net2"). A transition's adjacency matrix is a triplet `[_,_,_]` of `Bag{Place}` terms . A net is a term of sort `Map{Tran,Imatrix}` (renamed `Net`), i.e., a set of entries SPSVERBc13SPSVERBa1. A `System` term is the juxtaposition of a `Net` and a `Bag{Place}`. *System* module `PT-EMU` efficiently formalizes the PT operational semantics.

```
mod PT−EMU is
 pr PT−SYS .
 var T : Tran .
 vars I O H S : Bag{Place} .
 var N N' : Net .
 crl [firing] : N S => N S + O − I if T |−> [I,O,H] ; N' := N /\ I <= S /\ H >' S.
endm
```

EOS Specification. An EOS net (module `EOS-NET`) is the juxtaposition of three sub-terms of *sorts* `Net`, `Map{String,Net}` (`NeTypeS`) and `Map{Tran,Map{String,Bag{Tran}}}` (`Syncmap`). The resulting term is of *kind* `[Sysnet]`, due to possible inconsistencies among sub-terms. The 2nd and 3rd sub-terms specify net-tokens' types (associated to place labels) and synchronizations, respectively. The three kinds of events in a EOS meet the following conventions.

1. System-autonomous: system-net transitions not occurring in `Syncmap` subterm.
2. Object autonomous: net-token transitions for which:
   `op synchronized : Syncmap Tran -> Bool`
   evaluates to `false`.
3. Synchronisations: implicitly defined by exclusion.

A term of sort `Map{Place,Bag{Bag{Place}}}` specifies an EOS (nested) marking as a map from system-net places to nested multisets of places: a `Bag{Place}`term indeed represents a net-token's marking. For example, the EOS marking in Fig. 1 is described by the term (`"net1"`, `"net2"` denote the net-token types):

```
p(1,"net1") |-> 1 . nil + 1 . (1 . p(1, "a1") + 1 . p(2, "b1")) ;
p(2,"net1") |-> 1 . 1 . p(1, "a1") ; p(3,"net2") |-> 1 . (1 . p(1, "a2") + 1 . p(2, "b2"))
```

A `Eosystem` term (module `EOSYS`) is the juxtaposition (`__`) of sub-terms `Sysnet` and `Map{Place,Bag{Bag{Place}}}`. Due to possible inconsistencies, `__`'s arity is kind `[Eosystem]`. As usual, we use a membership axiom to connote well-defined terms of sort `Eosystem`.

Operation Semantics. Two rewrite rules specify the firing of system-net transitions –including possible synchronizations– and net-token autonomous transitions. Both rely on the PT firing rule. There may be several *firing instances* for a system-net transition t in EOS marking m. An instance of t, as well as m, are terms `Map{Place,Bag{Bag{Place}}}`, namely, an instance of t is a sub-marking of m. The system-net firing rule builds on two operators:

```
op firemap : Eosystem −> Map{Tran, Set{Map{Place,Bag{Bag{Place}}}}}.
op firings : ImatrixT Map{Place, Bag{Bag{Place}}} Syncmap −>
   [Set{Map{Place, Bag{Bag{Place}}}}] .
```

`firemap` computes the enabled instances for *every* system-net transition. In the event of synchronization, instances are filtered according to the enabled, synchronizing nested transitions. The firing of an instance of t may be non-deterministic

(Sect. 1): `firings` calculates the multisets of net-tokens to distribute on t's post-set.

```
mod EOS−EMU is
  pr EOSYS .
  inc PT−EMU .
  var FM : Map{Tran, Set{Map{Place,Bag{Bag{Place}}}}} . *** firing map
  var TI : Entry{Tran, Set{Map{Place,Bag{Bag{Place}}}}} . *** t instance
  var NeFS : NeSet{Map{Place,Bag{Bag{Place}}}} .
  var FS : Set{Map{Place,Bag{Bag{Place}}}} . *** output firing set
  vars I O M M' : Map{Place,Bag{Bag{Place}}} . *** firing instance/marking
  vars N N' : Net . var T : Tran . var Ty : NeTypeS . var Sy : Syncmap .
  var Q : Imatrix . var S : String . vars J K : NzNat .
  vars B B' : Bag{Place} . var B2 : Bag{Bag{Place}} .
  rl [select] : I U NeFS => I . *** non−deterministic instance extraction
  crl [inst] : N Ty Sy M => N Ty Sy (M − I) + O if N' ; T |−> Q := N /\
    firemap(N Ty Sy M)[T] => I /\ firings(T |−> Q, I, Sy) => O .
  crl [aut] : SN (p(J,S) |−> K . B + B2) => SN (p(J,S) |−> (K . B − 1 . B) + 1 . B' +
    B2)
    if N (S |−> (T |−> Q ; N') ; Ty) Sy := SN /\
    not(synchronized(Sy, T)) /\ (T |−> Q) B => (T |−> Q) B' .
endm
```

Module `EOS-EMU` formalizes the Eos operation semantics. Rules `inst` and `aut` encode the firing of a system-net transition and an autonomous nested transition, respectively. `inst` relies on the auxiliary rule `select` which emulates the non-deterministic selection of an instance: We exploit the flexibility of rule conditions that may include matching equations and *other rules*. `inst` implements double non-determinism: first for selecting an instance of t then for choosing one of the associated output markings. Rule `aut` exploits the `synchronized` predicate and the PT firing rule. Module `SIMPLE-EOS` specifies the Eos in Fig. 1.

```
mod SIMPLE−EOS is
  inc EOS−EMU .
  ops net type1 type2 : −> Net .
  op netype : −> NeTypeS .
  op m0 : −> Map{Place,Bag{Bag{Place}}} .
  op eosnet : −> Sysnet .
  op sync : −> Syncmap .
  eq net = t(1,"sys") |−> [1 . p(1,"net1") + 1 . p(2,"net1")+ 1 . p(3,"net2"),
    1 . p(4,"net1") + 1 . p(5,"net2")+ 1 . p(6,"net2"), nil] .
  eq type1 = t(1, "") |−> [1 . p(1,"a1"), 1 . p(2,"b1"), nil] .
  eq type2 = t(2,"") |−> [1 . p(1,"a2") + 1 . p(2,"b2"), 1 . p(3,"c2"), nil] .
  eq netype = "net1" |−> type1 ; "net2" |−> type2 .
  eq sync = t(1,"sys") |−> ("net1" |−> 1 . t(1,"") ; "net2" |−> 1 . t(2,"") ) .
  eq eosnet = net netype sync .
  eq m0 = p(1,"net1") |−> 1 . nil + 1 . (1 . p(1,"a1") + 1 . p(2,"b1")) ;
    p(2,"net1") |−> 1 . 1 . p(1,"a1") ; p(3,"net2") |−> 1 . (1 . p(1,"a2") + 1 . p(2,"
      b2")).
endm
```

4   Eos Model of a Production Line

We consider a production plant with two production lines as an example. A similar scenario has been used as a case study in [5]. We have raw materials, two operations $t1$ and $t2$ working on pieces of those, and two *production lines* (robots), both being capable of performing $t1$ and $t2$. We assume that the two lines have different qualities, i.e., it is better to execute $t1$ on line 1 and $t2$ on line 2. During the execution, one of the two lines may get faulty (a double failure has a negligible probability).

The scenario is suited for a Eos model as we have two obvious levels: the production site and the production plans. The model is specified in the syntax of the Renew tool [12]. The system level (Fig. 2) describes the production lines, the execution life cycle, and the line dropout. Place $p0$ indicates normal operation: In this mode, transition $t0$ takes two tokens (i.e., the raw materials) from place $p1$ and activates the normal production plan by generating a net-token of type *plan* (label *x:new plan*). Place *production plan* is a side condition for the two transitions representing the lines. They synchronise via their labels with net-token matching transitions. Net-tokens describe the different production plans: the normal operation (Fig. 3) and two fallback managing line dropouts. The standard production plan (Fig. 3) specifies a synchronisation by which $t1$ will execute on line 1 and $t2$ on line 2. When the plan is finished, the corresponding net-token is withdrawn via synchronisation (label *do_t3()*). For simplicity, the scenario restarts (transition $t4$), what makes the main loop of the model live.

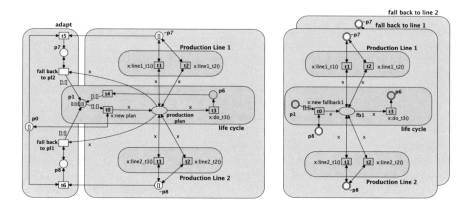

Fig. 2. The System Net modelling the Production Lines

We have the adaption part on the left of Fig. 2 (yellow nodes). Transitions $t5$ and $t6$ model the dropout of a production line: Upon a fault, the standard production plan is no more executable. Therefore, the transitions *fall back to production line 1/2* withdraws them. Places $p7$ and $p8$ indicate which line is off: In the case of line 1 ($p7$ is marked), we switch to the fallback plan 2 and vice versa. A fallback block has pretty much the same structure as the original one. The only difference is that it generates different net-tokens (labels *x:new fallback1/2*) in the two cases above. The fallback plans (not shown) look almost identical to the basic plan (Fig. 2) except for the labels, making *fallback 1(2)* to work only with line 1(2).

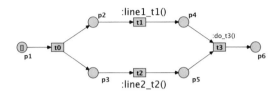

Fig. 3. Object Net modelling the standard Production Plan

The highly modular Eos encoding in MAUDE exploits the base net-algebra operator represented by the associative net juxtaposition (;).

```
fmod pline−NET is *** net structures used in the model
  pr PT−NET * (op emptyPbag to nil) .
  pr CONVERSION .
  vars S S' : String . var N : Nat . vars L L' : Bool .
  op l : Bool −> NzNat . *** boolean values denote the two lines/operations
  eq l(true) = 1 .
  eq l(false) = 2 .
  op str : Nat −> String . *** converts a number to a string
  eq str(N:Nat) = string(N:Nat,10) .
  op str : String Bool −> String .
  eq str(S,L) = S + str(l(L)) . *** concatenates S to the number of line L
  *** net−tokens' base modules
  op line : String Bool Bool −> Net . *** the 1st arg denotes the operation t_i,
      the 2nd line_j, the 3rd the net−token's label
  eq line(S,L,L') = t(l(L),str(S + ":line",L')) |−> [1 . p(1 + l(L),S), 1 . p(3 + 1(L
      ),S), nil] .
  op start : String −> Net . *** t0 (start−up): the arg is the net−token's label
  eq start(S) = t(0,S) |−> [nil, 1 . p(1,S) + 1 . p(2,S) + 1 . p(3, S), 1 . p(1,S)] .
  op assemble : String −> Net . *** t3 (assembler)
  eq assemble(S) = t(3,S + ":do_t3 ") |−> [1 . p(4,S) + 1 . p(5, S), 1 . p(6,S), nil] .
  *** net−tokens
  ops plan : −> Net .
  eq plan = start("plan") ; line("plan",true,true) ; line("plan",false,false) ;
      assemble("plan") .
  ops fallback : Bool −> Net . *** same structure as plan
  eq fallback(L) = start(str("fb",L)) ; line(str("fb",L),L,L) ; line(str("fb",L),
      not(L),L) ; assemble(str("fb",L)) .
  *** system−net's modules − labelled transitions represent synchronizations
  op pline : String Bool Bool −> Net .
```

```
  eq pline(S,L,L') = t(l(L),str(S + "_x:line",L')) |-> [1 . p(2,S) + 1 . p(6 + l(L')
      ,"-"), 1 . p(2,S) + 1 . p(6 + l(L'),"-") , nil] .
  ops sysnet lifecycle : -> Net . *** object- and system-net modules
  eq lifecycle = t(0,"x:new_plan") |-> [2 . p(1,"") + 1 . p(0,""), 1 . p(2,"plan")
      + 1 . p(0,"") , nil] ; t(4,"") |-> [1 . p(6,""), 2 . p(1,""), nil ] ;
      t(3,"x:do_t3 ") |-> [1 . p(2,"plan"), 1 . p(6,""), nil ] .
  ops adapt fbkcycle : Bool -> Net .
  eq adapt(L) = t(4 + l(L),"") |-> [1 . p(0,"") + 1 . p(6 + l(L),"-"), 1 . p(6 + l(L
      ),"") , nil] ; t(7,str(" fb_to_pl",not(L))) |-> [ 1 . p(2,"plan") + 1 . p(6 +
      l(L),""), 1 . p(6 + l(L),"") + 2 . p(1,""), nil] .
  eq fbkcycle(L) = t(0,str(" x:new_fb",L)) |-> [2 . p(1,"") + 1 . p(6 + l(not(L))
      ,""), 1 . p(2,str("fb",L)) + 1 . p(6 + l(not(L)),""), nil] ; t(3,str("fb",L) +
      "_x:do_t3 ") |-> [1 . p(2,str("fb",L)), 1 . p(6,""), nil ] .
  *** modular specification of EOS model's system-net (note the use of associative
      ';')
  eq sysnet = lifecycle ; pline("plan",true,true) ; pline("plan",false,false) ;
    adapt(true) ; adapt(false) ; *** adaptation sub-net
    fbkcycle(true) ; pline("fb1",true,true) ; pline("fb1",false,true) ; ***
        fallbacks
    fbkcycle(false) ; pline("fb2",true,false) ; pline("fb2",false,false).
  op K : -> NzNat . *** model's parameter (#raw pieces)
  eq K = 2 .
endfm

fmod PLINE-EOS is *** EOS specification of the production-line
  pr EOSYS .
  pr pline-NET .
  *** aliasing
  op netype : -> NeTypeS . *** net-token types
  op eosnet : -> Sysnet . *** EOS net-structure
  op sync : -> Syncmap . *** synchronizations
  op eosm0 : -> Map{Place,Bag{Bag{Place}}} . *** initial marking
  eq netype = "plan" |-> plan ; "fb1" |-> fallback(true) ; "fb2" |-> fallback
      (false) .
  eq sync = t(1, "plan_x:line1") |-> ("plan" |-> 1 . t(1, "plan:line1")) ;
    t(2, "plan_x:line2") |-> ("plan" |-> 1 . t(2, "plan:line2")) ;
    t(1, "fb1_x:line1") |-> ("fb1" |-> 1 . t(1,"fb1:line1")) ;
    t(2, "fb1_x:line1") |-> ("fb1" |-> 1 . t(2,"fb1:line1")) ;
    t(1, "fb2_x:line2") |-> ("fb2" |-> 1 . t(1,"fb2:line2")) ;
    t(2, "fb2_x:line2") |-> ("fb2" |-> 1 . t(2,"fb2:line2")) ;
    t(3, "x:do_t3 ") |-> ("plan" |-> 1 . t(3, "plan:do_t3 ")) ;
    t(3, "fb1_x:do_t3 ") |-> ("fb1" |-> 1 . t(3, "fb1:do_t3 ")) ;
    t(3, "fb2_x:do_t3 ") |-> ("fb2" |-> 1 . t(3, "fb2:do_t3 ")).
  eq eosnet = sysnet netype sync .
  cq eosm0 = p(1, "") |-> (2 * K) . nil ; p(7, "-") |-> 1 . nil ; p(8, "-") |-> 1 .
      nil ; p(0, "-") |-> 1 . nil .
endfm
```

Model Analysis. We give some evidence of formal verification with the intent of showing the benefits of an approach using `Maude`. Since adaptation should preserve the liveness of the production process, we focus on this kind of property. We use the base model-checking tool, namely the inline `search` command. We manage term wordiness using intuitive aliasing: `net` describes the system-net, whereas `eosnet` includes net tokens' description and synchronizations; `m0` and `eosm0` denote the initial marking of `net` –viewed as a PT system– and of the Eos model. One of the advantages of Eos is that we can separately analyse the two net levels.

```
search in pline-SYS : net m0 =>* X:System such that marking(X
    :System)[p(1,"")] + 2 * marking(X:System)[p(2,"plan")] +
    2 * marking(X:System)[p(2,"fb1")] + 2 * marking(X:System)
    [p(2,"fb2")]  + 2 * marking(X:System)[p(6,"")] =/= 4 .
search in pline-SYS : net m0 =>! X:System.
search in pline-EOS : eosys0 =>! E:Eosystem.
```

The first two `search`es operate on `net`'s abstraction as a PT-system. The first one verifies a state *invariant* which characterizes the production plan's life-cycle (assuming that p_1 initially holds four tokens): `search` has no matches, coherently with the fact that we seek a counter-example. Actually, it is a *structural* invariant holding for any initial configuration with $2 * K$ tokens in place p_1, but we cannot infer parametric results using simple model-checking. The second `search`, instead, verifies the presence of final states (deadlocks): Also in this case it has no matches. The last (again, unmatched) `search` is more significant because it proves that the whole Eos (including nested nets) is deadlock-free. Table 1 shows the performance of the last `search` as the model's parameter varies, on an Intel Core i7-6700 with 32 GB of RAM.

Table 1. Performance of `search` as model's parameter varies

K	# states	# rewrites	time (ms)
2	262	19801	38
5	2802	180539	453
10	13932	995724	3192
20	104007	19737494	56090
50	4186132	111564504	906253

5   Summary and Outlook

In this paper we have defined a `Maude` representation of nets-within-nets, more concretely: Eos. We are going to discuss some insights into our project.

What are the Strengths of the Approach? The `Maude` formalization presented here is an extension of (rewritable) PT specification [3]. Therefore a lot of code could be reused, which is beneficial for the implementation's reliability and efficiency as well.

The formalisation preserves central design issues of EOS: Specifically, the system-net and net-tokens have the same structure in `Maude`, which is coherent with EOS blueprint. This aspect is relevant when extending the architecture from the two-level case to an unbounded nesting of net-tokens in a marking [8].

Standard `Maude` facilities for formal verification (e.g., state-space search and model-checking) may be used with no additional costs. Additionally, the EOS firing rule is defined in a way that moving net-tokens around cannot be distinguished from moving ordinary tokens in PT. Therefore, we can easily define abstractions on the system's state (e.g., forgetting about the marking of net-tokens), which is essential to perform state space exploration efficiently.

Our approach fosters model extensions. A natural one would be the usage of inhibitor arcs. In our framework, this kind of arc is only a minor extension. Passing to Object-nets should be almost for free (using general Bags, with an arbitrary nesting, would be required). We may easily go one step forward, towards "rewritable" EOS, where the structure of both the system net and of net-tokens may change over time.

What was Complex? The most challenging aspect of the formalisation was the integration of the so-called *firing modes*. Roughly speaking, the firing rule of a system-autonomous event in an EOS collects the tokens of all net-tokens coming from system-net's places in the preset. When a system-net transition fires it distributes all these tokens on freshly generated net-tokens in the postset. The firing rule allows any possible distributions – an aspect which requires some tricky handling in `Maude`.

Limitations. The current formalisation fulfils the requirement that it provides a link to the world of programming. But we have to admit that like in any algebraic specification, terms describing EOS may be wordy, structurally complex and (consequently) difficult to read and manage. An aliasing mechanism (used in a naive way) might greatly help a modeller. Also, syntactic sugar would sweeten the approach. Of course, an automated translation from a high-level (graphical) description of EOS to the corresponding `Maude` module would be highly desirable.

Ongoing Work. In this paper, we were concerned with the `Maude` encoding of EOS. Our main motivation for this is to obtain a representation closer to the usual programming language world. Our intention is also to benefit from the advantages of a formal specification, i.e. the possibility to apply analysis techniques more easily. In the case of `Maude` the first idea is to apply state-space techniques, like LTL model checking. We also like to integrate structural PN techniques for EOS [10].

For the analysis of EOS, we need to struggle with scalability as the state space grows even worse than in PT nets. Possible approaches are the canonization of net-tokens [4] and the use of abstractions to obtain condensed state spaces. The latter can be expressed quite easily in `Maude` by adding additional equations on markings.

References

1. Bouhoula, A., Jouannaud, J.P., Meseguer, J.: Specification and proof in membership equational logic. Theoret. Comput. Sci. **236**(1), 35–132 (2000). https://doi.org/10.1016/S0304-3975(99)00206-6
2. Bruni, R., Meseguer, J.: Generalized rewrite theories. In: Baeten, J.C.M., Lenstra, J.K., Parrow, J., Woeginger, G.J. (eds.) Automata, Languages and Programming, pp. 252–266. Springer, Berlin (2003). https://doi.org/10.1007/3-540-45061-0_22
3. Capra, L.: A maude implementation of rewritable petri nets: a feasible model for dynamically reconfigurable systems. In: Gleirscher, M., Pol, J.v.d., Woodcock, J. (eds.) Proceedings First Workshop on Applicable Formal Methods, virtual, 23rd November 2021. Electronic Proceedings in Theoretical Computer Science, vol. 349, pp. 31–49. Open Publishing Association (2021). https://doi.org/10.4204/EPTCS.349.3
4. Capra, L.: Canonization of reconfigurable pt nets in maude. In: Lin, A.W., Zetzsche, G., Potapov, I. (eds.) Reachability Problems, pp. 160–177. Springer, Cham (2022). https://doi.org/10.1007/978-3-031-19135-0_11
5. Capra, L.: Rewriting logic and petri nets: A natural model for reconfigurable distributed systems. In: Bapi, R., Kulkarni, S., Mohalik, S., Peri, S. (eds.) Distributed Computing and Intelligent Technology, pp. 140–156. Springer, Cham (2022)
6. Clavel, M., Durán, F., Eker, S., Lincoln, P., Martí-Oliet, N., Meseguer, J., Talcott, C.: All About Maude - A High-Performance Logical Framework. LNCS, vol. 4350. Springer, Heidelberg (2007). https://doi.org/10.1007/978-3-540-71999-1
7. Hachicha, M., Halima, R.B., Kacem, A.H.: Formal verification approaches of self-adaptive systems: a survey. Procedia Comput. Sci. **159**, 1853–1862 (2019). https://doi.org/10.1016/j.procs.2019.09.357, proceedings of KES2019
8. Köhler-Bußmeier, M., Heitmann, F.: On the expressiveness of communication channels for object nets. Fund. Inform. **93**(1–3), 205–219 (2009)
9. Köhler-Bußmeier, M., Heitmann, F.: Liveness of safe object nets. Fund. Inform. **112**(1), 73–87 (2011)
10. Köhler-Bußmeier, M., Moldt, D.: Analysis of mobile agents using invariants of object nets. Electronic Communications of the EASST: Special Issue on Formal Modeling of Adaptive and Mobile Processes 12 (2009)
11. Köhler, M., Rölke, H.: Properties of object petri nets. In: Cortadella, J., Reisig, W. (eds.) ICATPN 2004. LNCS, vol. 3099, pp. 278–297. Springer, Heidelberg (2004). https://doi.org/10.1007/978-3-540-27793-4_16
12. Kummer, O., Wienberg, F., Duvigneau, M., Schumacher, J., Köhler, M., Moldt, D., Rölke, H., Valk, R.: An extensible editor and simulation engine for Petri nets: Renew. In: Cortadella, J., Reisig, W. (eds.) International Conference on Application and Theory of Petri Nets 2004. LNCS, vol. 3099, pp. 484–493. Springer-Verlag (2004)
13. Padberg, J., Schulz, A.: Model checking reconfigurable petri nets with maude. In: Echahed, R., Minas, M. (eds.) Graph Transformation, pp. 54–70. Springer, Cham (2016). https://doi.org/10.1007/978-3-319-40530-8_4

14. Stehr, M.O., Meseguer, J., Ölveczky, P.C.: Rewriting Logic as a Unifying Framework for Petri Nets, pp. 250–303. Springer (2001). https://doi.org/10.1007/3-540-45541-8_9
15. Valk, R.: Object Petri Nets: Using the nets-within-nets paradigm. In: Desel, J., Reisig, W., Rozenberg, G. (eds.) ACPN 2003. LNCS, vol. 3098, pp. 819–848. Springer, Heidelberg (2004). https://doi.org/10.1007/978-3-540-27755-2_23
16. Weyns, D., Iftikhar, M.U., de la Iglesia, D.G., Ahmad, T.: A survey of formal methods in self-adaptive systems. In: Proceedings of the Fifth International C* Conference on Computer Science and Software Engineering, C3S2E 2012, pp. 67–79. Association for Computing Machinery, New York (2012). https://doi.org/10.1145/2347583.2347592

Simultaneous OutSystems Integration to Different Instances of a System in the Context of a Single Environment

Borislav Shumarov[1]([⊠]) and Ivan Garvanov[2]

[1] ADVANT Beiten/University of Library Studies and Information Technologies, 90471 Nuremberg, Germany
borislav.shumarov@gmail.com
[2] University of Library Studies and Information Technologies, 1784 Sofia, Bulgaria
i.garvanov@unibit.bg

Abstract. For a REST integration in OutSystems to be implemented, the typical way is to use the platform's in-built capabilities: according to best practices, integration is being created within a dedicated module, with a single hard-coded target system base URL. However, a problem arises if the same Connector module should be used with different target systems' instances (having different base URLs) simultaneously from one OutSystems Environment. This paper proposes a minimalistic model for integration in the case of a typical token-based authentication, but a similar approach could be applied to other authentication methods. The proposed solution could be implemented not only right from the start of a project, but also in a setting where a typical existing OutSystems integration to a single instance of an external system is already in place and functioning (e.g. in production). The foundation of the solution lies in achieving a persistent mapping between an authentication token and a given target URL. This would allow for the URL to be defined at runtime, instead of the basic traditional design-time static approach, thus circumventing the platform's limitations.

Example implementation has been included as part of a Connector released on the OutSystems open-source market (Forge): https://www.outsystems.com/forge/component-overview/11958/xpressdox-connector

Keywords: OutSystems platform · low-code · REST integration · single environment · multiple instances

1 Introduction

Low-code is a broad term, but in general, it is used in the sense of application platforms, that facilitate software development via the means of visual workflows and models, instead of exclusively using written code in traditional programming languages (Java, C++, etc.). OutSystems is regarded as being among the leading enterprise low-code platforms on the market [1]. A typical OutSystems software factory consists of separate Environments, connected via a deployment pipeline. One Environment hosts many applications simultaneously, each comprised of 1 or more modules [2] and has only one configuration console (called "Service Center").

With the growth of a medium to large software factory in OutSystems usually a need arises for integration with external systems. A REST API has been the primary communication style of the Web since the mid-1990s, even shaping the World Wide Web's architecture [3]. Thus, REST integration is also one of the most widely used types of communication encountered in practice between an OutSystems application and an outside system.

The solution in this paper uses a setup of a REST API with a typical token-based authentication, where a username-and-password pair is provided in the first call to the system and a token is produced and used as an authentication mechanism in all the subsequent requests to the API, until the token's expiration. In the current context, an endpoint is defined as a set of a URL and an HTTP method, as defined by Wittern et al. [4].

One scenario observed in practice is the need for several OutSystems applications to communicate to the same instance of an external system via a single OutSystems REST-integration application module (conventionally named "Connector") simultaneously, all deployed within the same Environment. However, one problem arises, when there is a need for communication with different instances of an outside system, residing on different servers and thus having differing base URLs.

A possible complication (and one, encountered in practice) is that when a necessity for two (or more) simultaneous integrations with instances of the same type of an external system arises, it is usually only after the integration with the first system is already implemented and operational. With this, the in-built capabilities of the platform are already exhausted. The integration with a further instance becomes a cumbersome task. Thus, the need arises also for a standardized, minimally intrusive pattern for a solution, which requires none or a minimum amount of refactoring to be performed on the existing integration module, as well as on the consumer systems (Outsystems) and producer systems (the external system's instances).

In the context of OutSystems, the traditional way to implement a REST integration is by creating a module, which wraps the external REST endpoints. This is a use case, foreseen by the platform's designers, and is easily achievable. The out-of-the-box solution relies on the domain of the server to be set statically at design time, or via the configuration console (Service Center) after deployment [5]. This is currently strictly limited to one URL per environment per module and there is no straightforward approach, for it to be changed programmatically at runtime. Thus, it does not achieve the goal of allowing for parallel access from many consumer applications simultaneously.

This paper seeks to explore the possibilities for circumvention of the current OutSystems Platform's limitations for REST integrations with several instances of the same external system within a single OutSystems Environment. The solution should be "light", meaning as simple and quick to implement as possible, but at the same time, it should also optimize for a method with a refactoring footprint as small as possible. This is to be expected even in the case of an already existing integration module (Connector), where the aforementioned limitations have been noticed first after the Connector to a single target system's instance has already been brought in use and utilized heavily throughout an OutSystems Environment.

To find a solution to the problem, the first part of this paper explores known possible approaches while examining their characteristics with their corresponding advantages and disadvantages. The latter part describes in detail a possible solution that has been developed and successfully used throughout a real OutSystems software factory in a large enterprise.

2   Possible Solutions

As of the moment of writing, there are a couple of solutions, that the authors came across in the literature, in the practice, or could theorize. Some of these approaches have the potential to partially or entirely solve the current problem with a simultaneous connection to different instances of a system from one OutSystems Environment, but also come with drawbacks, limitations, and incapabilities or are entirely not possible currently, as described below.

2.1   A URL Passed Every Time with the Action

A solution, which is in some ways similar to the pattern proposed in this paper, has been observed in practice [6]. The base URL could be given as a parameter every time any Action from the Connector is being invoked and thus dynamically set similarly. As for the implementation cost of the Connector itself, in some cases, such a solution *could* be slightly easier to implement than our chosen solution, presented later in the paper. It would depend on the number of called endpoints, where an additional parameter (Base URL) should be added to each of them – in this case: to all of them. If the number is small (say less than 5) it may not justify the added cost of the design-time mapping, that our solution requires. However, additionally depending on the number of calls already made via the Connector throughout the whole Environment, it could require huge refactoring costs. The second drawback would be, that an additional input parameter to each and every REST Action contributes to the overall complexity.

2.2   Different Environments for the Different Integrations

A possible solution to the sub-problem of needing more than one domain set as the target URL for the same Connector module could be having more than one instance of the Connector deployed in more than one Environment. As the OutSystems limitation is one statically set Base URL per Environment, the problem won't be observed. In this case, every Connector would have the domain URL of the target system specified manually at design time. This would require no further development activity on the Connector whatsoever. The approach, however, is not without its drawbacks, 2 of which are: unnecessary fragmentation of the application landscape of the software factory, as well as higher costs associated. The latter stems from the fact, that any additional OutSystems Environment is usually separately licensed and lies on a separate piece of infrastructure. A third drawback is that this in turn would also increase maintenance costs and complexity, as well.

2.3 Clone of the Original Integration Connector's Functionality for Each of the Target System's Domains

Another similar solution would be to make an entire clone of the Connector for each new instance of an external system. Such a solution would mean duplicating, triplicating, etc. the resources for the operation and future maintenance tasks for the whole set of Connectors or different copies of the REST integration inside one Connector. It does not seem like a good solution at all, since the only reason for the existence of these copies would be the need for a single configuration parameter, thus strongly violating the widespread DRY (Don't Repeat Yourself) [7] principle.

2.4 Feature for Multiple URLs

The proposed solution would become obsolete in case OutSystems decide to include some functionality for handling this problem out of the box in a future release of the platform. This could be for example the option of adding more than one target URL for a particular REST integration. At the time of writing, there even exists a proposition for such a feature on the platform's community website [8]. However, it is still unclear, if such a feature would ever be introduced to the platform, thus paving the need for a different solution either way.

3   Proposed Solution

This paper proposes such a solution. The full code of the sample described here could be found as part of a Connector on the OutSystems Forge Marketplace [9]. It relies on the described setup and uses the authentication medium (token) as a means of identifying the current session of the invoker, thus mapping a session to a target URL, coming from the consumer (OutSystems application), as well. The authentication token is used in this case, but this is just a formality and a similar approach could also be implemented using a different medium, depending on the use case – for example, a pair of a username and a password; client id and client secret, etc.

First, instead of setting the base URL of the target system at design time, using the already described in-built OutSystems feature in Service Center, the base URL is passed as an input parameter to the wrapper Action for the authentication REST request from outside, along with the other necessary identification parameters. In our case, these are two input parameters: username and password. After a successful call to the target system, the token is extracted from the response of the external system's REST authentication endpoint and is saved in an output parameter, to be returned to the caller of the Action when it later reaches the end of its execution, as expected. But before that happens, the token is hashed (for security purposes), so that it could be persisted as an entry to the database in a table (OutSystems Entity), specifically created for this purpose. The entry contains a mapping between the base URL of the target system and the hashed token. This hashed token is then used before each subsequent request to the system as a unique key to search against, in order to fetch the base URL necessary to identify the target system. In the end, after a specified timeout, the entry in the database table is deleted. This last step is not strictly necessary, but it was implemented as means of some level of security (someone retrieving older authentication tokens), privacy (when were calls to the system made), and reducing the storage needed for the whole implementation in the long run.

Our GET request for the authentication token (Fig. 1) receives the necessary parameters, according to the API specification, but important is the parameter named "Runtime-BaseURL". It is a string, containing the target system's URL and using it dynamically for a run-time call to the corresponding system. There are a couple of options to pass it to the integration Action: as an additional URL, a body, or a header parameter for the GET request. An additional header parameter has been subjectively chosen as the least intrusive option among the 3, but another implementation would not constitute a fundamental difference in the overall method.

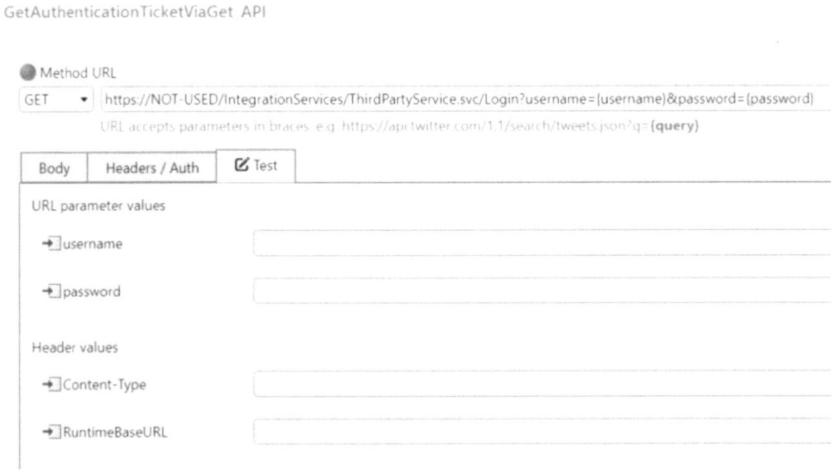

Fig. 1. REST call for obtaining authentication ticket: implementation.

Upon an endpoint invocation, the header parameter is set to the value of the corresponding base URL for the target system at run-time.

The Entity, where we store the mappings ("URLMap") has its sole purpose to contain a key-value pair between the hashed authentication token (used as a key) and the base URL for the target system's instance (as value). The token could also be set as a primary key, due to its uniqueness. For achievement of better performance, according to best practices, an index is being created for the token (key) column to improve the query speed, when used as a column to search against.

For the REST integration, an ordinary "OnBeforeRequest" Action has been created, according to the OutSystems documentation [10]. As the name implies, it executes always before an HTTP request to the target system is attempted. Further explained in the documentation, it receives the original data for the request as an input parameter, which in OutSystems is in the form of a particular structure. It has the headers and the URL query parameters as two sets of lists of key-value pairs, where the respective parameters reside. Then, a structure of the same type is returned from the Action and used to continue the normal flow of the integration call. The event handler executes whenever any of the integration Actions has been called before the actual call to the target's endpoint has been made.

Figure 2 depicts the OnBeforeRequest event handler for the REST integration. It filters the URL query parameters in a search for the corresponding parameter, containing the token string (in our case: "authenticationTicket"). If it has not been found, which is always the case during the first call to the system (whose role is to get that token for future use), the list with headers for the request is searched through for a parameter, corresponding to the base URL to be set. As already mentioned, in normal circumstances this is to be expected only during the first call to the system (the one, requesting an authentication token). When found, the base URL is set in the output structure for the value of the "BaseURL" parameter.

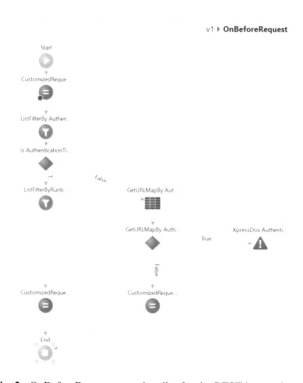

Fig. 2. OnBeforeRequest event handler for the REST integration.

In the case, that a token is found, we can conclude, that this is in the context of any of the regular (non-authenticating) REST calls to the target system. Then we proceed by reading the token from the input parameter, computing its hash, and using it as a key to finding the corresponding row in our database, containing the BaseURL set previously. If an entry has been found, we proceed by setting the "BaseURL" property of the output structure to the one, found in our database and the REST call continues as normal.

Functionality has been implemented as a process, triggered every time upon creation of a new URLMap Entity in the database. Its task is to wait for a pre-configured amount of time and to delete the corresponding row in the table afterward.

To demonstrate the simplicity of usage from a consumer application's point of view, the signatures of the first call Action to the external system - the authentication Action (Fig. 3) - and one of the possible subsequent call Actions (Fig. 4) are presented. The proposed solution adds only one further input parameter to only one of all the implementations (Fig. 3). The only change in the signature of an Action is the Base URL parameter, which has to be provided only once per authentication session: at the start, without overcomplicating initial implementations, as well as a possible refactoring of any consumer OutSystems applications in the Environment. For existing consumer applications already using Actions from the Connector, only the addition of a value for this new Base URL parameter is needed and only in one place: upon calls to "GetAuthenticationTicketViaGet" at the start of each authentication session. Any calls to other Actions from the Connector would require zero refactoring effort, as their signature stays exactly the same as before.

Fig. 3. Signature of the authentication Action for a new session.

Fig. 4. Signature of a normal Action.

4   Possible Modifications

The demonstrated solution is for a REST integration, but conceptually similar approaches could be used for other integration methods: e.g. SOAP. The underlying principle of the solution remains the same: providing the URL only on the first call to the external system via some form of a differentiating input parameter at run-time, thus escaping hard-coded platform limitations.

The solution in this paper could be modified in several different ways. One possible alternative to the custom-made map Entity in the database and the complexities of the hashing and automatic cleansing of the database entries after a specified time could be creating a Traditional Web Application in OutSystems and using a Session Variable for the current session. Further research can be made if there won't be any blockers and how exactly such a goal could be accomplished. However, the solution would require a Connector, created as a front-end Traditional Web Application, but without using any of the other functionalities, specific to this type of application type and not used in the way it is usually designed for. The current best practices are to create a Connector as a Service module [11], without front-end specific functionalities or the boilerplate, coming along.

Continuing the line of thought about such a modification, if from OutSystems they decide to include Session Variables as an integral part of a Service module type, this could make the task easier and more straightforward to implement. Additionally, perhaps further use cases could also benefit from a temporary out-of-the-box Session Variable in the context of a Service module type.

5   Conclusion

For a REST integration with an external system in OutSystems to be implemented, the traditional solution would be to use the platform's out-of-the-box capabilities. According to existing best practices, an integration is created within a Service module, with one hard-coded target system base URL per OutSystems Environment, which works well for the usual use case. However, a problem arises if the same Connector module is meant to be used simultaneously with several target systems' instances, which are defined by different base URLs. This paper proposed and demonstrated a light approach, which can be implemented in new integrations, as well as when such a need arises for already existing integrations later in the lifecycle of a system. The solution has been demonstrated for a traditional token-based authentication method and a REST API, but the principle could be used for other authentication mechanisms on one side or other connection methods (SOAP, etc.) on the other.

The first part of the proposed solution is the creation of a custom HTTP header, for it to be used for the passage of the base URL from consumer system to target system. The second part is the persistence of the mapping between an authentication token, produced by the target system and the already passed target base URL of that system. It is stored in a local database table for a specified time. The approach allows for the base URL to be delivered only once, upon the first authentication, and then to be retrieved and dynamically set as a target at runtime by each subsequent request with the already provided token, thus working around the platform's limitations. The full code of the sample described here could be found as part of a Connector on the OutSystems open-source marketplace [9].

References

1. Madrinha, J., Leitao, L.: Is Outsystems Portugal's Latest Unicorn?. https://www.forbes.com/sites/forbesinternational/2018/07/13/is-outsystems-portugals-latest-unicorn/. Accessed 25 Dec 2022
2. OutSystems: Platform Runtime | Evaluation Guide. https://www.outsystems.com/evaluation-guide/platform-runtime/. Accessed 25 Dec 2022
3. Taylor, R.N., Medvidovic, N., Oreizy, P.: Architectural styles for runtime software adaptation. In: 2009 joint working IEEE/IFIP conference on software architecture european conference on software architecture, pp. 171–180 (2009). https://doi.org/10.1109/WICSA.2009.5290803
4. Wittern, E., Ying, A.T.T., Zheng, Y., Dolby, J., Laredo, J.A.: Statically Checking Web API Requests in JavaScript. In: 2017 IEEE/ACM 39th international conference on software engineering (ICSE), pp. 244–254 (2017). https://doi.org/10.1109/ICSE.2017.30
5. OutSystems: Configure a Consumed REST API at Runtime. https://success.outsystems.com/Documentation/11/Extensibility_and_Integration/REST/Consume_REST_APIs/Configure_a_Consumed_REST_API_at_Runtime. Accessed 22 Feb 2022
6. Gomes, A.: how to covert REST API base URL to SITE property | OutSystems. https://www.outsystems.com/forums/discussion/40371/how-to-covert-rest-api-base-url-to-site-property/. Accessed 21 Feb 2022
7. Hunt, A., Thomas, D.: The pragmatic programmer. Addison-Wesley Professional (1999)
8. Birlogeanu, A.: Idea - Base URL for consumed REST services configurable at runtime | OutSystems. https://www.outsystems.com/ideas/2939/base-url-for-consumed-rest-services-configurable-at-runtime/. Accessed 21 Feb 2022
9. Shumarov, B.: XpressDox Connector - Overview | OutSystems. https://www.outsystems.com/forge/component-overview/11958/xpressdox-connector. Accessed 17 May 2022
10. OutSystems: Consume REST APIs: Simple Customizations, https://success.outsystems.com/Documentation/11/Extensibility_and_Integration/REST/Consume_REST_APIs/Simple_Customizations. Accessed 22 Feb 2022
11. OutSystems: Use Services to Expose Functionality, https://success.outsystems.com/Documentation/11/Developing_an_Application/Reuse_and_Refactor/Use_Services_to_Expose_Functionality. Accessed 20 March 2022

D-AI-COM: A DICOM Reception Node to Automate the Application of Artificial Intelligence Scripts to Medical Imaging Data

Andrea Vázquez-Ingelmo[1]([✉]) [ID], Alicia García-Holgado[1] [ID],
Francisco José García-Peñalvo[1] [ID], Pablo Pérez-Sánchez[2] [ID],
Antonio Sánchez-Puente[4] [ID], Víctor Vicente-Palacios[5] [ID],
Pedro Ignacio Dorado-Díaz[2] [ID], and Pedro Luis Sánchez[3]

[1] GRIAL Research Group, Computer Science Department, Universidad de Salamanca,
Salamanca, Spain
{andreavazquez,aliciagh,fgarcia}@usal.es
[2] Biomedical Research Institute of Salamanca (IBSAL), Salamanca, Spain
[3] University Hospital of Salamanca. CIBERCV and Biomedical Research Institute of
Salamanca (IBSAL), Salamanca, Spain
pidorado@saludcastillayleon.es
[4] University Hospital of Salamanca and CIBERCV, Salamanca, Spain
asanchezpu@saludcastillayleon.es
[5] Philips Clinical Science, Amsterdam, The Netherlands
victor.vicente.palacios@philips.com
https://ror.org/02f40zc51

Abstract. Artificial Intelligence (AI) has proven to be useful in several fields. The medical domain is one of the fields that benefits from the application of AI methods to automate and ease complex tasks including disease detection, segmentation, assessment of organ functions, etc. However, applying these kinds of methods to the variety of data formats involved in health contexts is not trivial. It is necessary to provide technologies that enable non-expert users to benefit from AI applications. This work presents a platform that acts as a DICOM reception node with the goal of automating the application of AI algorithms to medical imaging data. This platform is set to ease the process applying AI to their DICOM images by making the whole process transparent and straightforward for users without AI-related or programming skills.

Keywords: Information System · Medical Imaging Management · Artificial Intelligence · Health Platform · DICOM

1 Introduction

Over the years, artificial intelligence (AI) algorithms have grown in popularity and expanded their use. The ability to apply them to different problems and contexts provides broad support in data-intensive scenarios. One of the domains in which these scenarios are commonplace is the medical domain, where data is constantly being generated.

© The Author(s), under exclusive license to Springer Nature Switzerland AG 2024
A. Rocha et al. (Eds.): WorldCIST 2023, LNNS 801, pp. 199–206, 2024.
https://doi.org/10.1007/978-3-031-45648-0_20

In this sense, AI algorithms are becoming increasingly important when analyzing medical data to reach insights and find hidden patterns [1]. However, another additional difficulty within field is the variety of potential data formats: electronic health records, clinical trials, or even images.

The analysis of medical imaging can support several tasks, such as disease detection, segmentation, assessment of organ functions, etc. [2]. Introducing AI algorithms in these tasks can increase the benefits derived from medical imaging analysis by automating complex activities with similar performance compared to human skills [3].

But although the benefits of applying AI to medical images is unquestionable, it is important to provide usable and transparent technologies that make the use of these methodologies accessible to non-expert users.

In this work, we present the first version of a platform (D-AI-COM) focused on applying AI algorithms to DICOM (Digital Imaging and Communication In Medicine) [4] images. This platform has two main objectives:

1. To provide an independent and configurable service to apply AI algorithms to DICOM images on demand
2. To support the integration of custom algorithms by fostering the flexibility of the platform's components

The rest of this paper is organized as follows. Section 2 describes previous works on assisted application of AI to medical images. Section 3 outlines the proposed architecture and interface of D-AI-COM. Finally, Sect. 4 discusses the proposal as well as the conclusions reached during the development of the architecture.

2 Background

This platform has been proposed in the context of the Cardiology Department of the University Hospital of Salamanca. In this context, two platforms have been previously developed to support the application of AI in the medical domain.

The first one, the CARTIER-IA platform [5, 6], was designed to support the unification of structured medical data and imaging data. This platform allowed users to upload their datasets and images and match them through DICOM identifiers, such as the patient's, study's, etc.

The CARTIER-IA platform also provided a DICOM viewer and editor, which enabled users to perform segmentations without leaving the platform, as well as apply pre-uploaded artificial intelligence algorithms to their data (Fig. 1).

On the other hand, the other platform, KoopaML, focused on the application of machine learning algorithms to medical data through visual means [7–9]. This platform has the goal of facilitating the application of these kinds of algorithms to novice users (Fig. 2).

The D-AI-COM platform aims at continuing to enhance the technological ecosystem by providing a transparent node that does not require any explicit user action to carry out the application of AI algorithms to medical imaging. As will be detailed in subsequent sections, the node will transparently apply the algorithms whenever it receives a new image through the DICOM protocol.

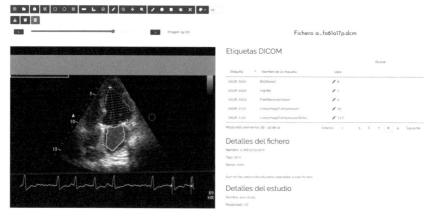

Fig. 1. The CARTIER-IA platform DICOM viewer and editor.

Fig. 2. The KoopaML platform workspace.

3 Proposal

This section details the proposed architecture and first version of the graphical interface for the D-AI-COM platform. The whole platform has been designed to focus on the flexibility and independence of the different modules.

3.1 Architecture

As introduced above, the platform must comply with different requirements specific to the context AI and medical images. This kind of scenario has a set of implications, among them:

1. Services must be continuously listening for incoming DICOM images
2. The application of AI algorithms is time- and resource-consuming, so they should be executed in the background and on-demand

In this sense, we designed the architecture outlined in Fig. 3. The platform consists of two separated but communicated services: a DICOM storage SCP (Service Class Provider) and a Django web application. The DICOM SCP is set to receive incoming images through the DICOM protocol, specifically it will listen for C-STORE commands.

To send an image to D-AI-COM, users need to configure the platform as a PACS (Picture Archiving and Communication System) by using the required parameters (IP address, port, and AE (Application Entity) title (Fig. 3).

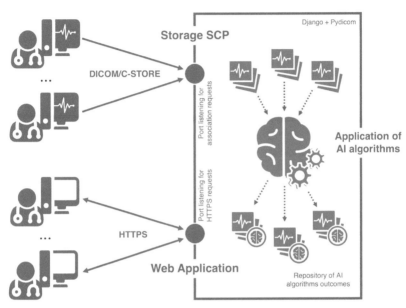

Fig. 3. Outline of the node's architecture. This node relies on two ports, one listening for incoming DICOM C-STORE requests, and the other listening for HTTPS requests for web application.

The workflow of the functionality is described in Fig. 4. The DICOM storage service is constantly listening for new associations. Whenever a new image is sent, it is analyzed to see if there are any issues regarding the DICOM protocol, and if it has been properly sent, it is stored in the node's filesystem.

The communication between the storage service and the web application is carried out through the relational Postgres database. Once the received image has been persistently stored, the SCP finishes the process by creating a new row in the database containing all the metadata (file name, patient ID, study ID, series ID, modality) and a reference to the path of the DICOM file.

Meanwhile, the Django web application is configured to monitor any new additions to the database, so when a new file has been received and stored through the SCP, a signal is triggered and processed by the web application, automatizing the whole process.

When the algorithms have been applied, a new image is generated with the results, which is also stored both in the filesystem and in the database (referencing the original DICOM file as a foreign key).

In this sense, the newly added file is inspected to see if any AI algorithm integrated in the platform supports its DICOM modality, and if it is the case, retrieves the scripts and executes them as a background process.

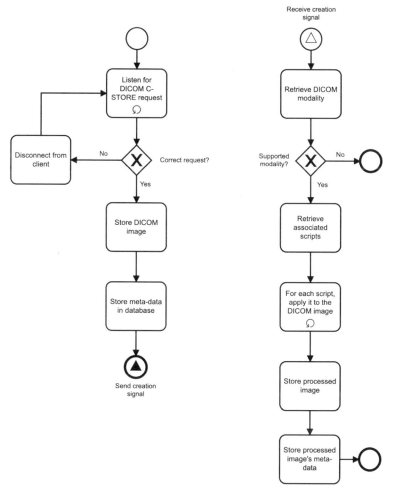

Fig. 4. Business process model of the workflow followed to receive and process incoming DICOM images. On the left, is the workflow for the DICOM storage SCP, and on the right, is the workflow of the web application.

3.2 Graphical Interface

A graphical interface has also been developed to ease the navigation and to provide access to the AI algorithms' outcomes. While the architecture provides the support to act like a PACS and to apply AI algorithms automatically, the interface focuses on making the processed DICOM images accessible and available for any user at the institution where the D-AI-COM node is deployed.

The first version of the interface enables users to navigate the received and processed DICOM images (Fig. 5). Users can add images to a favorite list to retrieve them faster.

In addition, the user interface allows privileged users to upload new AI algorithms, so the set of available algorithms can be modified without changing the codebase. To

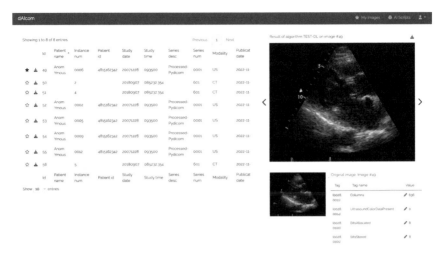

Fig. 5. First version of the graphical interface to navigate the results of the AI algorithms.

upload a new algorithm, it is necessary to provide some crucial data: the modality or modalities to which it can be applied, the model, and the python script that executes the model. Once the algorithm has been uploaded and validated, it is stored persistently in the database to be retrieved and executed when new images arrive.

4 Discussion and Conclusions

This work details a new platform for the automated application of AI algorithms to medical images. The approach taken to accomplish the development of the platform relies on a workflow with two main services:

1. The reception of DICOM images through an SCP
2. The application of AI algorithms to incoming images

This workflow has different benefits. First, the independence of the two services. The SCP does not depend on the web application and vice versa. In this case, these two services rely on the same database; the SCP is in charge of updating it with the received DICOM images and the web application monitors new additions to execute the corresponding algorithms on the fly. However, both services could work with their own databases as independent services.

Second, the workflows described in Fig. 4 are easily automatable, so the different events (such as receiving a new image or adding a new row to the database) can be traced and converted in signals that trigger subsequent actions. This method unburdens the web platform with time-consuming tasks, as they are executing in the background.

In this sense, the user does not need to perform any further action than sending the images to the D-AI-COM node through their imaging tools and accessing the web platform to get the results of the AI algorithms.

In addition, the possibility of adding new algorithms through the interface also increases the flexibility of the platform, as the scripts are not hard coded in the infrastructure and can act as interchangeable components.

The functionality of the node has been tested with real DICOM images, obtaining promising results. Relying on this node can make the process of applying AI methods more transparent and easier to lay users, as they only need to send the DICOM images through their own imaging management applications and consult or download the results through the interface.

Future research lines will involve the improvement of the graphical interface, as it is at its first stage of development and the user experience is crucial to enable users to understand the results. In addition, the D-AI-COM node will be integrated and tested in a real-world scenario to continue enhancing its features.

Acknowledgments. This research was partially funded by the Ministry of Science and Innovation through the AVisSA project grant number (PID2020-118345RB-I00). This work was also supported by national (PI14/00695, PIE14/00066, PI17/00145, DTS19/00098, PI19/00658, PI19/00656 Institute of Health Carlos III, Spanish Ministry of Economy and Competitiveness and co-funded by ERDF/ESF, "Investing in your future") and community (GRS 2033/A/19, GRS 2030/A/19, GRS 2031/A/19, GRS 2032/A/19, SACYL, Junta Castilla y León) competitive grants.

References

1. Rajkomar, A., Dean, J., Kohane, I.: Machine learning in medicine. N. Engl. J. Med. **380**(14), 1347–1358 (2019)
2. Litjens, G., Kooi, T., Bejnordi, B.E., Setio, A.A.A., Ciompi, F., et al.: A survey on deep learning in medical image analysis. Med. Image Anal. **42**, 60–88 (2017)
3. Liu, X., Faes, L., Kale, A.U., Wagner, S.K., Fu, D.J., et al.: A comparison of deep learning performance against health-care professionals in detecting diseases from medical imaging: a systematic review and meta-analysis. The Lancet Digital Health **1**(6), e271–e297 (2019)
4. Mildenberger, P., Eichelberg, M., Martin, E.: Introduction to the DICOM standard. Eur. Radiol. **12**(4), 920–927 (2002)
5. García-Peñalvo, F.J., Vázquez-Ingelmo, A., García-Holgado, A., Sampedro-Gómez, J., Sánchez-Puente, A., et al.: Application of artificial intelligence algorithms within the medical context for non-specialized users: the CARTIER-IA platform. Int. J. Interact. Multimedia Artific. Intell. **6**(6), 46–53 (2021)
6. Vázquez-Ingelmo, A., Alonso, J., García-Holgado, A., García-Peñalvo, F.J., Sampedro-Gómez, J., et al.: Usability Study of CARTIER-IA: A Platform for Medical Data and Imaging Management. HCII 2021: Learning and Collaboration Technologies: New Challenges and Learning Experiences, pp. 374–384, Virtual event (2021)
7. Vázquez-Ingelmo, A., Alonso-Sánchez, J., García-Holgado, A., Peñalvo, F.J.G., Sampedro-Gómez, J., et al.: Bringing machine learning closer to non-experts: proposal of a user-friendly machine learning tool in the healthcare domain. Ninth International Conference on Technological Ecosystems for Enhancing Multiculturality (TEEM 2021), pp. 324–329, Barcelona, Spain (2021 of Conference)

8. García-Peñalvo, F.J., Vázquez-Ingelmo, A., García-Holgado, A., Sampedro-Gómez, J., Sánchez-Puente, A., et al.: KoopaML: a graphical platform for building machine learning pipelines adapted to health professionals. Int. J. Interact. Multimedia Artific. Intell. (In Press)
9. García-Holgado, A., Vázquez-Ingelmo, A., Alonso-Sánchez, J., García-Peñalvo, F.J., Therón, R., et al.: User-Centered Design Approach for a Machine Learning Platform for Medical Purpose. HCI-COLLAB 2021, pp. 237–249, Sao Paulo, Brazil (2021)

Exploring a Deep Learning Approach for Video Analysis Applied to Older Adults Fall Risk

Roberto Aldunate[1], Daniel San Martin[2]($\boxtimes$)(iD), and Daniel Manzano[3]

[1] Centro de Innovación e Investigación Aplicada, Osorno, Chile
roberto@ceinina.com
[2] Departamento de Informática, Universidad Técnica Federico Santa María,
Valparaíso, Chile
daniel.sanmartinr@usm.cl
[3] School of Computing and Information Systems, University of Melbourne,
Melbourne, Australia
dmanzanoagua@student.unimelb.edu.au

Abstract. In some medical areas, video activity recognition has been used for patient rehabilitation, evaluating their performance doing some exercises to determine if they are correct or not. In this article we emphasize this approach applied on older adults' physical activity motivated by the problem caused by falls in this segment of the population. Furthermore, we have developed 8 Deep Learning models to classify different video recordings with the purpose of evaluating and determining how accurately those exercises are executed by people. This article is presented as a first step work, and taking into account this progress, good results were obtained considering the problem of the small number of samples, but addressed including the typical data augmentation techniques. The main results obtained from this work is that the models achieved between 71% and 89% of accuracy, depending on the exercise, and as a conclusion, it allows us to consider this approach to be a valid tool to address the problem of fall risk evaluation in older adults.

Keywords: Risk Falling · Older Adults · Deep Learning · *CNN-RNN Model*

1 Introduction

Falls on older adults are a serious threat to their health and well-being. In some countries like Chile, it is estimated that 1 in 3 older adults fall once or twice in a year [8]. More than 3 millions are treated in emergency departments annually for fall injuries. On the other hand, in 2014 the U.S reported a $28,7\%$ of adults falling at least once in a year according to the Center of Disease and Control Prevention, resulting in 29 million falls [3]. Health support for falls are most of the time expensive and restrictive, making difficult to provide a physically support for every older adult. In the worst scenarios, the damage caused by falls

A. Rocha et al. (Eds.): WorldCIST 2023, LNNS 801, pp. 207–218, 2024.
https://doi.org/10.1007/978-3-031-45648-0_21

may be deathly, and this important reason makes the self-assessment a very necessary tool in decrease the impact of this problem.

In the self-assessment line, mobile technology such as smartphones offer a potential solution to reduce the gap of access to fall risk assessment for older adults. Unlike other works, our focus is just evaluating this group of exercises by video recordings and not using any other devices like gyroscopes, accelerometers or analyzing any other physical activity. The main idea is the continuous tracing of older adults' physical activity, since it is the most effective way to prevent falls [11]. In the USA, almost 30% of the adults over 65 years old owned a smartphone in 2015. In 2020, the outlook in Chile reveals that 64% has a smartphone and 73% has a social network app such as Facebook, Instagram or Twitter [1]. This approach is deeply explored in the FONDEF ID20I10418 project [2], funded by the Chilean National R&D Agency (ANID), by means a smartphone application which allows older adults self-assess the falling risk. This FONDEF project represents the umbrella and provides the environment for the analysis presented in this article.

Due to the problem caused by falls in older adults, the main objective of this work is to develop a tool to assess the risk of falls using Deep Learning (DL) models. For this purpose, we analyzed 8 types of exercises for older adults, using machine learning techniques, specifically Convolutional Neural Network (CNN) and Recurrent Neural Network (RNN), to evaluate their physical activity through video recordings and classifying if these exercises are correct or not. One of the problems related with this project is the amount of training samples generated, so we include data augmentation techniques to increase artificially the number of samples. A visual validation of the fake data generated was performed, due to the chance of getting nonphysical examples. We emphasize that our approach only requires the use of videos and no other extra hardware requirements are necessary to assess the risk of falling.

This work is structured as follows. Section 2 presents related work, mainly in the use of Machine Learning. The description of the data, models and techniques used are presented in Sect. 3. Section 4 shows the results of this work, and finally, the conclusions and future work are presented in Sects. 5 and 6 respectively.

2   Related Work

A very important work is developed by [16], where the authors propose a Machine Learning Multi-class algorithm for fall detection using wearable sensors. They achieved almost a 99% of accuracy in their approach, showing that Machine Learning application could be a good strategy for risk falls problem. In addition, this work provides a summary of 24 studies where different algorithms were used to fall detection using wearable sensors.

Another interesting work is developed in [9], where authors proposed a RNN model. They used a pose model based on a combination of CNN and RNN, specifically a Long Short-Term memory ($LSTM$), which improves both the accuracy and efficiency. This model is applied to videos of any kind and not exclusively to the problem of fall detection.

In the same line of the previous work, [7] presents a *LSTM* model using the body joints for pose estimation. The aim is to learn the spatial correlation of human posture in order to achieve a *3D* approximation of the human spatial position.

The work in [12] presents a multi modal dataset with includes a collection of inertial sensor data from smartphones, smartwatches and earbuds, but also includes video with *2D* and *3D* pose estimation. This data is time-synchronized for people performing fully-body workouts. Notice that is not focused in older adults, but they used a *CNN*-based model for exercise recognition problem.

Although there is increasing work related to the Machine Learning approach for fall detection problem using sensors or general pose estimation using videos, but there is not enough effort tackling the problem using video/image processing to detect anomalies in standard exercises developed by older adults, similar to what health professionals do in clinical settings. Therefore, the idea of this work is to conduct research in this line in order to provide future applications that may help in the evaluation of the risk of falls in older adults.

3   Methods

3.1   Data Collection

The 8 exercises defined in the FONDEF ID20I10418 project [2] were collected on a daily basis over a period of several weeks. Those videos were tagged good or bad by human criteria. The exercises and their descriptions are:

1. Walking forward: Move 8 steps away from the telephone (4 meters approximately). Walk naturally towards the phone, taking small steps every 2 seconds.
2. Walking sideways: Stand sideways so that the left side of your body is in front of the phone. Walk forward from your position for 5 seconds. Turn around and walk in the opposite direction for 5 seconds. Turn again and walk for 5 more seconds.
3. Semi-tandem: Standing in front of the telephone, you must see your whole body on the screen. Take a half a step forward with your right foot. Hold the position and stare straight ahead.
4. Tandem: Standing in front of the phone, you should see your whole body on the screen. Position your left foot just in front of your right foot so that the heel of the front foot touches the toe of the back foot. Hold the position and stare straight ahead.
5. Visual Contact Time: Standing in front of the phone, you should see your whole body on the screen. Spread your feet shoulder-width apart and place your hand on your hips. Lean forward as far as possible. Turn on your hips, making circles with your trunk in a counterclockwise direction (to the left).
6. Standing on one leg: Standing in front of the phone, you should see your whole body on the screen. Stand on one foot only, raise the opposite foot (you must rest on the leg that gives you more security). Hold the position. If the raised foot touches the floor, lift it as quickly as possible.

7. Transition: Sit on a chair in front of the telephone without touching the back of the chair and with your arms stretched forward. After 5 seconds, stand up and sit down as fast as possible.
8. Four corners: Standing in front of the phone you should see your whole body on the screen. We are going to mark the 4 corners of a square in a imaginary way. Take step to your right and stay in place for 4 seconds. Step to the left and hold in place for 4 seconds. Step backward and hold in place for 4 seconds to return to the position.

We point out that unlike other generated datasets that only store the pose of the people, this one contains the complete video including the context of the image as well as the characteristics of the people's bodies.

3.2   Deep Learning

The classification models used by this work are based on Deep Learning (*DL*), because these kind of models are able to represent the world using a nested hierarchy of concepts [6]. Using the same approach presented by [10], we combine a Convolutional Neural Network with a Recurrent Neural Network in order to extract both spatial and temporal patterns included in the video data used by this work. Also, we take advantage of the pre-trained Convolutional Network to reduce the amount of computation by exploiting the knowledge acquired in similar task. These concepts will be briefly described below.

Convolutional Neural Networks (CNN). They are simply neural networks that use convolution in place of general matrix multiplication in at least one of their layers [6]. In this work, the *CNN* was designed to process videos and it provides automatic feature extraction, which is the primary advantage using this model. The input video is forwarded to a feature extraction network, and then the results are forwarded to a recurrent neural network. The *CNN* model used by this work is the *EfficientNet* due to the good results we achieved after testing different *CNN* models. *EfficientNet* is a type of *CNN* which uses a scaling approach, identifying and balancing carefully the network width, depth and resolution improving performance in terms of accuracy and efficiency [14].

Recurrent Neural Networks (RNN). They are the state of art algorithm to work with sequential data [6]. Due to its capability of remembering its inputs through an internal memory, thus makes them a perfect solution to resolve machine learning problems that involve sequential data. This work proposes the use of a Gated Recurrent Unit (*GRU*), a kind of *RNN* model composed by an update and a reset gate [5], in order to capture the dynamic of the frame sequences existing in the analyzed videos.

Transfer Learning (TL). As summarized by [4], and based on the stated by other authors, Transfer learning is a research problem that focused on storing knowledge acquired in one problem to be applied into a different but related problem. The use of *TL* is associated to problems with insufficient training data [13]. This work exploited this idea using a pre-trained *CNN* in order to reuse the knowledge obtained in image classification and get an new feature representation of the video frames. These frame representations are then used by the *RNN* model.

3.3   Augmentation

To get better results on video or image classification tasks, Deep Neural Networks (*DNN*) are trained on a massive number of data examples. In some areas, like medical imaging, data is scarce or expensive to generate, therefore, augmentation techniques are used to solve this problem [6]. Some of the techniques used in this work were: horizontal flip, pepper (augmenter that sets a certain fraction of pixel intensities to 0, hence they become black.), salt (augmenter that sets a certain fraction of pixel intensities to 255, hence they become white) and temporal variations.

3.4   Implementation

This work was implemented using Python and Tensorflow. The code was executed on a workstation with a NVIDIA GeForce RTX 3090 graphic card, to accelerate the training process by means of the Tensorflow compatibility with Graphic Processing Units (*GPUs*).

4   Results

4.1   Data

For the data generation process, we use the same idea for each exercises. Two classes were labeled as True or False, where True is the class with the exercises correctly performed and the False class use bad executed exercises. In addition, we complemented the False class using the other exercises that are not the one corresponding to the True class. Only in Tandem exercise we used Semi-tandem exercise as False class. Since we only had access to a few number of real examples, data augmentation techniques described in Subsect. 3.3 allowed us to increase the number of samples. The summary of the dataset created for this work is presented in Table 1.

4.2   Models

The models used in this work have the following characteristics:

- All the models used *EfficientNet-B7* for feature extraction, except for Walking forward exercise which uses *EfficientNet-V2M*.
- All the models were compiled using with the following configuration:

Table 1. Dataset split configuration for each exercise

Exercise	Correct class		Other	
	Train	Test	Train	Test
Walking forward	209	70	237	75
Waling sideways	300	60	330	60
Transition	270	78	278	78
Standing on one leg	300	48	286	54
Tandem	280	56	165	21
Semi-tandem	132	72	144	72
Visual contact time	165	35	171	36
Four corners	215	63	213	72

- Adam optimizer.
- Learning rate of 0.001.
- The loss function is categorical cross-entropy.
- 50 epochs, except for walking forward and Semi-tandem which use 20 epochs.
- Batch size of 32.
- The architectures of each model are detailed in Table 3 (see appendix).

Hyperparameter Optimization was used to get different model configurations, and the confusion matrices of the best results are presented from Fig. 1, 2, 3, 4, 5, 6, 7 and 8.

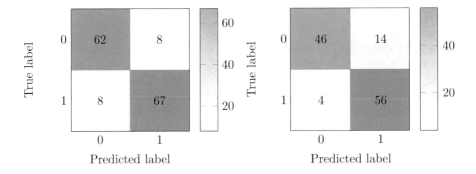

Fig. 1. Walking forward   **Fig. 2.** Walking sideways

The performance of the models are reported in Table 2. Regarding accuracy, the models obtained values between 71% and 89%. These differences can be explained by the diversity of exercises and the complexity required to identify movements that are very subtle, as in the case of the semi-tandem. The

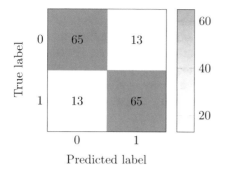

Fig. 3. Transition

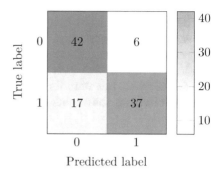

Fig. 4. Standing on one leg

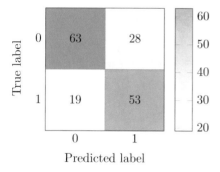

Fig. 5. Semi-tandem

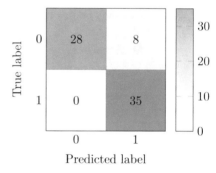

Fig. 6. Visual contact time

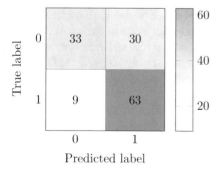

Fig. 7. Four corners

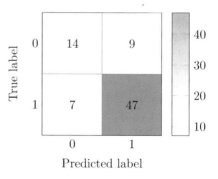

Fig. 8. Tandem

result observed in the four corners exercise can be attributed to the dynamics of the execution since it involved a much greater use of space compared to the other exercises. Nevertheless, these results are good enough as a first attempt to address this problem, and currently, we are working on improve these metrics in the near future. The data gathering is in an beginning process and still being too scarce. Also, raw data its actually imbalanced, some exercises have a greater amount of videos than others. Further work for data generation is needed.

Table 2. Summary of the results for each model

Exercise	Metrics			
	Accuracy	Precision	Recall	F1-Score
Walking forward	0.89	0.89	0.89	0.89
Walking sideways	0.85	0.80	0.93	0.86
Transition	0.83	0.83	0.83	0.83
Standing on one leg	0.77	0.86	0.69	0.76
Tandem	0.79	0.87	0.84	0.85
Semi-tandem	0.79	0.81	0.76	0.79
Visual contact	0.89	0.81	1.00	0.90
Four corners	0.71	0.68	0.88	0.76

5   Conclusions

In this study, we have proposed a Deep Learning approach for classifying 8 exercises recorded by video. For each exercise a model was built. The main idea of this proposal was to try the capability of neural networks to classify a correct exercise execution from those who are different or similar (poorly executed). The Neural networks architectures are simple and not all achieve a high accuracy. We utilized different video recordings, using just our own criteria for labeling each video as good or bad, and we did not have an expert for video review such as a kinesiologist. However the results obtained in this study showed that Neural Networks could help to complement this task using other approaches that are commonly used in this area like *Posenet* [15].

On the other hand, this work has many possibilities to be improved. First, it is necessary to have an expert in body postures. Second, the lack of videos on this type of exercise is a problem. It is necessary to collect more videos. In this study we used augmentation techniques to increase video amount. Third, video recording could be more standardized, we mean considering more specific instructions for exercise execution, for example for standing on one leg exercise we could label a video as good only if the exercise was executed with the right leg. In our case, due a lack of data, we flip some videos to increase the amount of video samples. Instructions about the camera positions could be another kind of standardization, for instance, the distance between person and the mobile, camera position to avoid going out of the picture frames.

As a final conclusion we can mention that the implemented method works to address the problem that leads to this project; we are able to obtain models that are able to determine with an accuracy of 71-89%, depending on the exercise, whether or not it corresponds to the evaluated exercise. Considering the results of the metrics for each test, we could also establish how well the exercise is performed.

6 Future Work

Because this article is first step, currently still in progress, we are aware that substantial improvements are required before the production version is released. One of the first needs is the acquisition and generation of more data. This is important because we are using two kind of deep learning models that require a significant amount of data to avoid problems associated with *overfitting*. Connected to the latter, the application that collects data already allows saving videos and only minor adjustments are needed to connect it to this work. Our idea consists of using the results of these models to complement the application that has already been developed and is associated with the project described in [2]. The target is calculating the fall risk index to which an older adult is exposed, after performing the exercises described in the Subsect. 3.1. These models will run as a service on a server and the app will get the evaluation using an *API*.

To make this work reproducible, in the near future we will include a repository with the model architecture and weights, to let the model available for any new research effort in this line. In addition, we want to provide a comparison between the existent models using wearable sensors to identify which tool is better or even to improve the results by complementing both ideas. We also expect feedback from health professionals to validate this tool as a complementary tool for fall risk assessment. Finally, the aim of this entire framework is obtain reliable results in order to address the problem of older adults risk falling.

Acknowledgment. The authors of this article want to thank the Chilean "Agencia Nacional de Investigación y Desarrollo (ANID)" and "ANID-Subdirección de Capital Humano/Doctorado Nacional/2019-21191017" - for the support in the development of FONDEF ID20I10418 project.

Appendix

Architectures of the models used in the work are presented in Table 3.

Table 3. Architecture of the models used for each exercise

Model	Architecture
Walking forward	– Input: Dense, 350×1280 units, linear activation – Hidden: • $2\times$GRU, 128 units, tanh activation, sigmoid recurrent activation • Dropout 0.4 • GRU, 128 units, tanh activation, sigmoid recurrent activation • Dense, 8 units, tanh activation – Output: Dense, 2 units, softmax activation.
Semi-tandem	– Input: Dense, 400×2560 units, linear activation – Hidden: • $2\times$GRU, 64 units, tanh activation, sigmoid recurrent activation • Dropout 0.4 • GRU, 64 units, tanh activation, sigmoid recurrent activation • Dense, 8 units, sigmoid activation – Output: Dense, 2 units, softmax activation.
Walking sideways	– Input: Dense, 350×2560 units, linear activation – Hidden: • $3\times$GRU, 128 units, tanh activation, sigmoid recurrent activation • Dropout 0.4 • GRU, 128 units, tanh activation, sigmoid recurrent activation • Dense, 8 units, sigmoid activation – Output: Dense, 2 units, softmax activation.
Visual Contact	– Input: Dense, 400×2560 units, linear activation – Hidden: • $3\times$GRU, 16 units, tanh activation, sigmoid recurrent activation • Dropout 0.4 • GRU, 16 units, tanh activation, sigmoid recurrent activation • Dense, 8 units, sigmoid activation – Output: Dense, 2 units, softmax activation.
Transition	– Input: Dense, 350×2560 units, linear activation – Hidden: • $4\times$GRU, 128 units, tanh activation, sigmoid recurrent activation • Dropout 0.4 • GRU, 128 units, tanh activation, sigmoid recurrent activation • Dense, 8 units, sigmoid activation – Output: Dense, 2 units, softmax activation.

(continued)

Table 3. (*continued*)

Standing on one leg	– Input: Dense, 350×2560 units, linear activation
	– Hidden:
	• 4×GRU, 256 units, tanh activation, sigmoid recurrent activation
	• Dropout 0.4
	• GRU, 256 units, tanh activation, sigmoid recurrent activation
	• Dense, 8 units, tanh activation
	– Output: Dense, 2 units, softmax activation.
Four corners	– Input: Dense, 350×2560 units, linear activation
	– Hidden:
	• 2×GRU, 16 units, tanh activation, sigmoid recurrent activation
	• Dropout 0.4
	• GRU, 16 units, tanh activation, sigmoid recurrent activation
	• Dense, 8 units, relu activation
	– Output: Dense, 2 units, softmax activation.
Tandem	– Input: Dense, 350×2560 units, linear activation
	– Hidden:
	• 2×GRU, 128 units, tanh activation, sigmoid recurrent activation
	• Dropout 0.4
	• GRU, 128 units, tanh activation, sigmoid recurrent activation
	• Dense, 8 units, sigmoid activation
	– Output:
	– Dense, 2 units, softmax activation

References

1. Adimark, G.: Microestudio: radiografía a los adultos mayores en chile 2020 (2020). https://www.gfk.com/es/prensa/radiografiaalosadultosmayores2020. Accessed 17 Nov 2022
2. Aldunate, R.: Fondef id20i10418: Self-assessment technology to prevent and reduce falls of older adults. Tech. rep, CEININA, Chile (2020)
3. Bergen, G., Stevens, M.R., Burns, E.R.: Falls and fall injuries among adults aged ≥ 65 years-united states, 2014. Morb. Mortal. Wkly Rep. **65**(37), 993–998 (2016)
4. Bozinovski, S.: Reminder of the first paper on transfer learning in neural networks, 1976. Informatica **44**, 291–302 (2020). https://doi.org/10.31449/INF.V44I3.2828. https://www.informatica.si/index.php/informatica/article/view/2828
5. Cho, K., van Merriënboer, B., Bahdanau, D., Bengio, Y.: On the properties of neural machine translation: encoder-decoder approaches. In: Proceedings of SSST 2014–8th Workshop on Syntax, Semantics and Structure in Statistical Translation, pp. 103–111 (2014). https://doi.org/10.3115/V1/W14-4012. https://aclanthology.org/W14-4012
6. Goodfellow, I., Bengio, Y., Courville, A.: Deep Learning. MIT Press (2016). http://www.deeplearningbook.org

7. Lee, K., Lee, I., Lee, S.: Propagating LSTM: 3D pose estimation based on joint interdependency. In: Ferrari, V., Hebert, M., Sminchisescu, C., Weiss, Y. (eds.) ECCV 2018. LNCS, vol. 11211, pp. 123–141. Springer, Cham (2018). https://doi.org/10.1007/978-3-030-01234-2_8

8. Leiva, A.M., et al.: Factores asociados a caídas en adultos mayores chilenos: evidencia de la Encuesta Nacional de Salud 2009–2010. Revista médica de Chile 147, 877–886 (2019). http://www.scielo.cl/scielo.php?script=sci_arttext&pid=S0034-98872019000700877&nrm=iso

9. Luo, Y., et al.: LSTM pose machines. In: 2018 IEEE/CVF Conference on Computer Vision and Pattern Recognition (CVPR), pp. 5207–5215 (2018). https://doi.org/10.1109/CVPR.2018.00546

10. Paul, S.: Video classification with a CNN-RNN architecture (2021). https://keras.io/examples/vision/video_classification/. Accessed 19 Nov 2022

11. Sherrington, C., et al.: Evidence on physical activity and falls prevention for people aged 65+ years: systematic review to inform the who guidelines on physical activity and sedentary behaviour. Int. J. Behav. Nutr. Phys. Act. **17**(1), 1–9 (2020)

12. Stromback, D., Huang, S., Radu, V.: MM-Fit multimodal deep learning for automatic exercise logging across sensing devices. In: Proceedings of the ACM on Interactive, Mobile, Wearable and Ubiquitous Technologies 4 (2020). https://doi.org/10.1145/3432701

13. Tan, C., Sun, F., Kong, T., Zhang, W., Yang, C., Liu, C.: A survey on deep transfer learning. In: Kůrková, V., Manolopoulos, Y., Hammer, B., Iliadis, L., Maglogiannis, I. (eds.) ICANN 2018. LNCS, vol. 11141, pp. 270–279. Springer, Cham (2018). https://doi.org/10.1007/978-3-030-01424-7_27

14. Tan, M., Le, Q.V.: EfficientNet: rethinking model scaling for convolutional neural networks. In: 36th International Conference on Machine Learning, ICML 2019 2019-June, pp. 10691–10700 (2019). https://doi.org/10.48550/arxiv.1905.11946. https://arxiv.org/abs/1905.11946v5

15. TensorFlow: pose estimation – tensorflow lite (2022). https://www.tensorflow.org/lite/examples/pose_estimation/overview. Accessed 19 Nov 2022

16. Zurbuchen, N., Wilde, A., Bruegger, P.: A machine learning multi-class approach for fall detection systems based on wearable sensors with a study on sampling rates selection. Sensors **21**(3) (2021). https://doi.org/10.3390/s21030938. https://www.mdpi.com/1424-8220/21/3/938

Organizational Models and Information Systems

Evolution of Fintech Companies in Guatemala from Prepandemic to Post Pandemic Period and the Options of Fintech Companies as Financial Provider for Projects

Stephen David Martinez-Mendez[1]([✉]), Fauricio Alban Conejo Navarro[2], and Janio Jadán-Guerrero[3]

[1] Maestría en Gestión y Dirección de Proyectos, Universidad Galileo, Ciudad de Guatemala, Guatemala
Davidm777@galileo.edu
[2] Universidad Latina de Costa Rica, San José, Costa Rica
fauricio.rogers@ulatina.net
[3] Centro de Investigación en Mectrónica y Sistemas Interactivos – MIST, Universidad Tecnológica Indoamérica, Quito, Ecuador
janiojadan@uti.edu.ec

Abstract. On the Pre-pandemic time most of the projects were traditional financial banking methods, that means the biggest companies had have receive 85%, leaving only the 15% for the Small and medium-sized enterprises (SMEs), in the pandemic period financial risk increased and trying to figure out how to survive the biggest companies update them technology for e-commers but SMEs did not have enough money to do it, for that reason they found a solution on fintech technologies. As a result, in the post pandemic era, fintech companies began to have a huge relevance not only as a technology supplier but also as financial provider. The main reason of this research is to find out if Fintech companies could be provider for projects to big companies and SMEs, the outcome shows that project managers of different industries see e-money as a secure payment method and with the fintech establishment on the economic scene they begin to feal more comfortable with the idea of using fintech companies as financial provider for projects in them companies. This research was carried out under the descriptive method, The tool selected to collect information was Google Forms, in this tool a multiple-choice survey was created, and email was sent with access to the survey. The results allow us to understand that they can be a permanent and reliable source to finance projects of different types of industries.

Keywords: Prepandemic · Pandemic · Post pandemic · Fintech · Projects · Financial

© The Author(s), under exclusive license to Springer Nature Switzerland AG 2024
A. Rocha et al. (Eds.): WorldCIST 2023, LNNS 801, pp. 221–228, 2024.
https://doi.org/10.1007/978-3-031-45648-0_22

1 Introduction

All projects are a temporary effort to create value through a unique product, service, or result. All projects have a beginning and an end [1]. This concept let us know how important projects for business development are for them diversity nature clouding be projects of construction [2]. Marketing, software, all kinds of maintenance, etcetera. But projects are not only important for private industries are also significant for countries development, non-governmental organization (NGO), International relationships, etcetera.

As we can see projects have a significant role in the economic growth [3]. Social and cultural. A successful project can be measure by key performance indicator (KPI). Such as, financial kpi to Amount of money allocated to the project, money flow, money reserve. Predictive kpi for sales estimate, return of investment, Clients. Communications kpi for Money earmarked for communication. Internal communication channels, external communication channels. Market kpi. As we can see kpi could be different for each project.

In Guatemalan economy, SMEs represent 99% of business section, creating 80% of employment and making the 40% of gross domestic product [4]. For all those reasons SMEs have of enormous importance in the economy performance. On the Prepandemic time big companies and SMEs use traditional work methods [5]. And suddenly everything changes with the pandemic situation. All companies form different industries needed to change from the old traditional work methods to a work from home or a hybrid method. Those changes required a technology and equipment investment [6]. For big companies that was not huge issue because they already have an information technology (IT) department however, SMEs had a hard time trying to update them technology to e-commerce, SMEs found in fintech companies the right solution and began to make deals with them.

Although SMEs are an important for the economic performance, they only have access to 15% of bank business credits [7]. Leaving the rest for big companies. All this without taking into consideration the high bank risk in Guatemala (see Table 1). That means Project financial option look toughest in post pandemic time for big companies and SMEs. For all those reasons born this research making the following question ¿Could fintech companies be a financial provider for projects?

Table 1. Bank risk by country

Country	Raking
Chile	3
Mexico	5
Panama	5
Peru	5

(continued)

Table 1. (*continued*)

Country	Raking
Uruguay	5
Colombia	6
Brazil	6
Guatemala	7
Costa Rica	8
El Salvador	8
Honduras	8
Paraguay	8
Argentina	9
Bolivia	9

Bank risk analysis for Latin American countries [8]. The lower number means less risk on the contrary higher number represent a huge risk. Lists order by name.

1.1 Fintech

The word Fintech refers to companies that focus on carrying out everything that a traditional bank does, but with a technological focus [9]. Fintech in Guatemala is a relatively new sector that was born on 2017, which makes it a sector little known but growing (see Fig. 1). it still presents several challenges to consolidate itself as a source of financing for projects [10]. Since its function within the Guatemalan market is more focused on the technological services sector. Due to the service approach, they already have a wide range of clients in their portfolio [11]. (see Fig. 2).

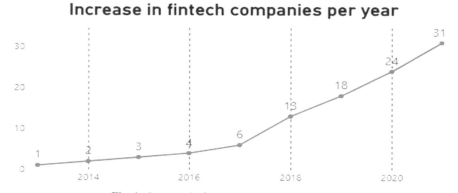

Fig. 1. Increase in fintech companies per year [12].

Service sectors of Fintech companies

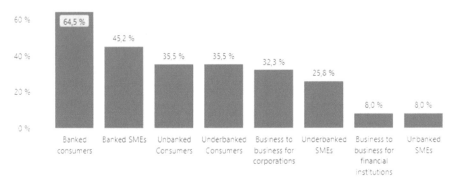

Fig. 2. Sectors of attention of fintech companies [12].

1.2 Wallet

It is a digital wallet that is used to manage digital money, there we can receive loan deposits, payments for services, transfers to other Wallets or bank accounts [13]. This technology is important as it is the future of financial transactions globally. This would be the way in which the Fintech would make the deposits of the credits provided by them, and in the same way it will be used for all types of transactions [14].

The impact of this technology has been so big that many companies focused on technology have created their own Wallets for their users [15]. Has been well received because it is only necessary to use an App on the mobile device to be able of money management and make financials transaction [16].

2   Methodology

This research was carried by applying the quantitative method, to understand the perception of different Project managers about Fintech companies as a provider for future projects under their leadership.

The tool that was selected to collect information was Google Forms; this tool has the option of a multiple-choice survey which were created. The questions were focused on knowledge management about the Fintech companies and the services that are provided such as the Wallet, and various questions to measure the confidence in Fintech companies.

Once the survey was created, it was complemented with the digital method of an email with access to the survey, to ensure that the project managers were the people who answered the questions. The survey was sent to 24 project managers who work for different industries and 20 of them answered all survey questions.

3   Result

The results of the questionnaire survey demonstrate the results from the project mangers that answered the digital survey. Below are the results of the answers gathered from a quantitative perspective the outcomes.

Every day fintech companies are scaling in different industries, making new agreements to supply technology, creating new business section, meanwhile fintech companies did a great job expanding their services not only to companies but also to people who cannot have an opportunity of have a bank loan because of credits issues and other situations, and they are part of fintech increase for that reason project managers start to know them and that knowledge will increase, as we can see on the figure below 50% of the project manager knew what a fintech is and 50% did not know. (See Fig. 3). This result clearly shows a clear lack of knowledge on this area as an emergent service.

Do you know what a Fintech company is?

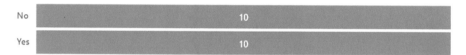

Fig. 3. From Google form survey

For those project managers who are already familiar with fintech companies, believe that fintech companies are a good choice of financial provider for projects, because fintech will offer you different payment arrangement options, as: They will give the digital money in a wallet, and they will only charge you the amount you spend, so if you only use the 90% of the loan, you do not need to pay for the full loan amount and that is a big difference with a bank that will charge you the full loan amount and project managers see these kind of agreements and believe that fintech companies have a big future, as we can see. (See Fig. 4). Comparing the results of Fig. 3 with Fig. 4 it shows the same consistency of project managers that know what a fintech is and same percentage for the ones that do not know about fintechs.

Would you use a fintech company as a financial provider for a project?

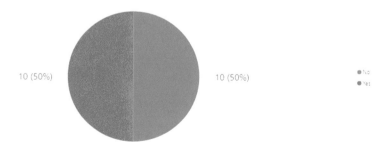

Fig. 4. From Google form survey

Most of surveyed Project managers of the survey have knowledge about the wallet technology with a total of 85%, due to their popularity and a result only 15% did not know about digital Wallets. The main reason that its popular is that the biggest tech companies developed their own wallet, you only need to go at play store, app store or a company website to get one, these wallets are secure, and people can use them in a grocery store, gas station, movie theater, restaurants, a lot of different places. As we can see. (See Fig. 5).

Do you know what a wallet is?

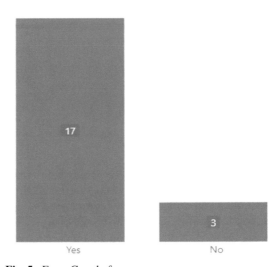

Fig. 5. From Google form survey

Although on Fig. 6 if competed with Fig. 5 it shows a contrast between the Project Managers that know about digital Wallets versus the projects managers that have utilized it. One of the most significant barriers is the technological backwardness, in Guatemala people prefer to use debit or credit cards because they feel that in many places they will reject a wallet as a method of payment, of course that situation is going to change year by year, but in this point in time some steps are made for wallet future in Guatemala. As we can see. (See Fig. 6).

Have you used a wallet service before?

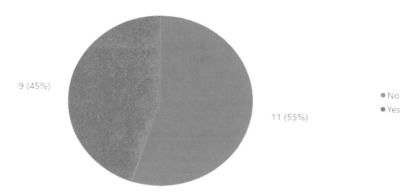

9 (45%)

11 (55%)

● No
● Yes

Fig. 6. From Google form survey

4 Conclusion

Like any new disruptive technology, Fintech still has a long way to go, this to become public knowledge on a large scale and be considered reliable by big companies and SMEs in all industries to be considered as financial provider for projects. They have in their favor that their services are increasingly used by SMEs, which means that as they create more commercial ties the trust between them will grow up allowing both parts to explore the opportunity for fintech companies of be financial provider for projects and seeing it, big companies will also explore the same opportunity.

The Wallet during the pre-pandemic period had little relevance in the market as a personal and business monetary management method, but in the pandemic time their relevance increased. The main reason was the quarantine period that we had to go through helped create a motivational confidence in this new money management technology is growing more and more. Many SMEs have benefited from the use of this technology for e-commers.

The purpose of this research is to know if Fintech can be a provider for projects, the results allow us to understand that they can be a permanent and reliable source to finance projects of different types of industries, SMEs are the sector that will be most benefited, because they are the ones that have the most commercial relationship with Fintech companies but also big companies can do it only giving them the opportunity of be financial provider for projects, which is why this financial sector represents an opportunity for growth and economic development for all parties involved.

To make this research the main adversity was the technological backwardness in Guatemala, the country has only one fintech association with only 31 companies, obtaining information about them was a challenge even though fintech companies are the financial future in Guatemala and the rest of the world, especially in this period where Business intelligence and artificial intelligence are having amazing results for companies.

References

1. Project Management Institute.: Project Management Body of Knowledge (PMBOK® Guide). 6th edn, Project Management Institute, Inc. (2017)
2. Kleimeier, S., Versteeg, R.: Project finance as a driver of economic growth in low-income countries. Rev. Finan. Econ. **19**(2), 49–59 (2010)
3. Cox, R.F., Issa, R.R., Ahrens, D.: Management's perception of key performance indicators for construction. J. Constr. Eng. Manag. **129**(2), 142–151 (2003)
4. Agexport. https://dataexport.com.gt/98-de-las-pymes-guatemaltecas-aceleraron-su-proceso-de-transformacion-digital/. Accessed 6 April 2022
5. Taylor, M., Murphy, A.: SMEs and e-business. J. Small Bus. Enterprise Develop. (2004)
6. Iriarte, C., Bayona, S.: IT projects success factors: a literature review. Int. J. Inf. Syst. Proj. Manag. **8**(2), 49–78 (2020)
7. Agexport. https://revista.dataexport.com.gt/2021/07/puede-el-credito-digital-impulsar-el-comercio-electronico-en-mipymes/?_ga=2.114397831.1099373343.1666577904-432373067.1666577904. Accessed 1 June 2021
8. S&P Global Ratings. https://www.spglobal.com/_assets/documents/ratings/es/pdf/2022-02-17-america-latina-panorama-bancario-por-pais-1-t-2022-bancos-sortean-la-tormenta-pero-nuevos-riesgos-los-acechan.pdf. Accessed 17 Feb 2022
9. Buchak, G., Matvos, G., Piskorski, T., Seru, A.: Fintech, regulatory arbitrage, and the rise of shadow banks. J. Financ. Econ. **130**, 453–483 (2018)
10. Lee, I., Shin, Y.: Fintech: Ecosystem, business models, investment decisions, and challenges. Bus. Horiz. **61**, 35–46 (2018)
11. Hellmann, T., Murdock, K., Stiglitz, J.: Liberalization, moral hazard in banking, and prudential regulation: Are capital requirements enough? Am. Econ. Rev. **90**, 147–165 (2000)
12. FintechGuatemala. https://www.guatemalafintech.com/_files/ugd/aaef0a_c06ec1d39a6a416ca1a62e8de9a01cc2.pdf. Accessed 1 Dec 2021
13. Rathore, H.S.: Adoption of digital wallet by consumers. BVIMSR's J. Manage. Res. **8**(1), 69 (2016)
14. Teng, S., Khong, K.W.: Examining actual consumer usage of E-wallet: a case study of big data analytics. Comput. Hum. Behav. **121**, 106778 (2021)
15. Leong, L.Y., Hew, T.S., Ooi, K.B., Wei, J.: Predicting mobile wallet resistance: A two-staged structural equation modeling-artificial neural network approach. Int. J. Inf. Manage. **51**, 102047 (2020)
16. Hassan, M.A., Shukur, Z.: Review of digital wallet requirements. In 2019 International Conference on Cybersecurity (ICoCSec), pp. 43–48 (2019)

Digital Business Transformation for SMEs: Maturity Model for Systematic Roadmap

Maximiliano Jeanneret Medina, Alessio De Santo, Philippe Oswald, and Maria Sokhn[✉]

HEG Arc, HES-SO, University of Applied Sciences Western Switzerland, Neuchâtel, Switzerland
{maximiliano.jeanneret,alessio.desanto,philippe.oswald, maria.sokhn}@he-arc.ch

Abstract. Digital maturity models (DMM) aim to measure the current level of a company's digitalization, as well as providing a path towards digital maturity. While many DMMs exist, several challenges surround their application for small and medium-sized enterprises (SME) that would ben the process on their own. The aim of this paper is to describe the development and the evaluation of a tool that assist SMEs with their digital transformation, from the maturity assessment to the setup of a systematic digital transformation roadmap. Following a Design Science Research approach, we elucidated the needs of SME managers regarding digital maturity through interviews (n=10). Then, we developed and evaluated a DMM integrated within an online tool through focus groups (n=3). We finally evaluated our tool in the context of six SME digital transformation initiatives. Our findings highlight that SME managers need prescriptions about digital maturity rather than a sole description of their organization' maturity stage. Evaluations show good results in terms of utility and validity. Moreover, a need for a more complete and specialized recommendation service was highlighted which pave the way for future research.

Keywords: digital transformation · maturity level · digital business · digital maturity model · digital transformation roadmap

1 Introduction

Small and medium-sized enterprises (SMEs), defined as organizations having fewer than 250 employees, a turnover of under EUR 50 million, or revenues of under EUR 43 million, are the backbone of the European economy [1]. Currently, 99 percent of European firms are SMEs [1] and the same pattern is observed in the Swiss Jura Arc region [2]. In a context of digital transformation, organizations need to re-think and possibly re-invent their business model to remain competitive [3]. Business leaders are not only responsible to grasp and embrace the possibilities of digital technology, but also to change their organizations as they integrate digital technologies [4]. However, SMEs are taking the digital turn

A. Rocha et al. (Eds.): WorldCIST 2023, LNNS 801, pp. 229–240, 2024.
https://doi.org/10.1007/978-3-031-45648-0_23

with delay [5]. In order to define a digital transformation strategy, managers must understand the current state of their organization and need an instrument that indicates possible areas of action [6]. Digital maturity models (DMM) disclose the dimensions that need to be designed in their descriptive role, and in their prescriptive function, they allow firms to identify the courses of action or competencies required to attain the desired level of maturity [6]. However, the use of a DMM is still constrained by a lack of validation [7] or suitability for SMEs [8]. In this study, we present the Design Science Research process of an online tool based on a DMM supporting both descriptive and prescriptive goals. Relying on previous literature regarding descriptive functionalities [6], we aim to enable SMEs to describe the courses of action required to attain the target degree of maturity through the prescriptive function. Following the Gregor and Hevner [9] recommendations concerning Design Science Research, this contribution is structured as follows: 2) Related work, 3) Method, 4) Artifact Description, 5) Evaluation, and 6) Discussion.

2   Related Work

2.1   From Digitization to Digital Transformation

Although the terms digitization, digitalization, and digital transformation are sometimes used interchangeably in the literature, they should be differentiated. All are linked concepts that are generally associated with successive phases implying digital technologies [10–12]. When focusing on organizations, each phase conveys specific digital resources, organization structure, growth strategies, and metrics [11]. Digitization refers to the encoding of analog information into a digital format (i.e., into zeros and ones) such that computers can store, process, and transmit such information [11]. Hence, computers supplanted physical carriers such as paper, which led to further automation in job routines [10]. Then, networking technologies and more recently, miniaturization paired with ever-increasing processing power, storage capacity, and communication bandwidth, has brought the goal of ubiquitous computing closer to reality [10]. Such capabilities lead to digitalization which broadly describes the many-fold sociotechnical phenomena and processes of adopting and using digital technologies in individual, organizational, and societal contexts [10]. In this context, organizations enhance existing processes and provide new functionalities made possible by digital technologies [11]. Finally, digital transformation is defined as *"a process that aims to improve an entity by triggering significant changes to its properties through combinations of information, computing, communication, and connectivity technologies"* [13]. More specifically, it describes a company-wide change that leads to the development of wholly new business models [10,11]. While most definitions include technological, organizational, and social components [3], the key difference between digitalization and digital transformation is made by the profoundness of the change [3,11,13]. Even if digitalization and digital transformation have been a topic for information systems research for decades [14], the current wave differs from previous ones because it influences all aspects of customers' life [3]. Customers, getting more and more used to sophisticated digital technologies, now expect such kind of sophistication in every digital experience [14]. These rising expectations put pressure on company executives, presenting new challenges [15] but also new opportunities. However, SMEs would not necessarily have the means to have a department for driving the digital transformation and may need a tool to help them as a starting point. DMMs might provide information on the state of SMEs' digital transformation efforts and the next actions necessary to get there.

2.2 Digital Maturity Models for SMEs

As mentioned by Pöppelbuß and Röglinger [16] *"maturity models represent theories about how organizational capabilities evolve in a stage-by-stage manner along an anticipated, desired, or logical maturation path"*. Maturity models serve descriptive, prescriptive, and comparative purposes, while their basic purpose consists in describing stages and maturation paths [16]. For a descriptive purpose, maturity models are used to describe a way to gauge an organization's competence in a certain area of expertise [7, 16, 17]. Specifically, a maturity model is used as a diagnostic tool where the current capabilities of the entity under investigation are assessed with respect to given criteria [16, 18]. For a prescriptive purpose, maturity models can also act as a roadmap for the transformation process [6, 7, 18] by enabling the identification of probable, predicted, or typical development pathways to the desired goal state [16]. Hence, specific and detailed courses of action are suggested for each maturity level of a process area [18]. Prescriptive functionality enables firms to describe the courses of action or skills required to achieve the target degree of maturity [6]. Finally, a comparative purpose allows for internal and external benchmarking [16].

Commercial vendors, consulting firms (e.g., [19]), and scholars (e.g., [6,12]) engaged a lot of effort to develop DMMs [7]. Such models are in line with the multidimensional character of the digital transformation [13] and often measure customer experience, operational processes, business models, and digital capabilities [7,8]. For instance, Berghaus and Back [6] developed a DMM in its descriptive function to show the characteristics through which digital transformation affects the organization and to develop maturity levels based on empirical data to derive a typical transformation path. Moreover, DMMs can be tailored to the size of the company [12] or to a particular application area [20]. However, the use of a DMM is still constrained by a lack of validation or suitability for SMEs [8]. This can be explained by the abstract nature of the digitalization *jargon*, the context-dependency of related concepts (e.g., society, company, individual), and the variations according to the perspectives on the topic (e.g., human, process, technology) which leads to ambiguity when it comes to the basic task of DMMs: measuring a company's level of digitalization [7].

3   Method

This research has been conducted in the Swiss Jura Arc region. The project of which this research is a part aims to : (1) inform regional SMEs about digital transformation, (2) provide them an online digital maturity self-assessment tool, and (3) provide an observatory of SME digital maturity. To conduct this study, we followed the Design Science Research approach [21] that addresses research through the design and evaluation of artifacts meeting identified business needs (see Fig. 1). Design Science Research has been identified as a suitable method for filling some of the gaps in research on maturity models [8].

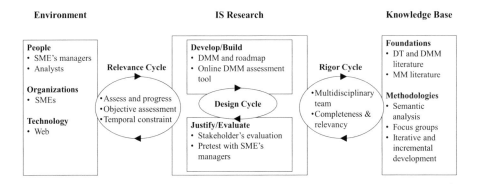

Fig. 1. Design Science Research Cycles (adapted from Hevner [22])

According to Hevner [22] the Relevance Cycle initiates Design Science Research with an application context that provides the requirements for the research. It defines acceptance criteria for the ultimate evaluation of the research results. To identify SMEs' needs, we conducted semi-directive interviews (n=10) involving ten stakeholders of the Swiss Jura Arc region. These stakeholders are regional economic and entrepreneurial associations, regional economic development departments, regional innovation hubs, technology-specialized clusters, IT companies, and academic institutions.

The Rigor Cycle connects the design science activities with the knowledge base of scientific foundations, experience, and expertise that informs the research project [22]. To do so, we performed a literature review about digital transformation and DMMs. The central Design Cycle iterates between the core activities of building and evaluating the design artifacts and processes of the research [22]. In our context, we built our DMM in two design cycles until meeting the SME objectives. To evaluate our DMM, we conducted three focus groups with regional SMEs. One researcher took the role of the moderator while another took the role of the observer. Each focus group lasted 2.5 h and was video recorded. We also evaluated the completeness and relevancy quantitatively based on a methodology for information quality assessment [23]. To evaluate the artifact obtained at the end of the second design cycle, we conducted a one-hour workshop with managers of regional SMEs. This workshop was part of a whole-day forum about digital transformation. Prior to the event, managers performed the digital maturity assessment of their company. During the workshop, two researchers presented the DMM (included in a online tool) and the maturity levels of the companies that have participated. After the event, we collected qualitative feedback about the utility of the artifact. Video records of focus groups and qualitative feedback were analyzed through thematic coding.

4   Artifact Description

4.1   Development Process

Relevance Cycle — Elucidation of Company Needs. Stakeholders compared different DMMs developed by industry or scholars. The Berghaus and Back DMM [6,24] was highlighted by SME managers since it encompasses and covers the majority of the identified maturity criteria. This DMM [24] includes nine dimensions, 26 criteria, and 65 indicators and has been validated by a rigorous process [6,7] conducted in a similar context to ours. More specifically, participants agreed with the multidimensional and staged view of digital maturity as well as its applicability in their context. However, participants raised the following limitations regarding the Berghaus and Back DMM initially retained: (1) managers needed a model that meet the double objective of assessing digital maturity as well as providing a plan for progress, (2) managers considered the assessment difficult and subjective (i.e. selecting a maturity level with a Likert scale), (3) managers have a limited amount of time to perform an assessment and mentioned that the model contains too many items, and (4) analysts found the maturity level calculation method unsuitable (i.e. inductive and too complex) to develop a digital observatory. The most notable feedback concerned the managers' need to have a baseline to be able to evaluate objectively the maturity of their organization. Based on these findings, the objectiveness and the temporal consideration were identified as predominant acceptance criteria.

Design Cycle 1 — DMM Adaptation. To provide more objectivity and reduce the number of items, we adapted the Berghaus and Back DMM [6]. The initial dimensions and criteria were kept. Indicators however were transformed into five statements each describing a maturity level per criteria (135 statements in total). For each criterion, the respondent chooses the statement that best fits their organization [16,18]. This approach is used in the context of digital maturity [12] and business processes [25]. Moreover, to provide rigorous statements within each criterion, we used semantic analysis to extract from each indicator a series of constituting primitives [13] and used those primitives to find specialized literature. Each statement is built upon 1) the digital maturity literature, and 2) specialized literature related to criteria. For instance, for the *Customer Interaction* criteria, we used the following digital maturity literature [4,12] as well as the following maturity model/framework literature [26,27]. At least two researchers and two SME managers of the project team performed an independent evaluation of each statement.

Design Cycle 2 — DMM Extension and Improvement. Participants suggested adding a criterion about cybersecurity and green IS. After deliberation with the project team, only cybersecurity was retained (IT4). The updated DMM with 27 criteria was accepted by stakeholders and SME managers. However, some managers found it difficult to position their organization within a maturity level. As a solution, they proposed to include help regarding the digitalization *jargon* as well as concrete examples. Hence, we create a set of almost

100 additional descriptions which accompany the 135 statements. For instance, to accompany the CX1-LV3 criteria (see Table 1), we provide the following help "*A newsletter is sent by email while infographics or videos are communicated through social networks.*" (see Fig. 1). To support the path to follow, we implemented an action plan — to be completed by managers — that contains for each criterion: an objective (pre-filled with the statement of the next maturity level) and an initial assessment of costs, time, difficulty, and priority. We finally implemented an online tool publicly available[1] that integrates our DMM with some SME's information inputs (e.g., state/region, economic sector, size) as well as the pre-filled roadmap.

In total, 15 managers from 14 regional SMEs representing 6 key regional economic sectors evaluated the tool. They evaluated the dimensions and criteria of the DMM as complete (M = 3.8, SD = 0.94) and relevant (M = 4.27, SD = 0.59) on a five-point Likert scale [23]. No significant weaknesses were detected, and two comments were negative and minor (i.e. add a criterion and a missing statement help). Participants evaluated different design alternatives to improve the online assessment experience. To answer each criterion, a slider has been chosen to convey a notion of progress when selecting a maturity level (see Fig. 2). The final version of the online DMM is concise to meet the time constraint of SME managers. Also, feedback from managers during the design cycles has been addressed.

4.2   Digital Maturity Model and Roadmap

Table 1 presents the resulting DMM of this study including nine dimensions, just as the Berghaus and Back DMM [6,24], and 27 corresponding criteria: (1) Customer Experience — CX1, CX2 and CX3 — Meets customer expectations for digital services and products that they take for granted, (2) Product Innovation — PI1, PI2 and PI3 — Addresses the importance of technological innovation in improving and creating offerings, (3) Strategy — ST1 and ST2 — Identifies the level of incorporation of digital within the organization's strategy, (4) Organization — OR1, OR2 and OR3 — Identifies the level of incorporation of digital at all levels of the organization, and no longer as a separate entity, (5) Process Digitization — PD1, PD2 and PD3 — measures the level of simplification, unification and automation of internal processes, (6) Collaboration — CO1, CO2 and CO3 — Meets the need to improve collaboration and communication between employees and teams, (7) Information Technology — IT1, IT2, IT3 and IT4 — Identifies the level of openness to new technologies and associated changes, (8) Culture & Expertise — CU1, CU2 and CU3 — Identifies the flexibility and agility of the information system for the success of digital projects, (9) Transformation Management — TM1, TM2 and TM3 — Identifies the flexibility and agility of the information system for the success of digital projects. Each criterion comprises five descriptive statements corresponding to a maturity level.

---

[1] https://digitalarchub.ch/.

Table 1. Digital Maturity Model.

Criterion	Level 1	Level 2	Level 3	Level 4	Level 5
Customer Interaction (CX1)	We interact with our customers through traditional channels only.	We interact with our customers through independent traditional and digital channels.	The content delivered to customers is appropriate for the digital channel.	Several interaction channels are complementary and partially integrated.	The customer experience is consistent and seamless across channels.
Analytics (CX2)	We do not collect customer data.	We collect customer data.	We analyze customer data.	We perform predictive analysis on customer data.	We perform prescriptive analysis based on customer data.
Data protection (CX3)	Personal data is not identified.	The collection and use of data is carried out in accordance with the recommendations.	We have developed a set of data protection procedures and processes.	We have a high level of expertise in data protection.	We follow best practices in data protection.
Business development (PI1)	We are not integrating digital into our business model.	Our activities are supported by digital technology.	Some aspects of our business model are complemented by digital.	Innovation based on digital technology helps to increase our revenues.	The development of our company fully and systematically exploits digital technologies.
Innovation capability (PI2)	There is no specific support for innovation.	We follow some practices adapted to innovation.	We follow a process adapted to innovation.	We have a set of tools and practices adapted to innovation.	Our company offers full support for innovation.
Customer integration (PI3)	We do not involve our customers in the development of digital innovations.	We involve our clients in an informal way.	We involve our clients in a formal and casual way.	We involve our clients in a formal and regular way.	Our customers are at the center of the digital innovation process.
Strategic innovation (ST1)	Our strategy does not involve digital innovation.	We have analyzed opportunities for digital innovation.	Digital innovation is integrated into our strategy.	Digital innovation plays a major strategic role.	We have developed a digital innovation program.
Digital commitment (ST2)	We have not yet committed resources to conduct digital projects.	We have planned digital projects.	Proofs of concept and prototypes based on digital technology have been completed.	A digital project of strategic importance has been completed.	Several digital projects have been successfully completed and this type of project is as important as the daily activities.
Digital team set-up (OR1)	We don't have a team dedicated to digital projects.	Our teams dedicated to digital projects are composed of employees of the same function.	Our digital project teams are cross-functional.	Our digital project teams are cross-functional and have a common vision.	Our teams dedicated to digital projects are efficient thanks to their optimal configuration.
Organizational agility (OR2)	We are not able to react to changes in our environment.	We have analyzed the changes in our environment.	We have taken some actions to adapt to the changes in our environment.	We can react quickly to changes in our environment.	The ability to reinvent ourselves is part of our DNA.
Partner network (OR3)	We have no active contacts in the digital field.	We exchange information with contacts active in the digital field.	We work with an active partner in the digital field.	We work closely with partners in the digital field.	We have a network of partners in the digital field with whom we work effectively.
Digital marketing communication PD1	We have no management of customer touch points.	We have integrated a few touch points.	We have integrated touch points consistently into the customer journey.	We evaluate our interactions with the customer and adapt them accordingly.	We continuously monitor, measure and improve our customer interactions.
Automation (PD2)	Process automation is not on the agenda.	We analyzed the potential for process improvement.	We have automated key processes.	We have automated a number of processes and regularly review their potential for improvement.	We have a proactive approach to process improvement through automation.
Data-driven business (PD3)	We do not collect or use any data.	We produce summary reports from the data collected.	We develop analytical models from the collected data.	We develop analytical models in a continuous process and on the basis of a common company infrastructure and processes.	Our business is data-driven, supported by an analytics strategy that is aligned with the company's strategy.

(continued)

Table 1. (*continued*)

Criterion	Level 1	Level 2	Level 3	Level 4	Level 5
Team work (CO1)	Our employees work in their own closed space.	Different collaborative tools are used by some teams.	Collaborative tools are used company-wide and in a standardized way.	Our employees share knowledge in an efficient, proactive and structured way through standardized collaborative tools.	The use of collaborative tools is continuously monitored, measured and improved.
Knowledge management (CO2)	We have limited digital skills.	We have an adequate level of digital skills for our business.	We have an adequate level of digital skills for our activities and are carrying out some actions to develop them.	We have a high level of digital skills and develop them regularly.	We have a high level of digital skills and have a proactive attitude to develop them.
Flexible working (CO3)	We do not promote flexible working.	Flexible working is possible for some employees.	Flexible working is widespread throughout the company and we have established a policy for this.	We promote various forms of flexible work.	Flexible working is part of the company's culture.
Agile project management (IT1)	We do not follow an agile project management approach.	Our digital projects follow some agile project management practices.	Our digital projects follow an agile project management method.	We are experienced in leading digital projects in an agile manner.	Our teams have an agile culture when it comes to digital projects.
Integrated architecture (IT2)	Our computer systems are isolated.	Work to open our systems is planned or underway.	Our systems are open internally and partly interconnected.	Our systems are interconnected and open internally and externally.	Our information system architecture is flexible, adaptable and scalable.
IT expertise (IT3)	We have little in-house IT expertise.	Our IT team is meeting the demands of our digital transformation with outside help.	Our IT team leads some of our digital transformation activities.	Our IT team leads most digital transformation activities.	Our IT team is a driver of digital transformation and is constantly experimenting with new ideas.
Cybersecurity (IT4)	Computer security measures are very limited.	Basic cybersecurity measures are implemented.	Comprehensive IT security measures have been implemented and are documented.	Cybersecurity governance guides cybersecurity activities.	We have optimal cyber security from a technical and organizational point of view.
Digital affinity (CU1)	Our employees are digital averse.	Some of our employees are open to digital technology.	Our employees generally have a positive attitude towards digital.	Our employees have a proactive and open attitude towards digital technology.	Our company has a digital culture.
Readiness to take risk (CU2)	We prefer not to take any chances.	Our digital projects are low risk.	Our digital projects involve moderate risk.	Managers are taking significant risks by implementing innovative digital solutions.	The culture of our organization encourages risk-taking and experimentation.
Error culture / No blame culture (CU3)	There is a general stigma attached to failure.	We tolerate failure when the risks are significant.	We accept failure as a natural part of experimentation.	We value and learn from failure.	We have a culture of right to make mistakes.
Governance (TM1)	We have no governance for digital transformation.	We have planned to develop a governance plan for our digital transformation.	Our digital transformation is guided by a clear roadmap, defined roles and processes.	Our digital transformation governance incorporates best practices.	Our governance for digital transformation is constantly evolving.
Performance measurement (TM2)	We don't measure our progress in digital transformation.	We plan to identify digital transformation goals and actions.	The objectives of our digital transformation are clearly defined and communicated within the company.	Achievement of digital transformation goals is reviewed periodically.	The effects of digital transformation are assessed along several dimensions and systematically.
Management support (TM3)	There is no specific support for digital transformation.	Management communicates the importance of digital transformation.	Management supports and encourages digital transformation.	Management is strongly involved in the digital transformation.	Digital leaders coordinate, facilitate and promote digital transformation activities.

After a self-assessment with the online tool, a manager obtains a summary of the SME maturity level, a short description, and a brief comparison with other companies over each criterion. Moreover, to initiate a change, an action plan is generated from the basis of the assessment. This personalized roadmap can be used as a starting point for a systematic digital transformation. Figure 2 presents the assessment and prescription process with a focus on one criterion.

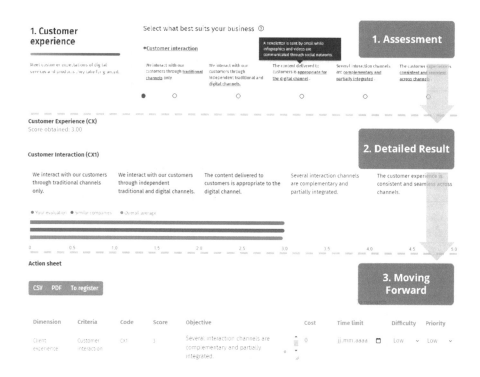

Fig. 2. Digital maturity assessment and roadmap tool focusing on the criterion CX1.

5   Evaluation

Six managers of different industrial SMEs participated in a digital transformation workshop and used our online tool as a starting point (further details are provided in Sect. 4.1). We precise that the results of this evaluation have been obtained on the final version of the tool.

All participants were generally satisfied with the workshop and the assessment of their organization with our tool. The assessment was performed in an average of 21.5 min, indicating respect for the temporal requirement. Some participants highlighted a difficulty to understand some maturity level indicators or mentioned indicators that were not applicable to their company. Moreover, some participants mentioned a need for a more specialized recommendation about how to be more digitally mature. Two companies have identified a need to assess the maturity of different departments in their organization which led us to propose a new service in our tool. Currently, this service is under development with two pilot enterprises.

6 Discussion

In this study, we presented the Design Science Research process [21] of an online tool for assessing the digital maturity of SMEs and defining action items for their transformation roadmap. Our findings highlight that SME managers need prescriptions about digital maturity rather than a sole description of their organization's maturity stage. Following this requirement as the main design constraint, we have moved towards an audit and roadmap approach. The designed DMM allows assessing the digital maturity of an SME through 9 dimensions comprising 27 maturity criteria. Each criterion includes five levels of maturity. Evaluations with SME managers during and after the design process has shown the utility and validity of our artifact. However, there is still a need for more specific recommendations regarding digital maturity [28]. Future works should investigate the action recommendation and its links with the digital maturity assessment. For instance, we could imagine providing personalized objectives based on the SME's characteristics for each criterion.

Our study is subject to several limitations. First, and considering the benefits of building our DMM on the dimensions and criteria proposed by Berghaus and Back [6,24] as well as the applicability to our context, we did not reconsider the dimensions of digital maturity. Secondly, the evaluation has been performed by a small number of SME managers. Thirdly, it should be mentioned that all artifacts were developed and evaluated in French, in the Swiss Jura Arc region, which could limit the results to this population.

We believe that our contribution bridge the sometimes dichotomous view between measuring digital maturity and providing an evolutionary path [6,7]. Moreover, the designed tool presents interesting implications for practice. It allows SME managers to self-assess their organization's digital maturity and initiate a transformation without requiring important or dedicated resources. We also believe that this tool may potentially lead to comparing SMEs' digital maturity in the long run, which would favor discussion and exchange for best practices. Hence one of the future works focuses on the Digital Maturity Observatory. This observatory will help browse the aggregated self-assessment data through filtered-based visualizations (i.e. digital maturity evolution over time, by sector, or region). By making this information available and updated, the observatory could serve as a reference in the management of the digital maturity of a region, stimulate SMEs and strengthen their competitiveness.

References

1. European Commission, Executive Agency for Small and Medium-sized Enterprises, Muller, P., et al.: Annual report on European SMEs 2020/2021 : digitalisation of SMEs. Publications Office (2021)
2. Federal Statistical Office, "Portrait des PME suisses, 2011-2020," tech. rep. (2022)
3. Reis, J., Amorim, M., Melão, N., Matos, P.: Digital transformation: a literature review and guidelines for future research. In: Rocha, Á., Adeli, H., Reis, L.P., Costanzo, S. (eds.) WorldCIST'18 2018. AISC, vol. 745, pp. 411–421. Springer, Cham (2018). https://doi.org/10.1007/978-3-319-77703-0_41
4. Peter, M.K., Kraft, C., Lindeque, J.: Strategic action fields of digital transformation: an exploration of the strategic action fields of Swiss SMEs and large enterprises. J. Strategy Manage. **13**(1), 160–180 (2019)
5. OECD, "The digital transformation of smes," In: OECD Studies on SMEs and Entrepreneurship, Paris: OECD Publishing (2021)
6. Berghaus, S., Back, A.: Stages in digital business transformation: results of an empirical maturity study. In: Proceedings of the 10th Mediterranean Conference on Information Systems (MCIS) 2016 (2016)
7. Thordsen, T., Murawski, M., Bick, M.: How to measure digitalization? a critical evaluation of digital maturity models. In: Hattingh, M., Matthee, M., Smuts, H., Pappas, I., Dwivedi, Y.K., Mäntymäki, M. (eds.) I3E 2020. LNCS, vol. 12066, pp. 358–369. Springer, Cham (2020). https://doi.org/10.1007/978-3-030-44999-5_30
8. Williams, C., Schallmo, D., Lang, K., Boardman, L.: Digital maturity models for small and medium-sized enterprises: a systematic literature review. In: International Society for Professional Innovation Management (ISPIM) Conference Proceedings, pp. 1–15 (2019)
9. Gregor, S., Hevner, A.R.: Positioning and presenting design science research for maximum impact. MIS Q. **37**(2), 337–355 (2013)
10. Legner, C., et al.: Digitalization: opportunity and challenge for the business and information systems engineering community. Bus. Inf. Syst. Eng. **59**(4), 301–308 (2017)
11. Verhoef, P.C., et al.: Digital transformation: a multidisciplinary reflection and research agenda. J. Bus. Res. **122**, 889–901 (2021)
12. North, K., Aramburu, N., Lorenzo, O.J.: Promoting digitally enabled growth in SMEs: a framework proposal. J. Enterprise Inform. Manage. **33**(1), 238–262 (2019)
13. Vial, G.: Understanding digital transformation: a review and a research agenda. J. Strat. Inf. Syst. **28**(2), 118–144 (2019)
14. Piccinini, E., Gregory, R.W., Kolbe, L.M.: Changes in the producer-consumer relationship-towards digital transformation. In: Wirtschaftsinformatik Proceedings 2015 (2015)
15. Matt, C., Hess, T., Benlian, A.: Digital transformation strategies. Business Inform. Syst. Eng. **57**(5), 339–343 (2015)
16. Pöppelbuß, J., Röglinger, M.: What makes a useful maturity model? a framework of general design principles for maturity models and its demonstration in business process management. In: Proceedings of the 19th European Conference on Information Systems (ECIS) 2011 (2011)
17. Becker, J., Knackstedt, R., Pöppelbuß, J.: Developing maturity models for it management. Business Inform. Syst. Eng. **1**(3), 213–222 (2009)
18. Maier, A.M., Moultrie, J., Clarkson, P.J.: Assessing organizational capabilities: reviewing and guiding the development of maturity grids. IEEE Trans. Eng. Manage. **59**(1), 138–159 (2012)

19. Gill, M., Vanboskirk, S., Freeman, E.P., Nail, J., Causey, A., Glazer, L.: The digital maturity model 4. 0, forrester (2016)
20. Canetta, L., Barni, A., Montini, E.: Development of a digitalization maturity model for the manufacturing sector. In: 2018 IEEE International Conference on Engineering, Technology and Innovation (ICE/ITMC)
21. Hevner, A.R., March, S.T., Park, J., Ram, S.: Design science in information systems research. MIS Q. **28**(1), 75–105 (2004)
22. Hevner, A.R.: A three cycle view of design science research a three cycle view of design science research. Scand. J. Inf. Syst. **19**(2), 87–92 (2007)
23. Lee, Y.W., Strong, D.M., Kahn, B.K., Wang, R.Y.: AIMQ: a methodology for information quality assessment. Inform. Manage. **40**(2), 133–146 (2002)
24. Berghaus, S., Back, A.: Gestaltungsbereiche der digitalen transformation von unternehmen: entwicklung eines reifegradmodells. Die Unternehmung **70**(2), 98–123 (2016)
25. Hammer, M.: The process audit. Harv. Bus. Rev. **85**(4), 1–14 (2007)
26. Saghiri, S., Wilding, R., Mena, C., Bourlakis, M.: Toward a three-dimensional framework for omni-channel. J. Business Res. **77**, 53–67 (2017)
27. Yrjölä, M., Saarijärvi, H., Nummela, H.: The value propositions of multi-, cross-, and omni-channel retailing. Int. J. Retail Distrib. Manage. **46**(11–12), 1133–1152 (2018)
28. Remane, G., Hanelt, A., Wiesboeck, F., Kolbe, L.M.: Digital maturity in traditional industries - an exploratory analysis. In: Proceedings of the 25th European Conference on Information Systems (ECIS) (2017)

Adopting Industry 4.0 and Lean Practices in Heavy Metalworking: Impact of Human Factors on Productivity

Cristiano Jesus[1,2,3](✉) ⓘ, Eduardo Pontes[1] ⓘ, Rui M. Lima[3] ⓘ, and Sérgio Ivan Lopes[1,2,4] ⓘ

[1] Advanced Production Systems Department, CiTin – Industrial Technology Interface Center, 4970-786 Arcos de Valdevez, Portugal
cristiano.jesus@citin.pt
[2] ADiT-Lab, Instituto Politécnico de Viana Do Castelo, Rua Escola Industrial E Comercial Nun'Álvares, 4900-347 Viana Do Castelo, Portugal
[3] ALGORITMI Research Center, School of Engineering, Production and Systems Department, University of Minho, 4800-058 Guimarães, Portugal
[4] IT – Instituto de Telecomunicacações, Campus Universitário de Santiago, 3810-193 Aveiro, Portugal

Abstract. Lean Manufacturing culture is enshrined as an industrial management policy. Moreover, production planning and manufacturing operations management tools, such as Work Sampling (WS), has long been successfully used. WS is known for its ability to support decision-making, especially regarding production capacity and delivery times. This paper presents the results of a pilot study developed in a heavy metalworking Small and Medium-Sized Enterprise (SME), located in the North of Portugal, that manufactures metal structures, and where WS has been implemented to assess human factors that impact productivity, such as physical disposition, fatigue, and stress, which have been estimated, based on biological markers acquired from workers while performing its working activities in the shop floor. This work proposes introducing the SME to Lean practices within the Industry 4.0 (I4.0) paradigm, as it is intended to move the WS tool from the traditional application for forecasting to a predictive and real-time tool that can be of great value for the Operation Manager, augmenting its capabilities to effectively manage relevant human factors that impact productivity, based on a dashboard fed with real-time data acquired from workers working at the shop floor. The paper shows that these factors have an impact on production performance, according to the initial hypothesis. The purpose of the presented research, based on the analyzed results, is (1) to develop an application model of Lean tools that privileges the predictive orientation in place of the preventive one, in a perspective that is assumed to be aligned with the pattern of Industry 4.0; and (2), to propose a way to introduce SMEs to the development circuit for Industry 4.0 with the first steps based on the consecrated good practices of industrial management remodeled to the pattern of Industry 4.0.

Keywords: Industry 4.0 · Lean Manufacturing · Ergonomics · Human Factors

© The Author(s), under exclusive license to Springer Nature Switzerland AG 2024
A. Rocha et al. (Eds.): WorldCIST 2023, LNNS 801, pp. 241–252, 2024.
https://doi.org/10.1007/978-3-031-45648-0_24

1 Introduction

It is not a new fact that industrial management tools are oriented towards maximizing the efficiency of production processes to provide resources both for the continuous improvement of the organization of operations and to eliminate, or reduce as much as possible, waste and work efforts that have no increased value for the business. The Lean tool framework has been thought out and developed for these purposes and is applied from the process design stage in, production planning, manufacturing monitoring, warehousing, and logistics. With the advent of Industry 4.0, both the technological resources to support these goals have increased, as well the opportunities for improvement; not only that, the main change provided by the digitalization of processes was the provision of the ability to reproduce in digital format, and in real-time, the current and concrete state of operations, something that substantially changes the way industrial management is done and how decisions are made. From a dynamic that had the main focus on planning and execution, as close as possible to the planned ideal, the orchestration of the factual is now also added. While in the first scenario, failures and defects are past occurrences and lessons to avoid in the future, in the second scenario, real-time monitoring and the larger quantity of measured variables can indicate circumstantial instabilities and poor performance factors that were previously invisible, so that engineers can act before problems happen, or can adjust the system so that performance improves.

Conversely, SMEs do not have the perception that the best practices and key factors in management concern them, nor that in their businesses there is some possibility of applying practices like Lean Manufacturing or Industry 4.0 because usually SMEs have limited resources to count on the help of consulting companies and, in addition, in many cases, the leadership lacks the long-term commitment that it is necessary [1]. Given this scenario, a possible approach is the development of competencies through executing projects strategically defined to promote experiences in specific fields [2].

This paper presents the pilot study conducted at a workstation in a heavy metalworking SME, a manufacturer of metal structures. The objective is to evaluate the impact of the workers physical disposition on productive performance. To this end, biological markers of fatigue and stress were analyzed throughout the workday of an operator for two months. The WS tool was used for this purpose, which allows the analysis of productive operations through direct observation and based on random sampling. This technique is useful for cases in which the production is not serial, as is this case. The aim is to identify the relevance of scientific research on the impact of human conditions on production forecasts, and on the application of Lean tools also for real-time management of the production system in action. With this study we intend to answer the following research questions: Q1: Can the evaluation of biological markers in tools such as WS make the assessment of production performance more accurate, and can the WS tool be adapted for the evaluation of data collected in real-time for production performance management? Q2: Can the adaptation of traditional best practices in industrial management to the Industry 4.0 model, essentially in the transformation of preventive approach tools to real-time predictive assessments, be a strategy to introduce SMEs for the adoption of Industry 4.0 practices?

The article was structured in such a way as to initially present the theoretical contextualization, in which we refer to recent publications that offer subsidies that support our proposal and that were selected from qualified journals in searches with keywords that we now also use in this article. In the next section, we explain the methodological approach and the research procedures adopted. The preliminary data from the pilot study are presented next, divided into general results from the assessment of the readiness of the company that guided the actions and results from the application of the lean work sampling tool. We end the paper explaining the limitations of the study and performing our critical analysis of the work performed which is in the conclusion section, in which we also present the answers to the research questions.

2 Theoretical Background

The set of philosophy, principles, and tools, identified as Lean, is usually evoked for industrial management programs that aim to analyze, diagnose, and eliminate processes that do not add value to the product, or that from the customer's point of view is waste. Materials or information that remain stored, in queues, or inspection, in addition to defects and errors, and workflow constraints, generate costs that, in theory, customers are unwilling to pay for [3]. Therefore, in the Lean perspective, the business strategy should focus on improving quality and service, while working on eliminating inefficiencies, and reducing time and costs [3].

A recent demand is the adaptation of businesses to the Industry 4.0 paradigm, which corresponds not only to the adoption of cyber-physical systems technologies, such as sensors, collaborative robots, big data analytics, and artificial intelligence but also includes the adoption of new and more effective management practices to deliver solutions involving both products and services. In this model, a manufactured product, for example, can be a service if its design is offered to the customer along with the service of adapting it to his requirements and 3D printing [4].

For Shahin et al. [5], Lean Thinking culture and its management model are important elements to prepare companies for the implementation of Industry 4.0 technologies and practices. Therefore, it corresponds to an essential stage of analysis, improvement, and rational organization of processes for the development of an adequate level of readiness, and thus being able to evolve to a more modern and dynamic configuration, as is the case of the adoption of Industry 4.0 practices.

According to Pereira & Sachidananda [6], and contrary to what common sense usually takes as an assumption, Industry 4.0 does not replace Lean Manufacturing as a model. This view is induced by the weight of the "industrial revolution" concept, which although true, is important to keep in mind that for the transformation provided by a revolution to be constructive and not destructive, organizations need to be prepared, or at least, initiated in scientifically recognized best practices.

Although Lean practices have existed since the 1960s, the term was first used to designate a systematized management model in 1988 [6, 7]. However, SMEs still suffer difficulties in their implementation, due to factors such as a lack of capital, knowledge, or qualified human resources. Access for these companies to management concepts, philosophy, and Lean tools, therefore, is still very restricted [3].

Concerning to Industry 4.0, SMEs are even further behind [8]. According to Jesus & Lima [9], SMEs have a large economic impact in European countries, and in Portugal,

99.9% of companies are in this category. However, while most large companies have process reconfiguration and digital transformation programs for Industry 4.0 in progress or the experimentation phase, SME managers do not see that such actions can be applied to their businesses. According to the authors, the importance of these companies is so significant that any movement they can pull off would already impact the economy.

Therefore, it is reasonable to consider that a roadmap for modernizing the business models and management practices of SMEs should include a phase of increasing the level of readiness of organizations by professionalizing them and developing results and performance-oriented culture [10] and then move on to digitalization [11, 12].

Dillinger et al. [8] evaluate and identify the Lean philosophy to leverage Industry 4.0 and propose an evolution strategy based on realization phases divided into three major stages: i) scoping, ii) planning and action design, and iii) implementation and evaluation. Whereas Veres [13] argues that the development of Lean Thinking in organizations should occur on the bases of education and training of the organizational mind in a program consisting of the phases of planning, training, development, and coaching. Both approaches can be seen naturally as complementary and effective in many cases. However, as already seen, this circuit is very difficult in SMEs for the reasons already listed and for the high incidence of improvisation and reactive posture very common in these environments.

Mofolasayo [14] suggest the introduction of tools such as Value Stream Mapping (VSM) [15, 16], Just in Time (JIT) [17], Heijunka, Total Productive Maintenance (TPM) [18], Visual Management (VM), Kanban, Single-Minute Exchange (SMED), 5S e Poka-Yoke, in a programmed and progressive way, implemented in a cycle of continuous improvement that involves the application adapted to the possible conditions, evaluation of the results, reflection on the principles of quality and waste, and then moving to a higher level of integration. Then, repeating the cycle so that a Lean culture is established naturally through the experience of new standards that by themselves drive the evolution.

Yilmaz et al. [19] consider the lack of adherence of SMEs to Lean management practices and Industry 4.0 due to, not only operational but also environmental and even social conditions. Similarly to Yilmaz et al. [19], Mofolasayo et al. [14] and also Marinelli et al. [20], Jing et al. [21], and Dossou et al. [22] suggest the use of Lean tools objectively to solve problems such as machine failures, low productivity, long lead times, long stops and waits, and others, while adopting typical Industry 4.0 tools such as RFID, IoT, Robotic, Big Data, simulation tools, for the automation of Lean processes. This approach has operations management in one vector, Lean tools in another vector, and the technology of Industry 4.0 in a third vector.

3 Methodology

The research proposal here presented shows the preliminary results of a pilot study conducted between August and October 2022 in the heavy metalworking industry, manufacturer of metal structures, located in Northern Portugal. The observation procedure was followed by critical analysis and supported by a literature review that substantiates the theoretical background. The results helped for clarifying the hypothesis and the research questions. The literature review was guided by findings identified from previous studies performed by the authors in the scope of Industry 4.0, the experience gained during the immersion process in the application environment, and previous experience. The device used to collect the operators biomedical data was *Polar H10 Heart Rate Sensor* and the software used for data analysis was *Polar H10 ECG Analysis.* It was observed one workstation and the activities of one worker. The worker who participated in the study signed a consent form where his data was safeguarded and only the monitored data would be used for research.

4   Preliminary Data

4.1   Assessment of Readiness

The aim of this pilot project is ultimately to introduce the SME under study into a development circuit for Industry 4.0 that aims at full digitalization of processes, integration of technologies and systems, and strategic management supported by predictive analysis with statistical data analysis, as well as the development of the innovation mindset. Given the current conditions of the SME under study, it is very important to highlight the goal of initiating the first steps in a development process that will be progressive, and the advancement will be determined at the same time by investments in technological modernization, reconfiguration of processes, and especially by the development of competencies at a pace determined by the return on investments. To get an understanding of what these first steps might be, we applied a readiness level assessment for Industry 4.0 [23].

We used as reference Sony & Naik [24] and Mittal et al. [25] to develop an understanding of "readiness" as a combined set of conditions and competencies for professional business and production management so that we consider strategic management spread over all decision levels, project management, process management, and continuous improvement as key factors. These conditions are structured to be able to leverage the progressive development of maturity in Industry 4.0. According to the Acatech model [26] "maturity" is the stage of evolution concerning the goals of Industry 4.0, considering the presence of the basic conditions (readiness) for this evolution.

4.2   Work Sampling

The fundamental objective is to evaluate the consistency of the hypothesis that real-time monitoring of data from actions performed at the workstation and from biological markers data indicating fatigue, stress, and others allows for decision-making in full compliance with the course of the process, as opposed to the traditional situation that only allows for forward-looking corrective and preventive actions.

For this study, a set of parameters related to Heart Rate Variability (HRV) have been considered [27]. Such parameters are directly obtained from an electrocardiogram (ECG), a signal obtained to assess the human hearts rhythm and electrical activity. Figure 1 illustrates an annotated example of an ECG signal obtained from a worker while working on the shop floor.

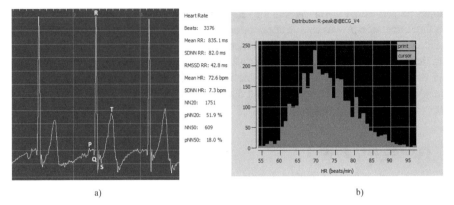

a)   b)

Fig. 1. a) ECG signal obtained from a worker on the shop floor (x-axis in seconds and y-axis in uV); b) Histogram and statistical analysis of the acquired data.

From the ECG signal we have computed the following parameters:

- meanRR, which measures the mean time between RR peaks in an interval series, cf. Fig. 1.
- meanHR, which measures the mean amount of heart beats per minute, h, cf. Fig. 1.
- SDNN(RR) measures the standard deviation of a RR interval series and is known as the "gold standard" for medical stratification of cardiac risk. SDNN(RR) values below 50 ms are classified as unhealthy, 50–100 ms have compromised health, and above 100 ms are healthy. It is supposed that training individuals to increase SDNN(RR) to a higher category could reduce their risk of mortality.
- RMSSD measures the root mean square of successive differences between normal heartbeats. It is obtained by first calculating each successive time difference between heartbeats in ms. Then, each of the values is squared, and the result is averaged before the square root of the total is obtained. The RMSSD reflects the beat-to-beat variance in HR and it is believed that lower RMSSD values are correlated with higher scores on a risk inventory of sudden unexplained death [27].
- pNN50 is the percentage of adjacent NN intervals that differ from each other by more than 50 ms and is closely correlated with peripheral nervous system (PNS) activity, which plays a key role in sending information from different areas of your body back to your brain, as well as carrying out commands from your brain to various parts of your body [27]. Even though recent studies provide some evidence that shorter intervals, under 5 min, can be used to infer these parameters with some reliability, to achieve more concrete data, those under 5 min were not considered.

The observation process was performed by monitoring the activities of a worker throughout a workday and by logging the data in a WS sheet, as depicted in Fig. 2.

The mathematical basis of WS is [28]:

$$PT_A = \frac{T_A}{OT} \times 100$$

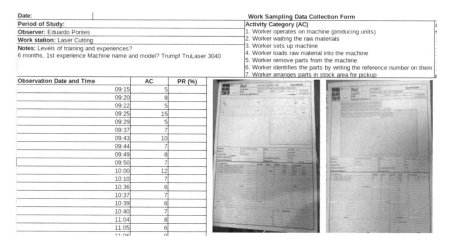

Fig. 2 - Work Sampling (part of).

where PT_A is the time percentage spent on a particular activity, TA is the time spent on the activity, A is the number of sample points, and OT is the time spent on all activities – the overall number of sample points. According to Robinson [29] the number of sample points is calculated as the followed way:

$$N = \frac{(1-p)Z^{2\lambda}/2}{pR^2}$$

where p is the time percentage spent on a particular activity, R is the level of precision, and $Z\lambda/2$ is the number of standard deviations for achieving the desired confidence level.

In the study, for the confidence level of 95% and margin of error 5%, the sample points are 385 measurements. Because of the characteristic of the activities, performed in a non-serial production line, we considered them as a measurement of the time (in minutes) spent on the activities. The observations took place for 8 days (2 days per week) for about 7 h per day, totaling 56 h, that is 3360 min. To consider that there are idle times or times unrelated to the productive operations, about 30%, of the resulting sample was sufficient. It was observed the activities of one operator of a sheet metal laser cutting machine.

The collection system had the problem of adherence to the workers skin, which sometimes generated noise in the signal, which was determined as an outlier for the study. Of the total data collected, around 5% were dismissed as such. The worker accused of discomfort when putting on the chest strap due to the difference in temperature of the device to his skin and, when in greater activity, of the heat caused by it that made him sweat more.

In the analysis, we considered the following approaches:

- When two or more of the parameters were below 60 ms in SDNN(RR), below 50 ms in RMSSD, and above 20% in pNN50 as a reference for an unhealthy momentary state.

- When one of the parameters was: below 30ms in SDNN(RR), below 25ms in RMSSD, and above 45% in pNN50 as a reference for an unhealthy momentary state.

Activities 1, 2, 6, 8, 9 and 15 can be considered mentally demanding and the rest more physically demanding. It was expected that physical activities could have a higher toll on the parameters monitored, and that was also considered. In this regard, after analyzing the collected data, these activities stood because their parameters were outside what is considered adequate from a healthy point of view, as shown in Table 1.

Table 1. - Highlighted activities for health concern.

Task	Mean RR (ms)	Mean HR (bpm)	SDNN (ms)	RMSSD (ms)	pNN50 (%)
T4–2022-09-14__16–53-00_16–59-00	567.6	122.7	177.8	181.8	46.4
T8–2022-09-14__17–55-00_18–00-00	509.4	138.6	182.6	223.6	65.2
T1–2022-09-21__13–00-00_13–06-00	742.7	81.7	82.4	46.4	20.3
T4–2022-09-21__11–30-00_11–35-00	678.3	89.1	52.4	33.5	4.3
T1–2022-09-21__17–35-00_17–40-00	530.6	137.8	210.2	196.8	55.8
T1–2022-09-26__16–27-00_16–35-00	762.1	80	96.5	47.9	20.4
T5–2022-09-26__10–10-00_10–16-00	712.8	84.7	53.4	29.3	4.8
T11–2022-09-26__11–01-00_11–22-00	806.2	75.2	75.8	46.2	22.4
T1–2022-10-10__13–30-00_14–04-00	805.9	75.4	91.1	47.9	21.5
T1–2022-10-10__15–42-00_16–00-00	853.9	71.1	90.5	47.3	20.7

From the mentally demanding, activities we will provide further information on:

- **T8–2022-09-14__17–55-00_18–00-00** in which the worker, performing a task that involves speaking with colleagues, was involved in a heated argument with the supervisor.
- **T11–2022-09-26__11–01-00_11–22-00** in which the worker was cleaning the laser cutting machine lenses and the worker himself described this activity as stressful due to the responsibility of handling such sensible and expensive material.

In these activities, it is clear how the mood and environment surrounding the worker took an effect on his well-being and work capabilities, which correlates with the monitored parameters.

Overall, once it is identified that one parameter or a pair of them exceeds the value considered for a healthy individual, action should be taken to take care of their well-being. From this study, we could conclude that taking this type of action would not affect the performance and productivity of the worker, but on the contrary, could increase them.

5   Statement of Limitations

The responses revealed a very low level of readiness and maturity, further evidenced by the observational process in which have been identified the almost exclusive practice of ad-hoc decision-making at all organizational levels (strategic, tactical, and operational) and the prevalence of manual actions in most operations. The computer-supported operations did not happen in an integrated way so there were several applications for the same process that required the manual intervention of employees for convergence. The WS option was made due to the productive systems characteristic of an exclusively non-serial, or "make-to-order" type. Moral, ethical, and social questions that the article completely omits. Moral, ethical, and social issues that may imply the implementation of the proposed solution are not the scope of this study. Recognizing the extreme importance of regulatory requirements for the implementation and possible social implications of digital transformation, this issue should be addressed in later stages, when the possible impacts emerge in future studies.

6   Conclusions

Supported by the literature, we used the strategy of implementing good management practices adapted to the Industry 4.0 model, so we applied the WS tool for the evaluation of production performance to add to the traditional method of time tracking of production operations also biomedical data of the worker. With the results obtained from this pilot project, we can answer the research questions as follows:

Q1: Can the evaluation of biological markers in tools such as WS make the assessment of production performance more accurate, and can the WS tool be adapted for the evaluation of data collected in real-time for production performance management?

This work consisted in structuring a method for applying WS to combine production operations with biomedical data. The device used to collect the operators biological markers data was *Polar H10 Heart Rate Sensor,* therefore, with the addition of new sensors and the development of digital interfaces and dashboards, it is perfectly feasible with the observed results to consider that WS can be a predictive analysis tool. At times when SDNN(RR), RMSSD, or pNN50 exceeded values considered adequate, decisions could have been taken on the spot if there were identified in real-time.

Q2: Can the adaptation of traditional best practices in industrial management to the Industry 4.0 model, essentially in the transformation of preventive approach tools to real-time predictive assessments, be a strategy to introduce SMEs to the adoption of Industry 4.0 practices?

The observed results and facts revealed that the digital transformation in SMEs takes place effectively and with fewer elements of uncertainty when framework conditions are achieved when improvised or immediate needs-driven practices are introduced into scientifically validated management models so that the basic competencies for evolution are also developed. By adapting traditional preventive guidance tools to predictive guidance, even if still in a very elementary way, we managed to develop the first steps of experimentation with the Industry 4.0 model in an SME.

In further work the WS tool will be integrated with sensors and a dashboard, and alternatives will be studied and discussed. Thus, a preliminary application of a dashboard will be shown as an example with features such as alerts, notifications, and others. Recognizing the extreme importance of regulatory requirements for the implementation and possible social implications of digital transformation, this issue should be addressed in later stages, when the possible impacts emerge in future studies.

Acknowledgment. This work has been done under the scope of "ErgoSafe5.0–Ergonomics and Safety for Industry 5.0", an R&D project funded and promoted by CiTin - Centro de Interface Tecnológico Industrial (http://www.citin.pt/en), in collaboration with the ADiT-Lab, Applied Digital Transformation Laboratory (http://adit.ipvc.pt/). C.J. and E.P. have been supported by operation NORTE-06–3559-FSE-000226, funded by Norte Portugal Regional Operational Program (NORTE 2020), under the PORTUGAL 2020 Partnership Agreement, through the European Social Fund (ESF). This work was partially supported by FCT – Fundação para a Ciência e Tecnologia within the R&D Units Project Scope UIDB/00319/2020.

References

1. AlManei, M., Salonitis, K., Yuchun, Xu.: Lean implementation frameworks: the challenges for SMEs. Procedia CIRP **63**, 750–755 (2017). https://doi.org/10.1016/j.procir.2017.03.170
2. Lückmann, P., Feldmann, C.: Success factors for business process improvement projects in small and medium sized enterprises – empirical evidence. Procedia Comput. Sci. **121**, 439–445 (2017). https://doi.org/10.1016/j.procs.2017.11.059
3. Zhou, B.: Lean principles, practices, and impacts: a study on small and medium-sized enterprises (SMEs). Annals Oper. Res. **241**(1–2), 457–474 (2016). https://doi.org/10.1007/s10479-012-1177-3
4. Tay, S.I., Lee, T.C., Hamid, N.Z.A., Ahmad, A.N.A.: An overview of industry 4.0: definition, components, and government initiatives. J. Adv. Res. Dyn. Contr. Syst. vol. **10**(14), 1379−1387 (2018)
5. Mohammad Shahin, F., Chen, F., Bouzary, H., Krishnaiyer, K.: Integration of lean practices and industry 4.0 technologies: smart manufacturing for next-generation enterprises. Int. J. Adv. Manufact. Technol. **107**(5–6), 2927–2936 (2020). https://doi.org/10.1007/s00170-020-05124-0
6. Clinton Pereira, H.K., Sachidananda,: Impact of industry 4.0 technologies on lean manufacturing and organizational performance in an organization. Int. J. Interact. Design Manufact. (IJIDeM) **16**(1), 25–36 (2021). https://doi.org/10.1007/s12008-021-00797-7
7. Palange, A., Dhatrak, P.: Lean manufacturing a vital tool to enhance productivity in manufacturing. Materials Today: Proc. **46**, 729–736 (2021). https://doi.org/10.1016/j.matpr.2020.12.193
8. Dillinger, F., Bernhard, O., Reinhart, G.: Competence requirements in manufacturing companies in the context of lean 4.0. Procedia CIRP **106**, 58–63 (2022). https://doi.org/10.1016/j.procir.2022.02.155
9. de Jesus, C., Lima, R.M.: Study of the Portuguese Challenges in the Context of European Union to Identify Adaptation Strategies for the Industry 4.0. In: Machado, J., Soares, F., Trojanowska, J., Ivanov, V. (eds.) Innovations in Industrial Engineering, pp. 25–35. Springer International Publishing, Cham (2022). https://doi.org/10.1007/978-3-030-78170-5_3
10. Bhadu, J., Kumar, P., Bhamu, J., Singh, D.: Lean production performance indicators for medium and small manufacturing enterprises: modelling through analytical hierarchy process. Int. J. Syst. Assurance Eng. Manage. **13**(2), 978–997 (2021). https://doi.org/10.1007/s13198-021-01375-6

11. Langlotz, P., Siedler, C., Aurich, J.C.: Unification of lean production and Industry 4.0. Procedia CIRP **99**, 15–20 (2021). https://doi.org/10.1016/j.procir.2021.03.003
12. Tiwari, M.: Fundamentals of lean journey. In: Lean Tools in Apparel Manufacturing, pp. 47–79. Elsevier (2021). https://doi.org/10.1016/B978-0-12-819426-3.00007-2
13. Veres, C.: Conceptual model for introducing lean management instruments. Proc. Manufact. **46**, 233–237 (2020). https://doi.org/10.1016/j.promfg.2020.03.034
14. Mofolasayo, A., Young, S., Martinez, P., Ahmad, R.: How to adapt lean practices in SMEs to support Industry 4.0 in manufacturing. Proc. Comput. Sci. **200**, 934–943 (2022). https://doi.org/10.1016/j.procs.2022.01.291.
15. Arey, D., Le, C.H., Gao, J.: Lean industry 4.0: a digital value stream approach to process improvement. Proc. Manufact. **54**, 19–24 (2021). https://doi.org/10.1016/j.promfg.2021.07.004
16. Banga, H.K., Kumar, R., Kumar, P., Purohit, A., Kumar, H., Singh, K.: Productivity improvement in manufacturing industry by lean tool. Materials Today: Proceedings **28**, 1788–1794 (2020). https://doi.org/10.1016/j.matpr.2020.05.195
17. Hemalatha, C., Sankaranarayanasamy, K., Durairaaj, N.: Lean and agile manufacturing for work-in-process (WIP) control. Materials Today: Proceedings **46**, 10334–10338 (2021). https://doi.org/10.1016/j.matpr.2020.12.473
18. Jadhav, P., Ekbote, N.: Implementation of lean techniques in the packaging machine to optimize the cycle time of the machine. Materials Today: Proceedings **46**, 10275–10281 (2021). https://doi.org/10.1016/j.matpr.2020.12.162
19. Yilmaz, A., Dora, M., Hezarkhani, B., Kumar, M.: Lean and industry 4.0: mapping determinants and barriers from a social, environmental, and operational perspective. Technol. Forecast. Social Change **175**, 121320 (2022). https://doi.org/10.1016/j.techfore.2021.121320
20. Marinelli, M., Deshmukh, A.A., Janardhanan, M., Nielsen, I.: Lean manufacturing and Industry 4.0 combinative application: practices and perceived benefits. IFAC-PapersOnLine **54**(1), 288–293 (2021). https://doi.org/10.1016/j.ifacol.2021.08.034
21. Jing, S., Feng, Y. and Yan, J.: Path selection of lean digitalization for traditional manufacturing industry under heterogeneous competitive position. Comput. Indust. Eng. **161**, 107631, Nov (2021). https://doi.org/10.1016/j.cie.2021.107631
22. Dossou, P.-E., Torregrossa, P., Martinez, T.: Industry 4.0 concepts and lean manufacturing implementation for optimizing a company logistics flows. Procedia Comput. Sci. **200**, 358–367 (2022). https://doi.org/10.1016/j.procs.2022.01.234
23. Jesus, C., Lima, R.M., Barretiri, L. and Lopes, S.I.: A Maturity model proposal with readiness level assessment for organizational processes improvement. In: 2022 17th Iberian Conference on Information Systems and Technologies (CISTI), jun. 2022, pp. 1–4 (2022). https://doi.org/10.23919/CISTI54924.2022.9820072
24. Sony, M., Naik, S.: Key ingredients for evaluating Industry 4.0 readiness for organizations: a literature review. Benchmarking: Int. J. **27**(7), 2213–2232 (2019). https://doi.org/10.1108/BIJ-09-2018-0284
25. Mittal, S., Khan, M.A., Romero, D., Wuest, T.: A critical review of smart manufacturing & Industry 4.0 maturity models: Implications for small and medium-sized enterprises (SMEs). J. Manufact. Syst. **49**, 194–214 (2018). https://doi.org/10.1016/j.jmsy.2018.10.005
26. Schuh, G., Anderl, R., Gausemeier, J., Wahlster, e W.: Industrie 4.0 Maturity Index. Managing the Digital Transformation of Companies (acatech STUDY). Munich: Herbert Utz Verlag (2017)
27. Shaffer, F., Ginsberg, J.P.: An overview of heart rate variability metrics and norms. Front. Public Health **5** (2017). https://doi.org/10.3389/fpubh.2017.00258

28. Škec, S., Štorga, M., Tečec Ribarić, Z.: Work sampling of product development activities. Teh. vjesn. 23(6), 1547−1554 2016. https://doi.org/10.17559/TV-20150606151030
29. Robinson, M.A.: Work sampling: Methodological advances and new applications. Human Factors Ergonom. Manufact. Serv. Indust. **20**(1), 42–60, (2010). https://doi.org/10.1002/hfm. 20186

Industrial Tourism Development in a Former Mining Area Using Dynamic Model Approach

Ionela Samuil[ID], Andreea Ionica[✉][ID], and Monica Leba[ID]

University of Petrosani, Petrosani, Romania
andreeaionica@upet.ro

Abstract. Tourism has become one of the most important industries and with the greatest potential for growth, offering products and services, investment opportunities but, above all, the creation of jobs, direct and indirect. Lately, in Romania, the tourism sector is perceived as an important catalyst for economic growth, offering great employment potential for both skilled and unskilled workers. Using this direction of development, both the growth of the national economy and the preservation of regional cultural heritage can be achieved. This sector, being predominantly a service sector, relies mainly on the workers in the system. Improving sales performance has a direct impact on increasing revenue margins and is a key driver of business strategy. Management's ability to optimize the distribution of the sales force for a given market niche creates the opportunity for the firm to achieve sales performance and, implicitly, revenue growth. This research uses dynamic modeling to simulate a sales model for an Industrial Theme Park, proposed to be developed in a mining perimeter in Romania. This model has the advantage of allowing the experimentation of alternative strategies for organizing the sales force and, as a result, the identification of that model that will lead to the achievement of the established performance indicators.

Keywords: Sales Force · Dynamic Modeling · Productivity · STELLA Architect

1 Introduction

Industrial tourism arose from the desire to preserve and present the industrial activities responsible for the development of some areas. This form of tourism represents a combination of specific leisure activities with a form of education. The concept has aroused the interest of many researchers, different business models of industrial heritage tourism sites being analyzed [1]. It is certain that the interest in industrial tourism is far from exhausted, the concept being in a continuous enrichment. This form of tourism can be helpful in the process of improving the image of the area where it is developed, becoming a source of income and job generators.

The context of the research is generated by the particularities of a mono-industrial area in Romania, the Jiu Valley, whose revitalization can be achieved through industrial tourism. The aim of the article is to propose a System Dynamics-based model for sales force planning for the Petrila Theme Park (a former mining perimeter). The expected

© The Author(s), under exclusive license to Springer Nature Switzerland AG 2024
A. Rocha et al. (Eds.): WorldCIST 2023, LNNS 801, pp. 253–263, 2024.
https://doi.org/10.1007/978-3-031-45648-0_25

results being the generation of sales force planning strategies to achieve the established performance indicators. The structure of the paper includes: Sect. 2 - Background (2.1. Sales force; 2.2. Using dynamic modelling to achieve increased sales force productivity), Sect. 3 – Research methodology, Sect. 4 - Results and discussions.

2 Background

2.1 Sales Force

In any organization that pursues profit, but especially in the field of services, the sales force is its most expensive and productive asset and can have a decisive impact on the economic performance of the organization [2]. Sales force sizing is a strategic management issue because it can overwhelmingly affect the organization's revenues and costs, resulting in profit or loss. Optimum dimensioning of the sales force provides agents with a suitable climate, they become challenged and motivated, and the interaction with customers is effective, thus achieving the profitability of the organization through sales. Under these conditions, sales compensation costs are reasonable, and the organization demonstrates its efficiency by increasing sales, profit, and achieving the established market share.

There are different approaches to solve the problem of optimal sizing of the sales force: [3] proposes a simple statistical procedure for estimating the optimal size of the sales force by estimating the costs induced by a salesperson's information gathering and processing; [4] uses the mixed-integer linear formulation and a heuristic algorithm to solve the sales force sizing problem; [2] proposes a mathematical model that can be used to determine the optimal size of the sales force within the organization.

Sales force productivity has been one of the most researched issues since the 1920s, although initially most research focused on the sales process and how managers can motivate salespeople to capitalize on current economic opportunities. After 1950, researchers' attention shifted to a better allocation, sizing and control of the sales force with an emphasis on operational research [5, 6] but without being able to provide answers and solutions to questions related to sales productivity. In the mid-1980s, the first research with a theoretical approach appeared, explanations were provided and normative guides were published regarding sales management [7]. In the last ten years, an intensification of research on sales force productivity can be noted, analytical models being used to analyze insufficiently researched problems [8–10] but also using the empirical approach to investigate different theoretical proposals and practices [11–14]. The potential of artificial intelligence and machine learning models in the process of increasing the productivity of the sales force is identified and researched [15].

2.2 Using Dynamic Modeling to Achieve Increased Sales Force Productivity

System dynamic (SD) modeling is used to frame and understand the dynamic aspect of various complex social and managerial systems [16, 17]. The dynamic modeling of the system is adapted to specific managerial phenomena and is carried out by mapping the structure of the analyzed system in order to facilitate the understanding of the behavior of the factors involved, and also to quantify the causal interactions, so as to obtain the simulation of the possible behaviors of the system, over time [18]. In particular, the SD modeling of sales force productivity allows highlighting all the variables associated with the phenomenon under observation. Feedbacks considered within the system can be positive (consolidated) or negative (balanced). While the former represents virtuous or vicious cycles related to a process of system growth or decline, the latter is associated with the situation where a lack of strategic assets (sales agents or financial resources) acts as a limitation in the process of increasing sales force productivity. Once the feedback loops are identified, the key variables related to sales force productivity are converted into stock and flow diagrams using a modeling and simulation environment. These charts allow decision makers to simulate the behavior of the sales force observed over time [17, 19]. Once the system model has been developed, calibrated, tested and simulated, if it behaves realistically, the inputs are changed to perform "what-if" analyzes of how short- and long-term outcomes would change in response to alternative strategic scenarios [20, 21].

Simulation is a valuable tool for entrepreneurs [17], allowing them to simulate years in the evolution when real experimentation is too expensive, this could be the case when launching a new business [22]. By simulating you can discover how complex systems work and where important levers are found, with this, time and space is compressed or expanded. Feedback loops, accumulations, time delays and nonlinear games to capture dynamic feedback processes, all are the focus of the holistic perspective, which, is taken by the dynamic systems. It focuses on all the relevant elements that contribute to strategy implementation and consequence awareness [17]. Moreover, by building the SD model, an opportunity to impose a commitment on stake holders (investors, business partners, collaborators) appears.

The strategic ideas from multiple actors in the model building process and the common understanding, is facilitated by SD. Reality is seen through all models, making them an imperfect representation of it [23], but, with the involvement of the stakeholders, both model accuracy and legitimacy can be improved while fostering the alignment of key players' mental model and group consensus on taken actions [24].

This group consensus, according to a strategic perspective, leads to an in-depth understanding of the analyzed system, through the potential causes and effects of the behavior of it [25, 26]. With this, a double-loop learning process is fueled by the stakeholder engagement, a management team learns from its past actions, while also changing the mental model of how the business works [27].

The theoretical logic, which is oriented to the description and exploration of how inputs produce outputs in a complex system of interconnected casual loops [28]., is also related to the features that made the simulation suitable for strategy design. When the theoretical focus is longitudinal, nonlinear, processual, or empirical data is difficult to obtain, the simulation is particularly useful, as noted by [22].

These are consistent with the pragmatic scientific approach needed to deal with research to increase sales force productivity [29]. As such, unlike other modeling and simulation approaches, the SD methodology can provide an understanding of how specific conditions can affect an observed phenomenon [30].

[31] uses SD techniques to model the sales and operations planning of one of the largest lime producers in Brazil using real data. After simulating different scenarios, analyzing the impact of internal company structure variables and the influence of external variables in the system, [32] demonstrates that the factor that most influences the system is the sales forecast.

3   Research Methodology

The literature review stresses the importance of considering the sales force productivity, the utility of SD and also, allows the identification of the premises that determined the main objective of the research, namely achieving a SD model for force sales planning. The study area identifies the particularities concerning the Jiu Valley revitalizing through tourism and values the Petrila mining industrial heritage, contributing to the objective of the research.

A Business Model (BM) CANVAS is presented and also, a simulation of the sales using STELLA Architect is generated. The results expected concerns the alternative strategies for using sales force to achieve established performance (Fig. 1).

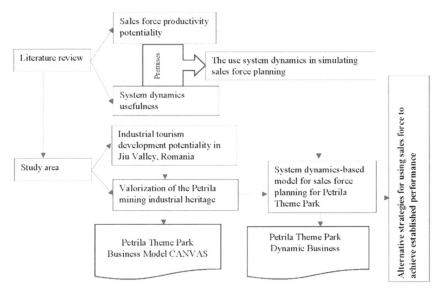

Fig. 1. Research methodology

A BM representation scheme can be seen as a tool to support the structural elements analysis of a business [32, 33], while a methodology based on the simulation of dynamic systems enables analyzes and provides appropriate data for strategy development in a flexible perspective on changes (internal and external) [34, 35]. Using a set of strategic assumptions and different scenarios through SD modeling predict the dynamic implications of strategies [36]. In real terms, entrepreneurs can explore these models and simulate scenarios (using the sales model, for example) and experience what might happen under a range of assumptions and different decision options [17, 37]. Dynamic modeling can be used as a strategy simulation tool to explore how strategies, decisions and others external events interact to generate long-term behaviors of key performance variables, and to explain why and how results are altered and unintended consequences appear. The theme park is treated as a complex system, for which STELLA Architect was applied to see the changes in the outputs for different cases. The dynamic simulation is supported by a mathematical model that integrates the input, output and state quantities to allow the study of different cases that may appear in the evolution of the system. Thus, the input sizes and the output sizes for the sales side were identified in order to determine how to cover the expenses related to the park as well as the remaining profit. However, in this model we also considered a simplifying hypothesis, that is, we took into account only the activities strictly related to the park, not taking into account the income obtained from space rentals.

4 Results and Ddiscussions

4.1 Business Model CANVAS Generation

From the multitude of approaches regarding the business model, we used the CANVAS board [38] which, although it is considered a rigid model, is highly appreciated because it allows the visualization of the business potential through the nine blocks arranged on the board (Fig. 2).

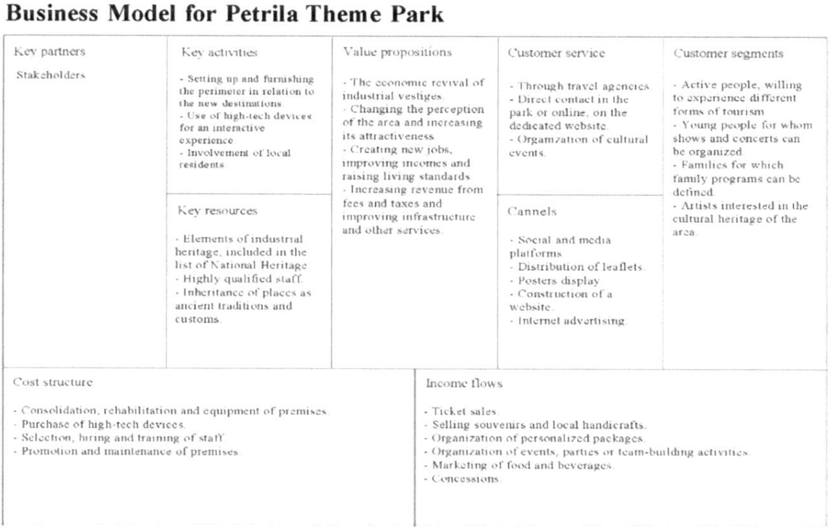

Fig. 2. Business Model CANVAS

Although it presents its advantages, it is limited because it does not allow highlighting the relationships and consequences that appear as a result of the interactions between different elements of the picture. Regarding the analyzed topic, the CANVAS board does not highlight the interdependence relationships that human resources in general and sales force in particular develop in the organization. This valuable asset of the company can be identified both in **key resources** in terms of employed people, in **cost structure** in terms of costs, in **income flows** in terms of revenues but also in **proposed value** in terms of profitability of the organization. In these conditions, the use of SD is required to identify the solution. Consequently, we looked into dynamic modeling because it allows the experimentation of alternative strategies without costs, until the identification of the final solution.

4.2 System Dynamics-Based Model for Sales Force Planning for the Petrila Theme Park

As a post-industrial city, the trend is to achieve the transfer of employed labor from production to services, given that its economic base has changed, and the role of tourism has become very important for the economy and development. We consider that industrial tourism is the only action compatible with the use of resources from the perimeter that would otherwise be lost. In order to develop activities specific to industrial tourism, as an alternative to industrial activities, research is carried out in the former Petrila mining perimeter, Jiu Valley. The idea of the Petrila Theme Park emerged as a necessity, but its development must be accomplished as an art, using both logic and emotion to achieve both efficiency and value. Today's hyper-competitive business environment requires tourism organizations to be agile and adapt quickly to the ever-changing environment. Since the research is based on an idea, and there is no empirical data for the analysis, we used estimates based on tourist flows recorded by the main similar tourism service providers in the market.

Internationally, the National Coal Mining Museum (Caphouse Colliery in Overton, England) must be mentioned. According to the information provided by the Museum management, approximately 150.000 tourists are registered annually. Likewise, the Wieliczka Salt Mine located in the Krakow metropolitan area, Poland is another successful example of industrial tourism. The mine, still functional, attracts over 1 million tourists annually. The Big Pit National Coal Museum, industrial heritage museum in Blaenavon, Torfaen, Wales, attracts around 130.000 tourists annually.

At the national level, the main industrial destinations are the salt mines that have succeeded in the transfer from industrial exploitation to tourist exploitation. Recently, the number of people who have chosen a tourist experience in the depths is significant. Thus, Salina Turda, included in the list of historical monuments since 2015, attracts around 400.000 tourists annually. Salina Praid has been operating as a treatment base since 1980 and records around 500.000 visitors annually. Salina Cacica becomes a tourist attraction in the middle of the 19th century, and the number of interested tourists is around 60.000. Salina Slănic - Prahova, the beginnings of the history of the salt mine date back to 1685, being of interest to about 170.000 tourists. And Salina Târgu – Ocna, which has the largest treatment base in the country, is of interest to approximately 90.000 visitors. Visiting fees for the presented tourist attractions are 30 - 60 RON.

For the proposed analysis, based on the flows of tourists registered by the main providers of specific tourist services, the sales objective is set at approximately 330.000

tickets annually, and the price of the access ticket at 50 RON, set by comparison with the access fees and the complexity tourist activities carried out in the previously presented establishments.

Of the total estimated tourist market, 30,000 are the number of those we consider as a safe tourist market, the other 300,000 being considered potential tourists. We applied a sociability coefficient of 20 and a fruitfulness rate of 1% to them.

As shown in Fig. 3, while the number of tourists and tickets sold positively influence the cash flow, the costs of the required work force reduce the corresponding stock (i.e. cash flow). This trend will be countered by identifying the level of sales productivity needed to increase income (cash flow).

The analyzed model assumes the existence of a number of sixteen employees, of which three people make up the management department, ten people make up the maintenance team and a sales team made up of three agents to whom we apply an initial sales productivity of 200 tickets / agent, in order to achieve a positive and constant cash flow over the 12 months under review. In this scenario, the price for an entrance to the perimeter is set at 50 RON, the sales revenue constituting the sales budget, and the expenses are those related to the workforce, representing the monthly wages paid, on average 4,000 RON / employee.

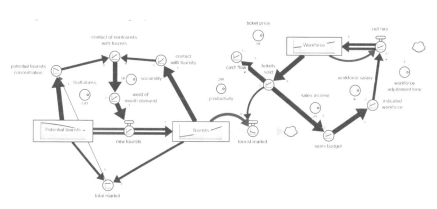

Fig. 3. Sales simulation using STELLA Architect

Using the tool allowed us to capture how cash flow is affected when productivity changes, all other indicators remaining unchanged. Three scenarios were run (Fig. 4), using different levels of productivity. The first scenario reflects an advantageous variant, when, with a productivity of 200 tickets / agent, we obtain an income stream that follows an upward trend. In the second scenario, using a productivity level of 160 tickets / agent, the resulting cash flow is linear. For the third scenario, a productivity level of 150 tickets / agent is considered, a situation in which the system becomes unstable (Fig. 5).

As we well know, productivity depends not only on the qualities and skills of the sales force, but also on the potential of the tourism market at a given time, conditioned by seasonality or accidental market fluctuations. The presented tool can be of real help in the process of adjusting the price elasticity according to the tourist demand curve.

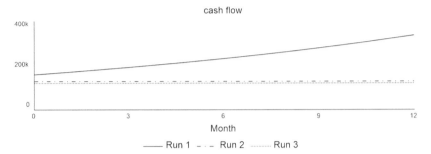

Fig. 4. Comparative cash flow evolution

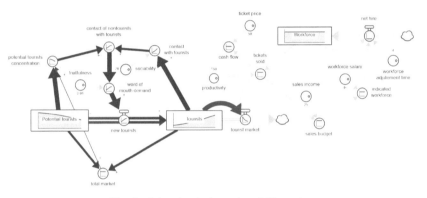

Fig. 5. Sales simulation at instability point

In reality, a number of other factors may appear to influence demand such as geopolitical changes, natural factors with a major impact or phenomena of the nature of the COVID-19 pandemic, which has affected us all, which may modify or influence the results obtained in the laboratory, factors that cannot be predicted but can be considered as probabilities in further model development.

5   Conclusions

The study on the optimal sizing of the sales force required within the Petrila Theme Park, highlights how a sales model can be built, using dynamic modeling, by experimenting with different scenarios until the most suitable one is identified. Thus, this type of approach creates the context for a "what-if" analysis that helps understand potential future scenarios.

The present article, through the obtained results, continues the research carried out [39, 40] proposing a business model for an Industrial Theme Park, to be developed in a former mining perimeter, with the use of SD for its construction. In this way, the proposed BM helps to gain a deeper understanding of how the business works and how it can create value for all parties involved, by highlighting the causal relationships between business system variables.

Thus, the model for planning the sales force for the Petrila Theme Park, based on SD, is designed with the aim of identifying the degree of involvement of the factors responsible for the process of increasing the cash flow, by increasing the productivity of the sales force. In general, when devising illustrative simulation models, the degree of realism is fundamental to effectively research the phenomenon under discussion. In particular, the degree of realism of the dynamic system model must be consistent with the purpose of the research, which in this case is to describe and explore the model for sales force planning and to simulate its effectiveness over time. As such, given that Petrila Theme Park is only a business idea and the sales force planning model is only a simulation tool used to determine the optimal sales force, it is not possible to validate the model by comparing the simulation result with a real, previous reference, constituting a limitation of the research.

The present research responds to a requirement expressed regarding sales force planning, aiming at validating the internal structure of the model so that future results are relevant and the model generates results comparable to other situations that fall within the context of the research.

References

1. Herman, K., Szromek, A. Naramski, M.: Examining the utility of a sustainable business model for postindustrial tourism attractions: the case of the European Route of Industrial Heritage. J. Heritage Tourism (2022)
2. Madhani, P.M., Optimal salesforce sizing and compensation cost: a mathematical approach. Compensation Benefits Rev. 1–8, (2018)
3. Darmon, R.: Optimal salesforce quota plans under salesperson job equity constraints, J. Adm. Sci. **18**, 87 (2001)
4. Salazar-Aguilar, M.A., Boyer, V. Sanchez Nigenda, R. Martínez-Salazar, I. A.: The sales force sizing problem with multi-period workload assignments, and service time windows, Central Europ. J. Oper. Res. **27**(1), 199–218, (2017)
5. Waid, C., Clark, D.F., Ackoff, R.L.: Allocation of Sales effort in the lamp division of the general electric company. Oper. Res. **4**(6), 629–647 (1956)
6. Semlow, W.: How many salesmen do you need? Harv. Bus. Rev. **38**, 126–132 (1959)
7. Basu, A.K., Rajiv, Srinivasan, L.V., Staelin, R.: Salesforce Compensation Plans: An Agency Theoretic Perspective. Market. Sci. **4**(4), 267–291 (1985)
8. Bhargava, H.K., Rubel, O.: Sales force compensation design for two-sided market platforms. J. Mark. Res. **56**(4), 666–678 (2019)
9. Jerath, K., Long, F.: Multiperiod contracting and salesperson effort profiles: the optimality of 'hockey stick, "giving up", and 'resting on laurels.' J. Mk Res **57**(2), 211–235 (2020)
10. Waiser, R. : Involving sales managers in sales force compensation design. J. Market. Res. **58**(1), 182–201, (20210
11. Daljord, O.S., M., Harikesh, N.S.: Homogeneous contracts for heterogeneous agents: aligning sales force composition and compensation. J. of Mkt. Resear. **53**(2), 161–182 (2016)
12. Tanjim, H., Shi, M., Waiser, R.: Measuring rank-based utility in contests: the effect of disclosure schemes. J. Mark. Res. **56**(6), 981–994 (2019)
13. Raghu, B., Hohenberg, S.: Self-selected sales incentives: evidence of their effectiveness, persistence, durability, and underlying mechanisms. J. Mark. **82**(5), 106–124 (2018)
14. Doug, C., Kim, B., Syam, N.B.: A practical approach to sales compensation: what do we know now? what should we know in the future? Found. Trends Market. **14**(1), 1−52 (2020)

15. Singh, R.R., Viswanathan, M., John, G., Kishore, S.: Do activity-based incentive plans work? evidence from a large-scale field intervention. J. Market. Res. **58**(4), 668–704 (2021)
16. Niladri, S., Sharma, A.: Waiting for a sales renaissance in the fourth industrial revolution: machine learning and artificial intelligence in sales research and practice. Ind. Mark. Manage. **69**, 135–146 (2018)
17. Forrester, J.W.: Industrial Dynamics. MIT Press, Cambridge, MA (1961)
18. Sterman, J.D.: Business Dynamics Systems Thinking and Modeling for a Complex World, McGraw-Hill Higher Education (2000)
19. Warren, K.: Strategic Management Dynamics. Wiley, Chichester (2008)
20. Ghaffarzadegan, N., Lyneis, J., Richardson, G.P.: How small system dynamics models can help the public policy process. Syst. Dyn. Rev. **27**(1), 22–44 (2011)
21. Zagonel, A., Rohrbaugh, J., Richardson, G.P., Andersen, D.F.: Using simulation models to address „what if" questions about welfare reform. J. of Policy Analysis and Manag **23**(8), 90–901 (2004)
22. Martin, E.G., MacDonald, R.H., Smith, L.C.: Policy modeling to support administrative decision making on the New York state HIV testing law. J. Policy Analysis Manag. **34**(2), 403–423 (2015)
23. Davis, J.P., Eisenhardt, K.M., Bingham, C.B.: Developing theory through simulation methods. Acad. Manag. Rev. **32**(2), 480–499 (2007)
24. Greenberger, M., Crenson, M.A., Crissey, B.L.: Models in the Policy Process: public Decision Making in the Computer Era. Russell Sage Foundation, New York (1976)
25. Vennix, J.A.M.: Group Model Building: facilitating Team Learning Using System Dynamics. John Wiley & Sons, Chichester, England (1996)
26. Bianchi, C.: Introducing SD modeling into planning & control systems to manage SMEs growth: a learning-oriented perspective. Syst. Dyn. Rev. **18**(3), 315–338 (2002)
27. Richmond, B.M.: The „strategic forum": aligning objectives, strategy and process. Syst. Dyn. Rev. **13**, 131–148 (1997)
28. Kim, H., MacDonald, R.M., Andersen, D.F.: Simulation and managerial decision-making: a double-loop learning framework. Public Adm. Rev. **73**, 291–300 (2013)
29. Torres, J.P., Kunc, M., O'Brien, F.: Supporting strategy using system dynamics. Eur. J. Oper. Res. **260**, 1081–1094 (2017)
30. Romme, A.G.: Making a difference: organization as design. Organ. Sci. **14**(5), 558–573 (2003)
31. Sastry, M.A.: Problems and paradoxes in a model of punctuated organizational change. Adm. Sci. Q. **42**, 237–275 (1997)
32. Santos, N.T., Santos, G.T., Silva, W.S., Ferreira, W.R: A system dynamics model for sales and operations planning: an integrated analysis for the lime industry, Int. J. Syst. Dyn. Appl. **9** (1), (2020)
33. Chesbrough, H.: Business model innovation: opportunities and barriers. Long Range Plan. **43**(2), 354–363 (2010)
34. Sosna, M.N., Trevinyo-Rodriguez, R., Velamuri, S.: Business model innovation through trial-and-error learning. Long Range Plan. **43**(2/3), 383–407 (2010)
35. Morecroft, J.: Strategic Modelling and Business Dynamics: a Feedback System Approach. Wiley, Chichester (2007)
36. Bianchi, C., Bivona, E.: Commercial and financial policies in family firms: the small business growth management flight simulator. Simul. Gaming **31**, 197–229 (2000)
37. Cosenz, F.: Supporting start-up business model design through system dynamics modelling. Manag. Decis. **55**(1), 57–80 (2017)
38. Bisbe, J., Malagueno, R.: Using strategic performance measurement systems for strategy formulation: does it work in dynamic environments? Manag. Account. Res. **23**(4), 296–311 (2012)

39. Ionică, A.C., Samuil, I., Leba. M., Toderas, M.: The path of petrila mining area towards future industrial heritage tourism seen through the lenses of past and present. Sustainability **12**(23), 9922 (2020)
40. Samuil, I., Ionica, A.C., Leba, M.: Business Model for Post-industrial Tourism from a System Dynamics Perspective, IEEE21, pp.1107–1113 (2021)

Implementing an Agile Project Management Methodology on a Minimum Viable Product Development

Danna Dias[1], Anabela Tereso[2(✉)], and Ingrid Souza[2]

[1] Master's in Engineering Project Management, University of Minho, Guimarães, Portugal
[2] ALGORITMI Research Centre/LASI, University of Minho, Guimarães, Portugal
anabelat@dps.uminho.pt

Abstract. In the past few years, the world has faced a rapid change in the digitalization of processes over several different economic sectors. But there are services where human contact is crucial, such as massage therapists, mindfulness guidance, yoga, therapy, etc. In those areas, digitalization and process automation are less important than human well-being. This project was developed in a not-for-profit Swedish startup that creates solutions, including services and products, to help well-being providers increase efficiency in their service. It focuses on selecting and implementing an agile methodology to support the software development of a Minimum Viable Product (MVP), with a restricted time and budget.

The research aims to answer the question: "Which agile method or framework best suits the studied startup environment, considering the restricted time and budget to develop an MVP and why?". To answer this research question, the research method used was action research. The selection, implementation, improvement, and adaptation of an agile methodology were done, considering the organization's needs, to support the delivery of a functional product that added value to the customer. The chosen methodology was Kanban, and through its improvement cycles, it was possible, with the help of the Product and Tech teams, to define a structure that would help them achieve their goals. It was noted that the methodology promoted excellent communication between the teams and aligned the customer's needs and the software development process. The project saved 14.5% of the budget and ended on time, considering the six months of planned development.

Keywords: Agile · Project Management · Software Development · Minimum Viable Product

1 Introduction

The Minimum Viable Product (MVP) concept comes from the startup environment and the product development literature. Lee and Geum [1] defend the importance of collecting feedback from the stakeholders before putting a product on the marked and also mention that the customer feedback on an MVP is crucial to avoid spending extra time and resources building less wanted product attributes. Conforto et al. [2, p.24] present the importance of "obtaining continuous feedback from the customer to respond to constant changes in requirements, needs, risks, new opportunities, and so forth".

The startup where this project took place has identified a lack of automation in the well-being services providers' administrative tasks, consequently wasting time they could use to help more people with activities that do not add value to their business or society. The work in this organization is divided into phases and during the first phase, finished in January 2022, the Product Team defined the Minimum Viable Product (MVP) concept. For the subsequent two phases, their solution's beta version (or MVP) would be developed, a platform to support the administrative tasks of well-being providers, such as yoga and meditation teachers. This research regards the second phase of work, with the duration of six months.

The startup environment is complex and uncertain, being tough to define the whole scope before the project starts, as it is traditionally done in project management [3]. The traditional approach "assumes that events are predictable and that all tools and techniques are well understood" [4], while the agile approach, according to Zasa et al. [5], aims to release different subparts of the output as soon as possible, at the end of each iteration, to collect feedback from the main project stakeholders.

Considering the complexity of the proposed project, the uncertainty around the solution that was defined, and the necessity of rapid responses to the stakeholders' constant feedback to guarantee the achievement of a product desired by them, the chosen approach for this project is agile. The agile methodologies also allow the possibility of developing the scope while developing the product [6]. In the scenario presented, it is non-viable to predict the project scope from start to end and create a robust project plan before starting it, once it was needed to understand the customers' needs and validate the solutions while developing them. Another point is that the organization has implemented, in the last phase, a Kanban Board to manage the tasks visually and some Scrum events, such as Sprint and Retrospective meetings, to create an agile space of decision-making within the team and to promote flexibility in the management of the deliverables. The Scrum events and Kanban Board do not represent a full implementation of either the Kanban Method or Scrum Framework. Still, it was working correctly for its purposes. The organization was open to agile methodologies, but the previous work management was not enough for software development.

2 Research Methodology

The research strategy that supported this study was Action Research (AR). It starts with a context and a purpose definition, followed by a research question: "Which agile method or framework best suits the studied startup environment, considering the restricted time and budget to develop an MVP and why?".

After defining the context, the AR is followed by cycles of research, and each cycle is composed of diagnosing, planning, acting, and evaluating action [7]. With the last cycle evaluation results, a new process, made of the same steps, starts.

The results of each action research cycle were validated by the researcher and the team working on developing the project to decide the study's next steps. It was crucial to the project that the studied team understands the importance of the applied project management methodology and collaborates for its success. It was also relevant to consider what can be implemented in the studied context and what is not valid to implement.

Throughout the action research cycles, qualitative and quantitative data were collected using the mixed method. The qualitative data considered in this research was the literature reviewed, the information discussed during the meetings with the group, and the feedback the team gave, which supported the evaluation and improvements on the proposed method to be implemented. The literature reviewed was the input to the action plan for the first cycle of the action research, supporting the diagnosis and defining the purpose of the study. It also helped determine which methodology best fits context of the organization, given what could be found in the academic literature.

The second and subsequent action research cycles were supported by team validation and feedback, to assess the impact of the actions on the success of the project. The result of each cycle was the input for the next one, adapting things, when necessary, to guarantee the success of the agile project management methodology implementation for the MVP creation.

For the quantitative data, the study analyzed the budget and the established time of delivery to determine if the chosen project management methodology helped the team to deliver the project on time and within budget.

3 Methodology Selection and Implementation

3.1 Selecting the Agile Methodology

Scrum is the most famous agile methodology in software development, followed by Kanban [8]. Based on this, the Kanban Method and the framework Scrum were chosen to be presented to the teams. An evaluation was then carried out to decide which one would be better accepted. The Scrum framework is well structured, with all the roles, time-boxed events, and a predetermined duration, going from two to four weeks [9]. On the other hand, the Kanban Method is not a predefined process or a template, as stated by Anderson [10]. Still, he defends that the team "should be empowered to evolve their unique process solution that obviates the need for such services and require a new set of tools" [10, p.44]. Anderson and Carmichael [11] suggest a meeting structure to implement and determine a feedback loop in the system, such as Kanban Meeting and Replenishment. Still, their time and frequency are determined by the team.

Despite the Tech Team was tiny, being one full-time and two part-time developers, the time for developing the MVP was only six months. The project is based on an open-source tool, which reduces its complexity in terms of development and time. The chosen methodology was Kanban. The decision was taken in agreement with the teams.

Before this project, the Tech Team only worked in the support and system administration, so the development process will also be implemented along with the beginning of the MVP building. Considering that, it will not be possible to measure the difference between the implementation of Kanban and the previous work, as it did not exist. The results will take into consideration the budget, schedule, and the final structure of the Kanban Method after the improvements.

3.2   Implementing the Kanban Method

The Kanban implementation was based on STATIK (System Thinking Approach to Introducing Kanban) defined by Anderson and Carmichael [11]. Considering the steps of STATIK:

- In Step 0, identify service, the software development process, was the service identified;
- In Step 1, understand what makes the service fit for the purpose of the customer, the service fit is to deliver a solution for the providers of well-being services to help them improve the management of administrative tasks and free up their time;
- For Step 2, understand the sources of dissatisfaction with the current system, it was not possible to find these sources as the software development process would be implemented in the organization at the same time as the Kanban method;
- For Step 3, analyze the demand, the demand was analyzed with the context of the organization, the technology chosen and the size of the team;
- In Step 4, analyze the capability, before the beginning of the project, the Tech Team and the Product Team met to align the deliverables of the next phase. They also analyzed if the team was enough to deliver what the Product Team had planned for the phase;
- In Step 5, model the workflow, the software development process was defined as presented in Fig. 1 below.

Fig.1. Software Development Process

The Tech Team is responsible for developing and testing the feature. After done and deployed, the Product Team tries the feature in the testing environment. When the feature is approved, it is delivered to the customer.

In Step 6, discover the classes of services, it was decided that it wouldn't have, at that moment, a differentiation in the classes of services. All the cards/tasks would be considered a standard work routine for developing the software.

In continuity with the STATIK for planning the implementation of the Kanban Method, as suggested by Anderson and Carmichael [11], in Step 7, they indicate the design of the Kanban Board. In the Kanban Board, the cards flow from the left to the right, so the card enters this board in the "Backlog" column, replenished by the Product Leader. The cards at the top of the "Backlog" column are the priority for the next two weeks of work. When the card is ready to be developed, which means it is a priority in the alignment between Tech and Product teams, and has the design tested and approved, and the card has enough information to be worked on by the Tech Team, it moves to the "To Do". The "To Do" column is the "Point of Commitment" of the Kanban Board. After moving to this point, the task must be delivered (Fig. 2).

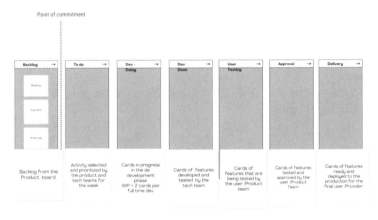

Fig. 2. Kanban Board for the Software Development Process

3.3 The Kanban System

The Kanban System regards the roles, events, and the Kanban Board. The Kanban Board was already presented in the previous topic. This topic will be divided into Kanban Roles, and Kanban Events. These following points will address how the Kanban System was designed to be implemented in the context of the research.

The proposed Kanban Board, roles and events were defined by the main researcher and presented to the teams, and with their agreement it was tested and improved with the cadences of the Kanban.

3.3.1 Kanban Roles

Anderson and Carmichael [11] define that two roles are needed for implementing and operating Kanban. They also defend that the tasks and responsibilities related to the roles are more important than the job title. The duties and people assigned to the position are the following (Table 1):

Table 1. Kanban Roles

Role	Responsibility	Responsible
Service Request Manager	- Understand the customer needs; - Transform customer needs into requirements to support software; - Create the backlog prioritize the work; -Test and approve the features	Product Leader
Service Delivery Manager	- Facilitating meetings and guaranteeing the Kanban Method is understood and followed by the teams; - Collect data and unblock the development team	Agile Project Manager

3.3.2 Kanban Events

Anderson and Carmichael [11] point out that the Replenishment Meeting and the Kanban Meeting are the baselines for implementing the Kanban Method. Other than that, also based on Anderson and Carmichael [11], the events implemented to support the Kanban implementation, in the context of this research, were the following:

- Kanban Meeting: the objective of this meeting is to visualize the Kanban Board, unblock tasks, and guarantee that the work is flowing through the process. The meeting would happen twice a week and last 15 min. The meeting was facilitated by the Agile Project Manager and has the participation of the Tech Team;
- Delivery Planning Meeting + Replenishment: Before this meeting, the Product Leader is responsible for creating the cards in the "Backlog" column, based on the roadmap and priorities defined for the project. During the meeting, the cards for the week would be moved from "Backlog" to "To Do", so the development could start. This meeting would happen biweekly, on a Monday, and was planned to be one-hour long. The Product Leader, the UI/UX designer, the Agile Project Manager, and the Tech Team attended this meeting;
- Service Delivery Review + Retrospective: In this meeting, the Product Leader would review, together with the Tech Team, whether the work planned for the previous two weeks went well and explain changes or possible malfunctions in the software to approve it or not. After that moment, a Retrospective Meeting was held, regarding the last two weeks of work. The Retrospective was done on a Board, with the columns "Well", "Not so Well", and "New Ideas". This meeting would happen biweekly, on Fridays, and was planned for 90 min.

4   Improvements

As explained before, the Kanban Method promotes the Kanban cadences as cycles of constant improvement in the system [11]. In this case, the most important event for feedback and improvement is the Retrospective Meeting.

To find a structure for the methodology that could work well in the context of the organization and help the teams to deliver the project on time and on budget, five improvement cycles took place during the research, as presented in Fig. 3 below. In this chapter, a summary of what has changed during each cycle will be presented.

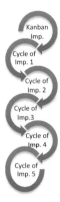

Fig. 3. Kanban cycles of improvement

4.1 First Cycle of Improvement

This first retrospective meeting was crucial to show the first impressions of using the Kanban Method and how the teams were feeling regarding managing the tasks. The two teams pointed out that they could see Kanban's transparency and that the Tech Team's collaboration has improved. Another point is that they no longer needed to ask someone else what's next, because they could visualize it on the Kanban Board.

In the not-so-well, the teams pointed out the need to improve the visualization of the system, as it was challenging to identify blocked cards. To solve that, in the new ideas, the teams suggested implementing new tags in the cards to signal when a card is blocked or when a card is ready to be tested by the Product Development Team.

One more point for improvement was to "improve synchronization of the Tech Team". The Tech Team identified that they communicated better during the meetings but did not try to reach out to the others, when they were working on the development tasks, so they decided to change it.

4.2 Second Cycle of Improvement

Right after the first Retrospective Meeting, the actions discussed during the meeting were implemented, and all of them became part of the process. The tags were implemented in the Kanban Board, and the Tech Team members started to have more one-on-one meetings and pair programming to sync the communication.

Regarding the methodology, the teams mentioned Kanban's transparency again. They also noted that the meetings were helping the work to flow.

Regarding the development process, the group mentioned the issue of going to the Review Meeting without all the tasks finished on the development side being tested by the Product Team, because the Tech Team had not deployed them. The teams decided to improve the process by determining that all tasks done by Thursday would be deployed, so the Product Team could test them before the Review Meeting that happens on Fridays.

4.3 Third Cycle of Improvement

Two relevant decisions were made in this meeting. The first one was to wait for the platform to be tested by a customer before developing details. The second one was to schedule an extensive review to prioritize what comes next and estimate what would be delivered in the first big release. To achieve the objective established at the beginning of the project, of having a functional MVP to start testing with the customers, it was crucial to be careful not to develop features that would not deliver value to the final customer.

During the meeting, it was decided that from that moment on, the Retrospective Meeting would happen monthly, not biweekly anymore. As the Kanban System was already integrated into the teams' routine, they decided that the process worked well, and no significant improvements were identified.

4.4 Fourth Cycle of Improvement

The Tech Team significantly improved and delivered good work in this cycle, especially the two developers mentioned. The teams used the Retrospective Meeting as an opportunity to acknowledge them.

On the "New Ideas", the teams suggested that the Tech Team could create a Roles and Responsibilities document to address the lack of clarity regarding the ownership of the tasks and responsibilities.

They identified that the time of the prereleases were very loose. To solve it, the Tech Team would define on Wednesdays what would be released on Thursdays. Other than that, the ideas implemented at the beginning of this cycle improved the work and would continue as part of the Kanban System.

4.5 Fifth Cycle of Improvement

In this retrospective, it was possible to notice the considerable improvement in the communication and alignment between the two teams. The prioritization of the work has also improved. They also mentioned the tremendous impact of the method and classified the work as agile and adaptable.

The teams did not give suggestions regarding the Kanban structure in the last two Retrospective Meetings. But in this cycle, the teams decided that it would be better to plan weekly with the last week's review on Mondays. And from that moment on, every Monday, the teams would have a meeting starting with the last week's review and then plan the work for the next week. The retrospectives would continue to happen monthly, and the Kanban Meeting would be on Wednesdays and Fridays.

4.6 Changes on the Kanban Structure

In the last improvement cycle, it was decided that the Kanban Planning would happen weekly and alongside the Review Meeting. So, every Monday, the two teams would start the meeting by reviewing what was delivered the previous week and planning and prioritizing the work for the following week.

Kanban Meetings continued to happen twice a week, as decided in the first improvement cycle. With the tests for what would fit best the teams' availability, it was established to have a fixed time to happen every Wednesday and Friday, as the teams noticed that the participation of everyone from the Tech Team was fundamental for the week's status.

In the beginning, the Retrospective Meeting happened every two weeks to adjust the Kanban structure to the teams' reality. After three cycles, the teams decided to change it to happen every month, and it continued to add value and improvements to the teams' processes. The communication and the Kanban System were already well established, reducing the need to meet biweekly.

Regarding the Kanban Board, some changes happened to improve the visibility and to attend to the changes in the software development process. The first change was adding the "In Review" column between "Dev–Doing" and "Dev–Done". So, a card only would move to the "Dev–Done" column when it was developed and approved by the Tech Team. The second change was the creation of tags to facilitate the visualization, prioritization, and management of the tasks.

5 Conclusions

The selection and implementation of the agile method was successfully done. It involved applying a new concept and a new process in a short period to achieve the project goals. Including the two teams since the definition of the method, hearing their needs, and taking into consideration the organization and people contexts, were crucial factors to answer the research question: "Which agile method or framework best suits the studied startup environment, considering the restricted time and budget to develop an MVP and why?".

The Kanban Method helped the improvement of the communication between the two teams involved, and it was clear since the project started, as well as the visual aspect of it. The board organization made the tasks flow at a good pace, considering the time we had to deliver the project.

The retrospective meetings were shown to be fundamental to the improvement of the methodology, being also the feedback needed for the Action Research, as a way of finding out how to make the method fit in the teams' context and help the delivery of the platform on time and on budget.

Regarding the time and budget, the project was delivered under budget, with an economy of 14.5%, and it had a duration of 6 months of work, as planned. There are already customers testing the platform, as it is already functional.

The flow of communication created, and the structure of meetings provided by the Kanban Method helped the teams to deliver the platform on time and on budget. With the Kanban System, we could constantly analyze and improve the process and manage the changes, aligning everyone involved in the project.

The flexibility of the Kanban Method fit nicely with the teams, as after each retrospective, we could improve the process to attend to the teams' needs. After three retrospectives, we had a structure that was working well, and it was already part of the teams' routine. After that, only minor changes were made to the Kanban System. The teams adapted well and will continue using the method to improve the platform in the next development phase.

The main contribution of research is to have proved that the Kanban method is effective to manage software development projects. Also, as we did not find any literature regarding the development of an MVP using the Kanban method, this work may contribute to this area of knowledge and be useful for other researchers and practitioners doing similar applications.

Acknowledgement. This work has been supported by *FCT – Fundação para a Ciência e Tecnologia* within the R&D Units Project Scope: UIDB/00319/2020.

References

1. Lee, S., Geum, Y.: How to determine a minimum viable product in app-based lean startups: Kano-based approach. Total Qual. Manag. Bus. Excell. **32**(15–16), 1751–1767 (2021)
2. Conforto, E.C., Salum, F., Amaral, D.C., Da Silva, S.L., De Almeida, L.F.M.: Can agile project management be adopted by industries other than software development? Proj. Manag. J. **45**(3), 21–34 (2014)
3. PMI: A guide to the project management body of knowledge (PMBOK ® guide). Project Management Institute. (PMI, Ed.) (6th ed.). Newtown Square, Pennsylvania 19073-3299 USA (2017)
4. Bergmann, T., Karwowski, W.: Agile project management and project success: a literature review. In: International Conference on Applied Human Factors and Ergonomics, pp. 405–414, July 2018
5. Zasa, F.P., Patrucco, A., Pellizzoni, E.: Managing the hybrid organization: how can agile and traditional project management coexist? Res. Technol. Manag. **64**(1), 54–63 (2020)
6. Beck, K., et al.: Agile Manifesto – Principles (2001a). https://agilemanifesto.org/principles.html. Accessed 9 Dec 2018
7. Coughlan, P., Coghlan, D.: Action research for operations management. Int. J. Oper. Prod. Manag. **22**(2), 220–240 (2002)
8. Sastri, Y., Hoda, R., Amor, R.: The role of the project manager in agile software development projects. J. Syst. Softw. **173**, 110871 (2021)
9. Schwaber, K., Sutherland, J.: The scrum guide. Scrum Alliance **21**(1) (2020)
10. Anderson, D.J.: Kanban: Successful Evolutionary Change for Your Technology Business. Blue Hole Press (2010)
11. Anderson, D.J., Carmichael, A.: Essential Kanban Condensed. Blue Hole Press (2016)

Application of the PM² Programme Management Methodology to the Portuguese Project Management Observatory

Vânia Novo[1], Anabela Tereso[2]([✉]), Paulo Sousa[2], Sofia Ribeiro-Lopes[3], and Pedro Engrácia[4]

[1] Master's in Industrial Engineering, University of Minho, Guimarães, Portugal
[2] ALGORITMI Research Centre/LASI, University of Minho, Guimarães, Portugal
{anabelat,paulo.sousa}@dps.uminho.pt
[3] Master's in Engineering Project Management, University of Minho, Guimarães, Portugal
[4] APOGEP, Lisboa, Portugal
pedro.engracia@apogep.pt

Abstract. The Portuguese Project Management Observatory (PPMO) is a non-profit organization and an initiative of the Portuguese Project Management Association (APOGEP). The PPMO was created in 2020 to serve the knowledge needs of project management, focusing on the status and evolution of project management in Portugal. The PPMO is being managed as a portfolio divided into programmes of projects since late 2021. To manage these programmes, the European Commission's methodology – PM² Programme Management (PM²-PgM) methodology – is being used, which was launched, in 2021, to the public as a free-to-use methodology. The methodology integrates artefacts, which were made available in 2022. The PM²-PgM is an extension of the PM² Project Managemen Methodology. It is easy to implement, customizable and has the best practices and tools adapted to the management of programmes. This paper shows how the methodology was implemented and customized to better suit the context and needs of the PPMO's "Creation of Knowledge Programme".

Keywords: PM² Programme Management Methodology · PM²-PgM · Programme Management · Portuguese Project Management Observatory

1 Introduction

The Portuguese Project Management Association – APOGEP, was created in 1994, and is a member of the International Project Management Association – IPMA. APOGEP works in six strategic areas: i) certification of project managers, project management consultants and trainers; ii) certification of organizations (IPMA Delta model); iii) registration of training entities according to the IPMA-APOGEP model; iv) organization and promotion of events in project management; v) support to the creation of a community of project managers; and vi) dissemination of the main news related to project management [1].

A. Rocha et al. (Eds.): WorldCIST 2023, LNNS 801, pp. 274–283, 2024.
https://doi.org/10.1007/978-3-031-45648-0_27

The Portuguese Project Management Observatory (PPMO) was created in 2020, as a strategic initiative of APOGEP. The PPMO is a non-profit organization and was created as a "satellite" of APOGEP "world" with the aim to assist on the achievement of some of the primary objectives of this association, such as the creation and dissemination of knowledge about the status and evolution of project management in Portugal.

The PPMO is constituted by a volunteer group of project managers, students, and academics from University of Minho and 17 other Portuguese universities. It is being managed as a portfolio composed by two programmes: i) Creation of Knowledge programme; and ii) Dissemination of Knowledge programme. In this paper the Creation of Knowledge is the focus, which is being managed, using the PM^2 Programme Management (PM^2-PgM) methodology created by the European Commission and launched in 2021. Since it is a recent methodology, there is a lack of scientific papers and information about it. Thus, the management of the PPMO, applying PM^2-PgM methodology, will be a contribution to the project and programme management knowledge.

The Creation of Knowledge programme started in late 2021 and is expected to be finished by 2025, when all the projects are completed. The programme is composed by eight projects that create knowledge in the project management field, including scientific papers, master theses, and PhD dissertations. At the moment, the initiating and planning phases of the programme are finished, and the programme is in its executing phase. The closing phase was already initiated due to the necessity of analysis of the whole PM^2-PgM methodology for a master thesis developed in this context.

The research questions aimed to be answered in this scientific paper are: Q1 – *"How to apply the PM^2-PgM methodology in the PPMO's Creation of Knowledge programme?"*; and Q2 – *"Does the application of the PM^2-PgM methodology create advantages to the management of the programme?"*. The expected results of this research are: R1 – PM^2-PgM methodology implementation in the Creation of Knowledge programme of PPMO; R2 – recommended improvements on the PPMO management; and R3 – a lessons learned process.

In Sect. 2 it is presented the literature review. Section 3 presents the research methodology adopted. The description of the PM^2-PgM methodology is presented in Sect. 4. The results and discussion are in Sect. 5. Finally, in Sect. 6 appear the main conclusions of the study.

2 Literature Review

This section presents a brief literature review concerning the topic of study (programme management) and the PM^2 programme management methodology.

2.1 Programme Management

There are many definitions for programme management. Initially, programme management was defined as a series of structures and processes used to coordinate multiple projects that constitute the organization's business strategy [2]. However, it is essential to differentiate between projects and programmes [3] since programmes are sets of projects with related goals and are interdependent [4]. Hence, according to the European Commission methodology [4], programme management is the "process of applying knowledge, skills and actions to a programme to achieve programme's objectives and benefits". In more detail, programme management is defined as an integrated structure that coordinates and manages a set of interrelated projects, aligning and allocating resources to achieve the expected benefits that would not be achieved if projects were managed individually [5]. Thiry [6] argues that the implementation of a model of programme management will benefit both the programme managers and the organization with the tools and practices that project management lacks. Additionally, Young et al. [7] concluded in their study that the biggest problem encountered was within the management of programmes where the idea of the individual impact of the projects on the strategic objectives was unclear. Thus, in the approval and initiation of programme management, it is crucial to align the individual projects with the overall programme and organization strategy to create value and achieve the expectations formulated [8, 9].

2.2 PM^2 Programme Management Methodology

The PM^2-PgM methodology was launched as an extension of the PM^2 methodology, in 2021, by the European Commission. This methodology is free-to-use and provides solutions targeted to programmes [10], allowing programme managers to manage and monitor their programmes more efficiently, with the most appropriate tools along the whole programme lifecycle [4]. The methodology includes the best programme management practices globally accepted. It comprises the methodology guide, a set of artefacts made available in 2022, and a set of effective mindsets in managing programmes. The artefacts are essential in each programme phase since they facilitate communication with stakeholders and provide a clear view of the programme goals. In addition, these documents work as a starting point to control and monitor the programme implementation [4].

It is important to note that the PM^2 Project Management methodology is recent and its extension focused on programmes (PM^2-PgM) is even more recent, hence the lack of scientific information on it, as shown in Table 1. Thus, the results obtained with the search of "PM^2", "project management methodology" and "program*" in different databases return the number of documents presented.

Based on these results, Fig. 1 was obtained, showing the publications per year. Note that the results in Google Scholar were only those that could be fully accessed (57 documents). The first publication was in 2014, and their number has grown until 2019. Between 2019 and 2020 the number of publications decreased, but between 2020 and 2021, it increased, reaching a maximum of 18. The numbers for Scopus (4) and WoS (2) are smaller.

Table 1. Research strings and obtained results

Databases	Research string	Link	Results
Scopus	(TITLE-ABS-KEY("PM^2") AND TITLE-ABS-KEY("project management methodology") AND ALL("program*"))	Scopus	4
WoS	(TS=("PM^2") AND TS=("project management methodology")) AND (ALL=("program*"))	WoS	2
Google Scholar	"PM^2" "project management methodology" "program*"	Google Scholar	65

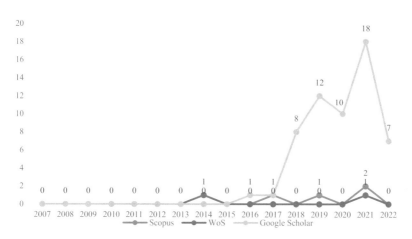

Fig. 1. Publications by year from 2007 to 2022

3   Research Methodology and Study Context

Action-research methodology [11] was applied to this research project, as methods, processes and techniques for programme management were proposed and applied, promoting change in the PPMO, with the involvement and cooperation of the researchers and the team.

This paper was made in the context of the Creation of Knowledge programme of the Portuguese Project Management Observatory. The PPMO is an APOGEP initiative focusing on improving activities and skills of Project Management (PM). PPMO links APOGEP with the academic community and includes project managers, students, teachers, professionals, and all interested in PM. The main objectives of the PPMO are [12]: (1) Conduct an annual report about Project Management in Portugal; (2) Stimulate the creation of knowledge in the area; (3) Aggregate the generated knowledge, particularly in Academia and Industry; and (4) Support APOGEP in the definition and dynamization of dissemination actions.

To be aligned with its objectives of creating and disseminating knowledge on the PM area, PPMO is organized as a portfolio of two programmes, with various projects, as shown in Fig. 2.

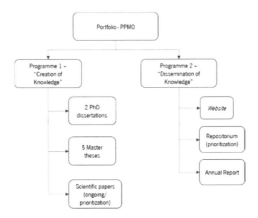

Fig. 2. PPMO's ongoing and prioritization projects

The PPMO team is composed of around twelve volunteers. Its organizational structure is shown in Fig. 3.

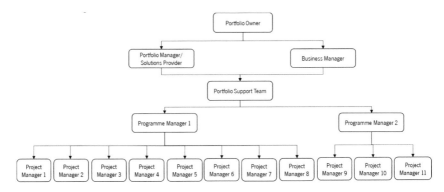

Fig. 3. PPMO's organizational structure

4 Description of PM² Programme Management Methodology

The PM²-PgM methodology includes a 61 pages' guide and 16 templates (the artefacts). The PM²-PgM methodology is customizable to any context and can be used and freely personalized to adapt to the needs of each context. This methodology assumes a top-down approach in which new projects are initiated once the executing phase of the programme has started. However, there are considerations for emergent programmes, where ongoing projects are included within the programme structure.

The methodology provides a governance structure, a programme lifecycle (divided as Initiating, Planning, Executing and Closing phase, with Monitoring and Controlling activities), guidelines to programme activities, artefacts, and orientations for their use, as well as the most effective mindsets. The different phases and artefacts assigned to each phase are represented in Fig. 4.

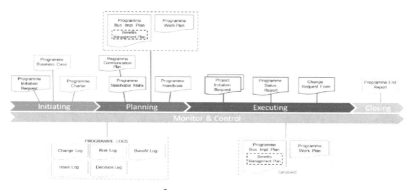

Fig. 4. PM^2-PgM Artefacts landscape [4]

The initiating phase focuses on identifying the programme's needs, problems, and opportunities. In this phase, the primary goals and expected benefits are defined, which is crucial for the programme's success. As presented in Fig. 4 the Programme Initiation Request (PIR), the Programme Business Case (PBC) and the Programme Charter (PC) should be developed to formulate the proposal better, evaluate the viability, and define the mandate and organization of the programme.

In contrast to projects, a programme's planning phase focuses more on delivering benefits and organizational change to achieve those benefits. The artefacts involved in this phase are the Programme Handbook (PH), Programme Stakeholder Matrix (PSM), Programme Communication Plan (PCP), Programme Business Implementation Plan (PBIP), and Programme Work Plan (PWP). Developing all these documents facilitates the planning of processes and responsibilities for those processes, the analysis of stakeholders, the creation of the communication plan, and the planning of the programme roadmap and deadlines for different deliverables.

In the executing phase there is more focus on the projects that constitute the programme and their coordination with the programme activities and goals. In this phase, project documents (such as Project Initiation Requests) are analyzed, and Programme Status Reports (PSR) are developed to assess the progress and status of the programme.

The closing phase aims to evaluate the programme, capture lessons learned throughout the programme lifecycle, formulate post-programme recommendations, and sustain benefits delivery. These activities should be documented in the Programme End Report (PER).

Regarding monitoring and controlling activities, many documents are available to develop from the beginning to the end of the programme. These artefacts aim to collect and analyze information about the programme status, performance indicators, control of incidents and unexpected deviations, to mitigate them and restore the programme. These artefacts are the Risk Log, Issue Log, Decision Log, Change Log and Benefit Log.

5 Results and Discussion

The "Creation of Knowledge Programme" was an emergent programme where some projects were already ongoing at the time of the initiation of the programme, so it was essential to consider the considerations that the methodology has for emergent programmes. This way, it was possible to adjust the methodology to the programme better, as it is intended. Some adjustments occurred right at the very beginning, such as the assignment of the roles and responsibilities before the implementation of the methodology. However, more adjustments were needed, namely adjustments to the recommended processes.

In the programme's initiating phase, an overall schedule was developed for a holistic view of the ongoing projects and their status. Additionally, it was necessary to resort to the PSM in this phase, contrary to what the methodology recommends (using it in the planning phase). The filling of the artefacts exhibited important topics to analyze to fulfil the programme's existing needs. The PIR aggregated all the strategic goals of PPMO and the overall view of the programme, making a clear view of the idea adjacent to the programme execution. The PBC was significant in defining the programme scope and evaluating the alignment between the PPMO strategic and programme goals. The PC allowed to describe in detail all the programme characteristics and projects included, highlighting the benefits map where all programme benefits were aligned with the strategic goals, outcomes, and necessary resources.

The planning phase involves higher effort since it has the most detailed artefacts compared with other phases. In this phase, the PH should be developed, and it is where the tailoring of the PM^2-PgM process is documented. Figure 5 shows the adjustments made throughout the implementation of the methodology and development of the artefacts. Three of the sixteen artefacts have not been fully developed. However, two of those three were analyzed and incorporated the most relevant information in the PH, which serves as a summary of the different area plans. The PWP was important to aggregate all the expected results for each defined stage and assign deadlines to achieve those results.

The executing phase suffered the most changes. Firstly, because this was an emergent programme, where many projects were already ongoing, and secondly, due to the volunteer context of PPMO, which translated into restricted availability of team members and consequently greater difficulty in coordinating the projects, as expected by the methodology. Thus, it was only possible to develop one simplified PSR to draw some conclusions about the progress of the projects and issues that may have happened. Figure 6 shows more visually the progress status of the different projects.

The closing phase started with filling of the PER with the lessons learned up to that point. This phase will be finished as soon as all projects are concluded, and the executing phase is completed. Although PER is not fully developed, it is already possible to understand its importance to the successful evaluation of the programme and the fulfilment analysis of stakeholders' expectations. Additionally, this will serve as a future consultation document with all lessons learned and recommendations for future programmes.

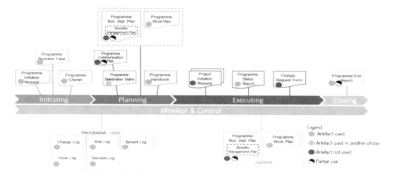

Fig. 5. Artefacts landscape and adjustments made

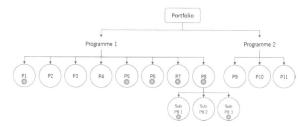

Fig. 6. Projects progress status

Monitoring and controlling activities occurred throughout the programme lifecycle supported by the artefacts provided by the PM^2-PgM methodology. All logs were prepared in the initiating phase and updated in subsequent phases, as needed. These documents helped identify in detail all issues, risks, decisions, benefits, and changes included in the programme for consultation and definition of corrective actions, if necessary. Apart from the artefacts provided, it was essential to add the Meeting Minutes from the start of the programme, to include important points discussed at meetings, relevant decisions, changes to be made and identify actions needed to solve any problems. In addition, these documents served as an agenda for future meetings, ensuring that time was optimized.

From the point of view of those responsible for applying the methodology, it proved to be intuitive, easy to use and easily adapted to the context of the programme. In addition, the artefacts helped to consider all important aspects of the programme, serving as a comprehensive guideline to manage all areas. Considering the volunteer context of PPMO and the low availability of its members, it was challenging to coordinate all the projects, resulting in an obstacle to the full implementation of the methodology. Therefore, only the projects' planning was analyzed, and a brief description of them was made, with information collected on the projects' progress at the study's beginning and end.

The application of the PM^2-PgM methodology in the PPMO made it possible to learn some lessons. First, the importance of prior knowledge of the PM^2-PgM methodology to better adapt and apply it, since some relevant aspects are mentioned superficially in the methodology guide. Another lesson is the importance of using the methodology at all levels (projects and programme), which was not possible in this case. In this way, it would be easier to obtain certain project information needed for programme management. Finally, the importance of using the Meeting Minutes, which are not included in the set of artefacts. These documents were important not only to register the decisions made in the meetings and all relevant information, but also to determine the topics to be discussed at subsequent meetings.

6 Conclusions

This paper aimed to answer the research questions: Q1 – *"How to apply the PM^2-PgM methodology in the PPMO's Creation of Knowledge programme?"* and Q2 – *"Does the application of the PM^2-PgM methodology create advantages to the management of the programme?"*. The expected results were: R1 – PM^2-PgM methodology implementation in the Creation of Knowledge programme of PPMO; R2 – recommended improvements on the PPMO management; and R3 – a lessons learned process.

To answer Q1, the methodology guide was analyzed, and its artefacts, which allowed a better understanding of the methodology's philosophy and all its activities and proposed tools. Applying the methodology to the Creation of Knowledge programme and following all its processes and guidelines enabled answering the first research question. However, the characteristics of the programme made it necessary to adapt and adjust the proposed process from the beginning, as many projects were already in execution, in contrast to the top-down approach that the methodology follows. All sixteen artefacts were analyzed. However, only thirteen were developed due to the programme's context and complexity. Furthermore, two of the three not developed (PCP and PBIP) were analyzed, and some parts were incorporated into the PH, which worked as a summary of several areas of the programme. Finally, the use of some artefacts was adapted in terms of the phase in which they were developed, namely the PSM, which had to be developed in the initiating phase together with the PC. All these adjustments led to the conclusion that the methodology can indeed be adapted to the PPMO program's needs.

To answer Q2, all results obtained with the application of the methodology and development of the artefacts were analyzed. Based on these results, it was possible to observe some advantages that the methodology brought to the programme, emphasizing the guidelines provided that ensure not to forget important aspects. Also, the PSR worked as an essential tool to check projects' progress and prove the impact of the low time availability due to other teams' professional occupations in them.

In summary, regarding the expected results, it was possible to implement the methodology in the Creation of Knowledge programme, record a process of lessons learned, and with that, and recommend improvements on the PPMO management.

References

1. APOGEP, Associação Portuguesa de Gestão de Projetos - Sobre Nós (2022). https://www.apogep.pt/sobre-nos. Accessed 3 Jan 2022
2. Partington, D., Pellegrinelli, S., Young, M.: Attributes and levels of programme management competence: an interpretive study. Int. J. Project Manage. **23**(2), 87–95 (2005). https://doi.org/10.1016/j.ijproman.2004.06.004
3. Artto, K., Martinsuo, M., Gemünden, H.G., Murtoaro, J.: Foundations of program management: A bibliometric view. Int. J. Project Manage. **27**(1), 1–18 (2009). https://doi.org/10.1016/j.ijproman.2007.10.007
4. European Commission, PM^2 Programme Management Guide 1.0 (2021). https://doi.org/10.2799/193169
5. Shehu, Z., Akintoye, A.: Construction programme management theory and practice: contextual and pragmatic approach. Int. J. Project Manage. **27**(7), 703–716 (2009). https://doi.org/10.1016/j.ijproman.2009.02.005
6. Thiry, M.: Combining value and project management into an effective programme management model. Int. J. Project Manage. **20**(3), 221–227 (2002). https://doi.org/10.1016/S0263-7863(01)00072-2
7. Young, R., Young, M., Jordan, E., O'Connor, P.: Is strategy being implemented through projects? Contrary evidence from a leader in New Public Management. Int. J. Project Manage. **30**(8), 887–900 (2012). https://doi.org/10.1016/J.IJPROMAN.2012.03.003
8. Artto, K., Dietrich, P.H.: Strategic business management through multiple projects. In: Morris, P.W.G., Pinto, J.K. (eds.) The wiley guide to project, program, and portfolio management, pp. 1–33. Wiley, New Jersey (2007)
9. Too, E.G., Weaver, P.: The management of project management: a conceptual framework for project governance. Int. J. Project Manage. **32**(8), 1382–1394 (2014). https://doi.org/10.1016/J.IJPROMAN.2013.07.006
10. Kourounakis, N.: An Overview of PgM^2 Programme Management - PM^2 Alliance (2020). https://www.pm2alliance.eu/forum/an-overview-of-the-pgm2-programme-management/. Accessed 19 Feb 2022
11. Saunders, M., Lewis, P., Thornhill, A.: Research methods for business students, 8th edn., vol. 195, no. 5. Pearson, New York (2019)
12. OPGP, Sobre Nós (2021). https://observatorio.apogep.pt/sobre-nos. Accessed 3 Jan 2022

The Relationship Between Digital Literacy and Digital Transformation in Portuguese Local Public Administration: Is There a Need for an Explanatory Model?

José Arnaud[1]([✉]) [iD], Henrique São Mamede[2,3] [iD], and Frederico Branco[2,4] [iD]

[1] Instituto Politécnico do Cávado e do Ave, Barcelos, Portugal
jarnaud@ipca.pt
[2] INESC TEC, Porto, Portugal
jose.mamede@uab.pt, frederico.branco@utad.pt
[3] Universidade Aberta, Lisbon, Portugal
[4] Universidade de Trás-os-Montes e Alto Douro, Vila Real, Portugal

Abstract. We cannot neglect digital literacy because it is undeniable how much technology is part of our lives. Ignoring it and the tools and services it provides us, which greatly facilitate the human experience, is simply a mistake. Recognising the importance of digital literacy, primarily due to the digital transformation in Portugal, it will be necessary to have technological skills to overcome some limitations. Information and Communication Technologies are seen in this environment as a factor that can contribute, on a large scale, to the inclusion of individuals with a digital literacy deficit, both in the Portuguese Local Public Administration and in society in general. The growth of digital transformation causes almost all jobs to need digital skills and participation in society. It takes digitally intelligent employees who know not only to use but also innovate and lead to new technologies because digital transformation may not be successful without that capacity. Thus, it is pertinent to develop, propose and validate an explanatory model that improves the relationship between digital transformation in Portuguese Local Public Administration and the digital literacy of its employees.

Keywords: Digital Literacy · Digital Transformation · Explanatory Models

1 Introduction

Some authors, such as [1], define digital literacy as the ability to locate, organise, understand, evaluate and analyse information using technology. The same authors claim that digital literacy is knowing how to communicate with computers and digital technologies. Walton [2] defines digital literacy as the ability to find, evaluate, use, share and create content using the Internet and Information and Communication Technologies (ICT). The assertive use of digital technologies improves academic, personal and professional development. Based on these authors, we can say that digital literacy is a vital competence in

A. Rocha et al. (Eds.): WorldCIST 2023, LNNS 801, pp. 284–291, 2024.
https://doi.org/10.1007/978-3-031-45648-0_28

the knowledge society. At the same time, it can constitute a barrier to personal development and social integration in case it is absent or underdeveloped in the population. The definition of digital literacy changes according to different authors because new and technological innovations change how people use technology and perform tasks. Digital literacy implies that the user can use the information constructively.

Thus, digital literacy is an essential element, both from the point of view of economic development and from the social point of view, establishing a border between info-included and info-excluded. Castells [3] reinforces this premise, as it states that the reality of info-exclusion goes beyond simple access to the Internet, namely, all the consequences that this access entails in itself and how access is carried out in the case it happens. This means that digital literacy and the respective competencies should not be based and focused on a simple instrumental use but on a perspective framed in a social practice [4].

As we are in the age of digital transformation, the digital empowerment of people is critical. Roberto [5] refer that digital technologies have implemented changes in how we live, learn and work, making it logical and coherent to include new teaching and learning resources on a digital basis in society. Many people may not follow the digital transition's development, which can sometimes imply conflicts. Digitally transforming Portuguese Local Public Administration may imply that people must be prepared and equipped with tools and, in this way, be the main drivers of this transformation. Thus, it is possible to rethink the processes associated with this transformation and use the added value of technology to implement the respective implementation. Thus, it is essential to reflect on the current state of digital skills in our society, more specifically, the digital skills of employees in the Portuguese Local Public Administration, and to propose reforms and structural measures that are the factual basis of sustainability.

Nowadays, digital transformation is a reality in several economic sectors, and in some cases, it is the basis for transforming businesses. The connection of digital technologies between organisations has increased an enormous amount of structured and unstructured data in the field of knowledge management, causing the business transformation, optimising resources and increasing value to those involved. In this context, organisations seek a more significant competitive environment in digital transformation, implying new challenges for knowledge management and supporting transformation.

This article is structured as follows: the introduction in Sect. 1 addresses the theme of digital literacy, digital transformation and their relationship; the purpose and research questions in Sect. 2, where the main objective is focused on being solved and where each review question is specified; Sect. 3 presents the theoretical framework and literature review reporting the exploratory work in preliminary literature review and Sect. 4, the research methodology to be used is presented; Finally, in Sect. 5 the authors conclude that better digital literacy will improve performance and efficiency in organisations.

2   Purpose and Research Questions

A study by [6] closely connects with human capital development. Most managers/decision makers do not perceive the human asset as a driving force but as the most significant barrier imposed by digital transformation, considering the low level of skills and capabilities required by digital transformation. Digital transformation will be a process of changing mindsets regarding the use of technology, as people are at the centre of digital transformation and are responsible for its success. It is, therefore, essential to make people part of the process, assuming them as a critical factor of change; otherwise, people are not part of the solution but part of the problem. Thus, it is essential to carry out digital transformation in Portuguese Local Public Administration, making the country competitive through its central pillar: people.

We intend to verify if there is any relationship between the digital literacy of employees of the Portuguese Local Public Administration and if this influences, in some way, the digital transformation in the organisation, developing an explanatory model that can solve or mitigate this relationship.

2.1   Objective

The objective related to reality is located in the area of digital literacy as a focus of incidence in implementing digital transformation. The problem to be solved is how the level of digital literacy of employees in the Portuguese Local Public Administration conditions impacts the implementation of digital transformation. From the systematic literature review, it will be possible to understand what was discussed about the problem. It will be crucial to know the view of employees of the Portuguese Local Public Administration on the subject, identify employees' limitations in using technologies, and understand the reasons that lead to difficulties in using them.

Thus, the objective will be to develop an explanatory model that allows for mitigating the problems detected, favouring the adoption of measures aimed at combating digital literacy among employees of the Portuguese Local Public Administration.

2.2   Research Questions

Given the previous discussion, the research question will be: Does the digital literacy of employees in the Portuguese Local Public Administration impact digital transformation with efficient and effective models that support this connection? As a sub-question: What is the relationship between digital transformation in Portuguese Local Public Administration and the digital literacy of its employees? The research will focus on improving the digital literacy of employees in the Portuguese Local Public Administration during digital transformation, developing an efficient explanatory model. We end by quoting a paragraph that can be read in the document of the Action Plan for the Digital Transition of Portugal and which is particularly relevant in this context: "The Portugal Digital Action Plan is the engine of transformation of the country, which aims to accelerate Portugal, without leaving anyone behind, through the digital empowerment of people, the digital transformation of companies and the digitisation of the State". [7].

3   Theoretical Framework and Literature Review

It is imperative to define the theoretical framework and the literature review to demonstrate the importance of this research and the problem it intends to solve. A preliminary literature review was carried out in the Google Scholar and Scopus databases published between 2001 and 2022. For the selection and evaluation of the study, an analysis was carried out on 74 articles related to digital literacy, with the transformation and with the Portuguese Local Public Administration. Some articles were discarded due to not being related to the topic. In Fig. 1, we can see the number of articles per year.

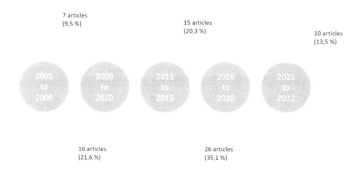

Fig. 1. Articles included in the study

The selection of articles was based on their content, having been categorised by author, title, year of publication, theoretical approach, context and relevant information for future research to contribute to knowledge and research on digital literacy and digital transformation.

In the literature review, it is found that the most popular concept of digital literacy belongs to Paul Gilster. Gilster [8] states that digital literacy requires a profound change in how we view the concept of literacy and that it forces us to rethink how we locate, evaluate, organise, read and write information. Gilster defines digital literacy as: "the ability to understand and use information in multiple formats from a wide variety of sources when it is present via computers." [8].

For [9], the concept of digital literacy is the ability of people to perform digital actions in the field of work, education, leisure and other aspects of everyday life. Digital literacy will depend on the personal situation, a lifelong learning process. In this sense, Martin presents three levels for developing digital literacy: digital competence, digital use and digital transformation (see Fig. 2).

Figure 2, [9] addresses digital literacy only at levels 2 and 3. Level 1, digital competence, is a requirement for a predecessor of digital literacy but cannot be described as digital literacy [10].

According to [11], all the people in an organisation related to a digital transformation are fundamental to how this transformation takes place, whether due to the simple need to acquire technical skills or the desirable involvement in taking advantage of the full potential of technologies adopted so that they can participate in the inevitable organisational transformation in these processes. The authors also refer to the problem of an eventual division between older collaborators and less integration in the changes in contrast to the new ones, more adapted to the technologies being adopted.

In recent years we have witnessed growing technological development, and organisations that do not keep up with this development may experience moments of uncertainty,

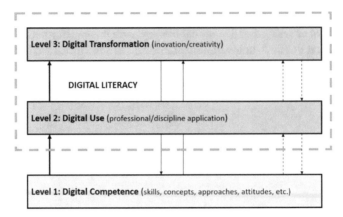

Fig. 2. Digital Literacy Levels (adapted from [9])

which will drive digital transformation [12, 13]. The definition of digital transformation can be quite broad, ranging from the profound transformation of organisational activities to the advanced use of digital technologies. However, there are some common aspects that organisations should consider when implementing digital transformation: operational processes, customer experience and business models that allow data reprogramming and standardisation, as well as decentralisation and process autonomy [14]. Other definitions emphasise the paradigm shift in all aspects of life, making change happen due to the introduction of technology. This transformation accelerates the cultural shift that requires new skills, providing people with new skills and knowledge [15].

Thus, and more specifically, digital transformation can be defined as "the most profound and accelerating transformation for business activities, processes, competencies, and models to leverage the changes of digital technology and their impact in a strategic and prioritised way" (p. 723) [13]. The focus of digital transformation is on changes within organisations and their transformations at different levels, including strategy, people, leadership culture and technology [16]. People, processes and technology are the three most essential vectors in digital transformation, but if the digital literacy of the people involved is scarce, it could jeopardise this transformation. The research will focus on people, specifically on the digital literacy of employees in the Portuguese Local Public Administration.

4 Research Methodology

The Design Science Research (DSR) methodology will support the research work. The option for DSR is because it is an adequate methodology for developing research of a technological nature, proposing the production and development of artefacts and the resolution of problems through them [17]. The investigation will result in the elaboration of an explanatory model, being the DSR the most appropriate methodology, as it supports the investigation intuitively and practically. However, it should be noted that the result will not focus on the presentation or implementation of a product or service but on something conceptual, such as an explanatory model [18]. Such a model constitutes the artefact foreseen in the DSR.

The DSR methodology can be used and applied in research on digital transformation. The knowledge needed to research the subject should involve paradigms in the science of behaviour and its design, which refers to the modelling of the artefact that will be developed. Behavioural science approaches research by developing theories and designs by developing and evaluating what was created to achieve the objective [19]. Thus, we can consider DSR as a research method that involves creating and improving something innovative that comes from a specific problem [17]. Since this research aims to find an efficient and effective explanatory model to improve the relationship between digital transformation and digital literacy, the DSR fits in intending to support the work to be carried out adequately.

Communication with public entities dealing with digital transformation and their employees will be crucial to understanding how this relationship can improve, paying attention to even the most minor details. The version of the DSR methodology will be the one proposed by Peffers [20], represented in Table 1. From this process, an explanatory model will be developed that will then be validated in previously selected local public entities. This process is developed based on six fundamental steps.

Table 1. Design Science Research applied to this work (adapted from [20])

Activity	Description
1. Identify the problem and motivation	Identify the focus of the study intended to be carried out and its value in responding to the problem. The resources needed in this step will be the problem's state of the art and the solution's relevance. To this end, it will be necessary to carry out a survey of state of the art (using an appropriate protocol) regarding digital transformation and digital literacy in Portuguese Local Public Administration and explanatory models that explain this relationship
2. Define the objectives for the solution	Specify the requirements (quantitative and/or qualitative) to present a possible solution and how it can be implemented. The resources needed at this stage will be the state of the art of the problem and the knowledge of possible solutions previously presented. At this stage, it is also essential to know already implemented solutions, if any, to understand their effectiveness and serve as a comparison
3. Design and development	Focus on the search for knowledge for the construction of the final artefact, which can be achieved by separating the main problems encountered during the research into something simple. The necessary resources understand the knowledge of the theory that can give rise to the solution. The artefact will be built at this stage, from its definition to its design and development. The artefact to be developed will consist of constructing an explanatory model

(*continued*)

Table 1. (*continued*)

Activity	Description
4. Demonstration	Demonstrate that the developed artefact solves all or part of the problem through a case study. Required resources include knowledge of how to use the artefact to solve the problem. In this phase, interviews/surveys will be carried out with employees of the Portuguese Local Public Administration, more specifically with employees from different municipalities, to verify and validate the artefact developed
5. Evaluation	Validate whether the results are efficient by focusing on methods that allow the feasibility of the final solution to be determined. If it is necessary to improve the artefact, return to step 3 (design and development) or 4 (demonstration). At this stage, it is essential to enhance the understanding of the research carried out, as it allows those who have access to the artefact to understand and learn from what is presented. Construction activities should be evaluated considering the artefact's applicability and generalisation [17]
6. Communication	Presenting the final solution through writing a doctoral thesis. It is vital to disseminate the results obtained to researchers and professionals in the field, and the work must be published in journals or conferences related to the research area. This way, publications will be submitted to conferences and scientific journals for dissemination throughout the work

This view of DSR, based on Peffers, is simple and meets the investigation needs for this case, providing a practical and problem-oriented process, as its steps range from problem identification to its demonstration and communication. Something useful will be built, as this is the purpose of the DSR methodology.

5   Conclusion

With this study, we can contribute to the digital literacy of Portuguese Local Public Administration employees by improving the relationship with digital transformation. This contribution would mitigate the current digital literacy deficit in the Portuguese Local Public Administration.

It will be verified that the digital transformation process in the Portuguese Local Public Administration is a reality and that the improvement of the digital literacy of its employees will lead to better performance and organisational efficiency. Better digital knowledge is an intangible and precious asset in an organisation.

Thus, it is concluded that using technology in conjunction with digital literacy improves efficiency and facilitates the competence and innovation of Portuguese Local Public Administration. We also add that having excellent digital literacy leverages the skills and knowledge of Portuguese Local Public Administration employees with the support of ICT.

References

1. Tabusum, S., Saleem, A., Sadik, M.: Digital literacy awareness among arts and science college students in tiruvallur district: a study. IJMSR **2**(4), 61–67 (2014)
2. Walton, G.: Digital literacy: establishing the boundaries and identifying the partners. New Rev. Acad. Librariansh **22**(1), 1–4 (2016)
3. Castells, M.: A Sociedade em Rede. Lisboa: Fundação Calouste Gulbenkian (2012)
4. Freitas, M.: Letramento digital e a formação de professores. Educação em Revista **26**(3), 335–352 (2010)
5. Roberto, M., Fidalgo, A., Buckingham, D.: De que falamos quando falamos de infoexclusão e literacia digital? Perspetivas dos nativos digitais. Observatório (OBS*) J. **9**(1), 43–45 (2015)
6. Herceg, I., Kuc, V., Mijuskovic, V., Herceg, T.: Challenges and driving forces for Industry 4.0 implementation sustainability, vol. 12, no. 10, pp. 1–22 (2020)
7. Portugal Digital – Plano de Ação para a Transição Digital de Portugal (2020). https://www.portugal.gov.pt/gc22/portugal-digital/plano-de-acao-para-a-transicao-digital-pdf.aspx. Accessed 24 Jan 2022
8. Gilster, P.: Digital literacy. Wiley Computer Pub, New York (1997)
9. Martin, A.: DigEuLit – a European Framework for Digital Literacy: a Progress Report. Journal of eLiteracy, Vol 2. https://citeseerx.ist.psu.edu/viewdoc/download?doi=10.1.1.469.1923&rep=rep1&type=pdf. Accessed 24 Jan 2022
10. Martin, A.: Digital Literacy and the "Digital Society". In: Lankshear, C., Knobel, M., (eds.) Digital literacies: concepts, policies and practices, pp. 151–176. Nova Iorque: Peter Lang (2008)
11. Nadkarni, S., Prügl, R.: Digital transformation: a review, synthesis and opportunities for future research. Manage. Rev. Quar. **71**, 233–341 (2020)
12. El Hilali, W., El Manouar, A.: Towards a sustainable world through a SMART digital transformation. In: Proceedings of the second international conference on networking, information systems and security NISS 2019. (2019)
13. Hamidi, S.R., Aziz, A.A., Shuhidan, S.M., Aziz, A. A., Mokhsin, M.: SMEs maturity model assessment of IR4.0 digital transformation. In: Proceedings of the 7th international conference on kansei engineering and emotion research 2018 KEER, pp. 721–732 (2018)
14. Junge, A.: Digital transformation technologies as an enabler for sustainable logistics and supply chain processes – an exploratory framework. Brazilian J. Oper. Product. Manage. **16**, 462–472 (2019)
15. Jeladze, E., Pata, K.: Smart, digitally enhanced learning ecosystems: bottlenecks to sustainability in Georgia. Sustainability **10**(8), 2672 (2018)
16. Heilig, L., Lala-ruiz, E., Voß, S.: Digital transformation in maritime ports: analysis and a game-theoretic framework. Genomics **18**(2–3), 227–254 (2017)
17. Hevner, A.R., March, S.T., Park, J., Ram, S.: Design science in information system research. MIS Q. **28**(1), 75–105 (2004)
18. Gregor, S., Hevner, A.R.: Positioning and presenting design science research for maximum impact. MIS Q. **37**(2), 337–355 (2013)
19. Hevner, A. R.: A three cycle view of design science research. Scandinavian J. Inform. Syst. **19**(2), Article 4 (2007)
20. Peffers, K., Tuunanen, T., Rothenberger, M.A., Chatterjee, S.: A design science research methodology for information systems research. J. Manag. Inf. Syst. **24**(3), 45–77 (2007)

Assessing Poor Adoption of the eID in Germany

Philipp Liesbrock[1] and Eriks Sneiders[2(✉)]

[1] German Federal Foreign Office, Berlin, Germany
[2] Department of Computer and Systems Sciences, Stockholm University, Stockholm, Sweden
eriks@dsv.su.se

Abstract. Personal e-services guarded by eID remain foreign to public organizations and citizens in many countries. In this context, Germany is an interesting case to study. While much of the e-government adoption research focuses on developing countries, Germany is a technologically advanced and wealthy country, close to the Nordic countries with widely accepted eIDs for private and public e-services. Still, Germany experiences rather poor adoption of its eID. Because most public services in Germany are provided through municipalities, we asked the municipalities why they rarely provide e-services with an option for citizens to identify themselves with the eID. The result is ten reasons of poor adoption of e-services and the eID, and a background story.

Keywords: eID · Electronic Identification and Authorization · Adoption of eID · Adoption of E-services

1 Introduction

Electronic identification and authentication (eID for short) is essential for secure and personal e-services. eID may be physical, such as a chip in an ID card and a card reader attached to the computer, or a mobile app. Popularity of eID follows the availability of e-services that require eID [1] because "with a lack of a digital service context, the eID becomes completely useless" [2]. Who develops and promotes eID also matters. In countries where eID is developed and driven by public organizations, the number of use cases remains low [3], and so does the popularity of eID. The Nordic countries, on the contrary, have popular privately run eIDs. BankID in Sweden, which is owned by a number of banks [4], was used by 98.7% of the population in the age group 21–50 to unlock public and private e-services in 2019 [5].

In the context of eID, Germany is an interesting country to study. Germany has the fourth largest economy in the world and a high standard of living. Germany hosts the headquarters of world-wide known companies such as Volkswagen, SAP, Bosch and BioNTech. Germany pursues a federal High-Tech Strategy [6]. German eID is included in the ID card and is available to every citizen. Nevertheless, only 6% of German citizens had made use of their eID in 2019 [7].

There exists research regarding adoption and non-adoption of public e-services by German citizens [8], therefore we explore the other side – why German public sector

A. Rocha et al. (Eds.): WorldCIST 2023, LNNS 801, pp. 292–301, 2024.
https://doi.org/10.1007/978-3-031-45648-0_29

organizations (PSOs) rarely provide e-services with an option for citizens to identify themselves with eID. It is important to understand the stance of PSOs because PSOs, not the citizens, provide public services and decide to use or not to use eID.

In Germany, public services are delivered to citizens mostly through municipalities [9], which means that there are many small local public service providers as opposed to a few large ones that cover entire Germany or a federal state. Therefore PSOs in this study are municipalities, and the results reflect the point of view of the municipalities.

Further in this paper, the next section describes our research method. Section 3 presents the results of the interviews with the municipalities. Section 4 summarizes the reasons of poor adoption of e-services and eID in Germany, as well as verifies these reasons with relevant previous research. Section 5 concludes the paper.

2   Method

In order to find out why German PSOs, primarily municipalities, rarely provide eID-powered public e-services, we interviewed knowledgeable employees at the municipalities. In order to make a representative sample of municipalities, purposive sampling was applied first. Key regions of Germany were identified, diverse in culture, economy, history, and political background. Municipalities in the metropolitan areas of Berlin, Hamburg, Munich, Leipzig and Cologne were targeted as urban areas with high population density. Municipalities in North-Rhine-Westphalia were targeted as rural, sparsely populated conservative areas [10], while municipalities in Mecklenburg-West Pomerania were targeted as rural, sparsely populated but left-leaning areas [11]. Municipalities in rural Saxony were targeted because extreme political right had become the leading political force there [11]. Municipalities in Bavaria were targeted as the wealthy ones [12], whereas those in the western coal region as those among the poorest ones [13].

After the purposive sampling, snowball sampling was used where respondents suggested new potential respondents. Eventually 212 municipalities were contacted; only seven agreed to be interviewed. Three interviewees were from Bavaria, one from an urban area, one from the western coal region, one from rural North-Rhine-Westphalia, and one from rural Saxony. Although the number of respondents is not high, their backgrounds are different, therefore we consider the sample representative for Germany.

The semi-structured interviews were conducted online, mostly via Zoom, in German language. All interviews were recorded and converted into text by a voice transcription tool, then translated into English by a translation tool, then manually checked and corrected where necessary. Thematic analysis was applied to the interview texts. The interview guide, the codes of the thematic analysis, and detailed description of the data collection and analysis process are described in a report [14].

3   PSOs' Opinion About E-services and the eID

The analysis of the interviews has identified four themes, presented in Sects. 3.1–3.4.

3.1 eID Drivers and Constraints

The attitudes of the respondents towards implementing the eID are categorized into subthemes in the following subsections.

Legal Requirements Dictate What and How E-services are Delivered. All participants highlighted the importance of legal requirements and potential changes in the law regarding e-service delivery and the implementation of the eID. In this context, the German Online Access Law (OZG) is a central piece of jurisdiction. According to the OZG, by the end of 2022 at the latest, the federal government, the federal states, and the municipalities must offer a predefined number of public services online via administrative portals and connect them to build a portal network. The OZG, therefore, functions as a change management driver and mandates the implementation of new solutions. One interviewee summarizes how the OZG is driving the digitalization of public services: "Through the Online Access Act, [...] the direction is quite clearly [...]: What is possible on the Internet should also be made available to citizens."

There is also a strong connection between the legal complexity of services and the complexity of digitalizing the services. A public service whose legal basis is straightforward and well-formulated is easier to digitize than a public service with an extensive legal background and various conditions and possible outcomes. Therefore, simplifying the respective law and making it more accessible is suited to achieve more digitalization of public services. "The simpler and clearer the law is formulated, the easier it is to digitize it. [...] our credo is that at the federal level as well as at the state and local level, the law simply has to be simplified so that it can be digitalized throughout."

The legal complexity of certain services does not vanish with the digitalization of the services. Instead, the citizens are confronted with a complexity that the administration absorbs in analogue applications. This explains why municipalities are reluctant to digitize services with a more complex legal background. They fear that citizens will struggle with the application procedure and make incorrect submissions which the administration must sort out, which increases the workload. This phenomenon was stressed by two interviewees: "The fact really is, online applications lead to more work in the administration, because they simply contain more errors. But that is not the fault of the citizens, it is simply our complex legal system."

However, the legal basis of public services is also an opportunity to increase the availability of public e-services further. A legal mandate requiring municipalities to offer certain e-services removes the discretional element. This kind of "legally forced digitalization" will ultimately generate more e-services for citizens.

Laws and guidelines do also enhance the inter-organizational cooperation between PSOs. In Germany, the OZG is guided by the so-called "Einer für Alle" (EfA) – one for all – principle. Each PSO should digitalize and share its services so that other PSOs do not have to create the web infrastructure from scratch. This reduces expenses, time, and resource usage. The fundamental idea of EfA is that PSOs collaborate and share the workload rather than individually re-develop each e-service.

The OZG recognizes several security levels for e-services, where only the highest security level legally requires the eID as an identification method. Most e-services do not require the eID and can be conducted with an unverified user account using an email address and a password.

One interviewee, who would only implement the eID for public e-services if it is legally mandated, does, however, criticize the fact that the eID was not made a legal obligation for certain private services, for example, opening a bank account or registering a mobile number: "The eID is too little known. It is too little used. In Sweden, in Scandinavia, no one would give such an answer because it is used every day. The Austrians use their card every day. In Germany, we never managed to take that step."

Role of Civil Servants. In addition to the legal requirements, individuals in the administration have been identified as both drivers and constraints for the implementation of e-service delivery in conjunction with the eID. Individuals can be the deciding factor for the success or failure of eID and e-services. In this context, the PSO management and leadership, and their attitude towards e-services and the eID play an essential role in motivating employees through setting goals and allocating resources. The lack of motivated employees will result in the lack of e-services because engaged civil servants are those who create such services.

Furthermore, one must consider the different attitudes and beliefs that may co-exist. For example, the leadership of a municipality may be very supportive of the eID implementation for e-services. At the same time, the desk officers who deal with customers daily may see no value in it. "If you were to ask the head of our authority, […] he would immediately say that it is good and important. […] Many case workers still see this as an additional burden or have reservations in general."

Case workers do not see any additional value in e-services for their work. On the contrary, e-services initially translate into an increased workload as digitizing public services is complex and breaks well-established routines. Case workers fear that their job could become obsolete once all services are digitized. They aim to have a convenient and stress-free work environment. The eID, which only complicates matters in their eyes, is therefore not desired or promoted.

Because German law dictates that the eID is only mandatory for e-services with the highest security requirements, PSOs have a certain level of discretion when implementing the eID for e-services. Therefore, it is crucial to convince the regular municipal desk officers of the potential eID benefits. Still, even the people who work with public e-services daily rarely use their eIDs: "I have the ID card app on my mobile phone. I need things exactly every two or three years when I apply for a new certificate for Elster and then even I have to think about what the PIN was again? That is too little, that is far too little."

Administrative Resources and Structures is another subtheme that helps explain why German PSOs rarely provide public e-services with eID. This subtheme goes beyond the influence of specific individuals in the administration and focuses on the broader existing organizational framework in which the municipalities in Germany operate.

One structural problem that the interviews revealed is the lack of eID marketing. There are no resources allocated for marketing of the eID, no one is explicitly charged with the duty to promote the eID for e-services: "We have no distribution. We have the best product with all the background information, but we don't have a single salesperson, neither at federal, state or municipal level."

Another structural problem concerning e-services is the difficulty of transferring administrative knowledge into accessible digital public services. Face-to-face, the office workers are able to steer the application in the right direction; online, citizens are expected to submit a completely digital application with all the relevant information provided: "The colleagues who sit there [...] know exactly which questions to ask, where to ask the questions. [...] When I now put the citizens in front of the online applications, we notice more and more how difficult it is to translate our specialist know-how that we have in the administrations, and what the legislator wants, into the language that is close to that of the citizens', so understandable language, that there are no mistakes."

Not all municipalities can develop their e-services because they lack the required human and financial resources. Nonetheless, more substantial involvement of private sector solutions is not considered. Instead, resource-scarce municipalities depend on the solutions that other PSOs have already created. Consequently, such resource-scarce municipalities can only copy what others have developed and cannot decide whether the eID should be used or not. Therefore, more capable PSOs strongly influence the landscape of e-service and eID.

If a department does not have e-services available, it does not necessarily mean they are uninterested or unwilling. Lack of personnel or financial resources is still an important aspect to consider.

3.2 PSOs' Interpretation of Citizens' Perceptions

This theme explores how PSOs interpret the feelings and attitudes of citizens towards e-services and the eID as an identification and authentication method. It is important to stress that this theme only reflects PSOs' perception of their citizens attitudes, not necessarily the actual attitudes that the citizens have. However, PSOs' perceptions are significant because they shape which e-services and how e-services are made available to the citizens.

PSOs have an impression that most citizens would use e-services with the eID but lack the knowledge to do so. The eID is not self-explanatory, the citizens need a short hands-on introduction what and how they can do with the eID. A good usable application would be self-explanatory and would not require additional guidance. Still, no one cares to make the eID more user-friendly or to explain how to use it.

PSOs believe that citizens perceive the eID as a hurdle. Citizens associate the eID with a particular data-sensitive procedure they rather want to avoid, especially because it is rarely needed. "My approach is to avoid it [the eID] as much as possible because it is a hurdle that I want to skip. And [...] the spread is not particularly high. [...] Many don't have it. Many find it too much of a hassle to set up and get used to it. Many have an increased inhibition threshold, because they say: 'Oh, that's somehow highly sensitive, my data is there and so on. I don't really trust that.'" Citizens do not use the eID regularly, lack experience with the eID, and do not trust it. The eID remains a foreign IT artefact to them. Consequently, as citizens do not trust and do not like the eID, PSOs do not enforce it either.

The eID is not user friendly. Citizens are familiar with authentication solutions from the private sector that deliver a smooth and hassle-free user experience, and expect the same level of usability from the public sector solutions, and are easily frustrated if the public solutions have deficiencies: "Administrative services are compared to online shopping. So, it has to be like Amazon." Even the eID mobile app is too cumbersome.

Therefore PSOs implement the eID mostly if it is mandated by law: "We try to keep it as low-threshold as possible. We have clarified with the lawyer that we would only use the eID function in cases where written form is really required."

3.3 Role of Alternative Identification and Authentication Methods

Even though the eID is the official government solution for secure identification and authentication online, there still exist alternatives for citizens. Previously, citizens could perform basic authentication for public services in person, via telephone, or email by just mentioning their alleged name. Today, this low-threshold identification and authentication still persists: "Citizens can call and make the application this way. And then we chose an electronic way, according to the motto: an email address is enough for us […]. We try to exclude the identification if it is not required by law." In most cases, what PSOs consider a low-security authentication equals to "no security" authentication. PSOs show a high level of trust in their citizens and do not take cybersecurity risks seriously. Potential fraud attempts are not considered an issue.

Furthermore, the eID being too user-unfriendly facilitates alternative authentication methods, such as username and password, which are considered more straightforward to use than the eID in its current shape.

Especially in smaller municipalities, it is still effortless for citizens to get the service by just coming into the office. There is no need for an online process. Many citizens also want personal contact with municipal employees. Direct face-to-face communication between citizens and PSOs is still a norm in Germany [15]. Offering a variety of alternatives to the eID is a "multi-channel strategy": citizens can make use of the eID, they can come in person or, for most services, simply use regular email.

3.4 Involvement of the Financial Sector

The respondents have repeatedly highlighted how adoption of the eID by banks, which are private organizations, would help boost its acceptance by citizens. The banking sector is of high relevance for many citizens, they deal with their financials frequently. It is a highly regulated sector where reliable fraud-prevention IT solutions are needed.

PSOs think that the eID, if regularly used for online banking login procedures, would make its occasional use for public e-services more accepted: "Austrians use their card every day. In Germany, we never managed to take that step. The exception in the Money Laundering Act was created for the banks."

It is nevertheless not entirely clear why private banks should use the eID for online banking login procedures if they have their own successful solutions: "But now online banking works, I have my app and code and such. I don't think they would change that and replace it with the eID, why should they do that?".

Instead, public e-services could adopt the login procedure from banks, like BankID is adopted by public e-services in Sweden, Norway, and Finland [4].

4   Reasons of Poor Adoption of E-services and eID in Germany

Although this research targets the German eID, it is difficult to separate poor adoption of the eID from poor adoption of the e-services. In Table 1, we summarize the reasons of the poor adoption according to seven German PSOs. The third column references the section where the reason is discussed, the fourth column shows the number of respondents that have expressed the reason.

Table 1. Reasons of poor adoption of e-services and eID in Germany.

	Reason	Section heading	Num
R1	PSOs receive negative feedback from citizens – eID is too difficult to understand and use	PSOs' Interpretation of Citizens' Perceptions	5
R2	Because eID is difficult to use, PSOs prefer other means of identification and authentication, such as username and password	Role of Alternative Identification and Authentication Methods	6
R3	There are not many e-services that require eID, therefore citizens rarely use them, and find using eID a hurdle	PSOs' Interpretation of Citizens' Perceptions	3
R4	PSOs do not assess cybersecurity risks and see no value in using strong identification and authentications methods	Role of Alternative Identification and Authentication Methods	2
R5	PSOs do not feel responsible for the success of eID, and they do not know anyone who is responsible	Administrative Resources and Structures Role of Civil Servants	5
R6	PSOs consider digitalization of public services as additional service offering, not as evolution of the services or a replacement to face-to-face service delivery	Role of Alternative Identification and Authentication Methods	3
R7	PSOs think that many public services are too difficult to digitalize, therefore they prefer face-to-face meetings with citizens	Legal Requirements Dictate What and How E-services are Delivered	5
R8	PSOs think that deploying e-services will make their work process more difficult	eID Drivers and Constraints	3
R9	Smaller PSOs lack financial and human resources to develop their own e-services	Administrative Resources and Structures	4
R10	PSOs are not used to co-operating with private suppliers and do not consider a private-sector delivered eID	Administrative Resources and Structures	2

Tsap et al. [16] have explored 39 publications and established twelve categories of eID acceptance factors from the citizens' point of view. Table 2 shows seven categories that are linked to the reasons in Table 1. The other five categories do not seem closely related to the reasons in Table 1.

Table 2. Categories by Tsap et al. [16] linked to the reasons from Table 1.

Category	Reasons
Ease of use	R1, R2, and R3
Complexity as "a difficult to understand mechanism of the system"	R1 and R3
Awareness as "seeing reason and purpose", "knowing how to use it", "comprehending"	R1 and R3
Functionality as "usefulness"	R3 and R4
Trust and *privacy concerns* stand close and interconnected	R3, R4, and even R10, where the latter is distrust and aversion of PSOs to co-operation with private e-service providers
Cultural and historical factors	R4, R6, and R10

Distel [8] has investigated adoption and non-adoption of public e-services in Germany from the citizens' point of view. The following factors facilitate citizens' inclination to use public e-services: (i) trust in the administration, (ii) perceived technical competence, and (iii) expectation of increased efficiency. We think that only "trust in the administration" is somewhat related to R3. After all, the reasons in Table 1 explain poor adoption of e-services, not high adoption levels. As to non-adoption of public e-services, according to Distel [8], (i) most citizens do not perceive any need for e-services because the conventional way seems easier and more convenient, which relates to R6 and R7. Furthermore, (ii) citizens often need personal consultation, therefore they skip e-services and visit the office instead, which relates to R7 and R8.

Eight of the ten reasons in Table 1 have been confirmed by [16] and [8]. Two reasons – R5 and R9 – have not been confirmed because R5 and R9 are specific to municipalities, whereas [16] and [8] have explored acceptance of eID and (non-) adoption of e-services from the citizens' point of view.

5 Conclusions

The study investigates why German public sector organizations (PSOs), in this case municipalities, rarely provide e-services with an option for citizens to identify themselves with the eID. Table 1 summarizes the reasons. The research results are descriptive in nature, we do not make any claims who and how should act upon these reasons.

In Germany, e-services are not considered useful because it is easy to walk into the local government office and manage the contact in the old-fashioned way. Many citizens prefer face-to-face communication, especially if they need personal consultation before they submit an application. PSOs think that many public services are too difficult to digitalize because the underlying laws are too complex. Furthermore, PSOs think that e-services create more work for case workers because of too many errors in online applications that the citizens must fill and submit on their own; face-to-face expert guidance through the application allows applying professional knowledge on the spot.

The eID is not considered useful either. PSOs have high level of trust in their citizens and disregard the risk of fraud; to state the name in an email is enough. Citizens do not like the current implementation of the eID because it is user-unfriendly, therefore PSOs avoid eID as much as possible. There exist alternative identification and authentication methods, such as username and password, if any ID security is needed at all.

Another problem of the eID is lack of marketing, lack of ownership. No one cares to create more use cases for the eID, not one cares to make it more user-friendly. A possible solution could be a private institution responsible for the eID development and promotion within a public-private partnership. However, PSOs seem to be reluctant to take this path. The eID is a data-sensitive artefact with heightened scrutiny. There are enough precedents in Germany where private companies operate in sensitive areas such as airport security or military equipment. Therefore, a general distrust and aversion towards private players can explain why PSOs do not involve a private sector entity to reorganize, own, and promote the eID.

It is not clear how long PSOs will be able to avoid cybersecurity risks. And when they cannot avoid the risks anymore, there should be a user-friendly and generally accepted eID in place. Most probably such eID will appear in a public-private partnership, perhaps in a partnership with banks that already have well-functioning eID routines in place [4]. Any bank can serve as an identity guarantee for its customers. It is not clear, however, who is going to put effort into developing such a partnership.

PSOs have suggested that citizens would like to use public e-services. Especially in the era of smartphones, that is a natural choice. It is not clear, however, which public services should stay face-to-face and which should move online.

Currently, the majority of public services are provided to citizens through local municipalities. We believe that establishing uniform e-service solutions that cover entire Germany or a federal state, rather than only one of the 11 014 municipalities, would make a huge cumulative efficiency improvement, as well as cost saving. And such a change requires a good eID.

References

1. Räckers, M., Hofmann, S., Becker, J.: The influence of social context and targeted communication on e-Government service adoption. In: Wimmer, M.A., Janssen, M., Scholl, H.J. (eds.) EGOV 2013. LNCS, vol. 8074, pp. 298–309. Springer, Heidelberg (2013). https://doi.org/10.1007/978-3-642-40358-3_25
2. Söderström, F.: Introducing public sector eIDs: the power of actors' translations and institutional barriers. Doctoral dissertation, Linköping University Electronic Press (2016)
3. Arora, S.: National e-ID card schemes: a European overview. Inform. Secur. Tech. Rep. **13**(2), 46–53 (2008)
4. Grönlund, Å.: Electronic identity management in Sweden: governance of a market approach. Identity Inform. Soc. **3**, 195–211 (2010)
5. Wemnell, M. Statistik BankID – användning och innehav. https://www.bankid.com/assets/bankid/stats/2020/statistik-2020-01.pdf. visited on 11 November 2022
6. High-Tech Strategy 2025. https://www.bmbf.de/bmbf/en/research/hightech-and-innovation/high-tech-strategy-2025/high-tech-strategy-2025.html. visited on 25 October 2022
7. eGovernment Monitor 2019. https://initiatived21.de/publikationen/egovernment-monitor-2019/. visited on 25 October 2022

8. Distel, B.: Assessing citizens' non-adoption of public e-services in Germany. Inform. Polity **25**(3), 339–360 (2020)
9. Kramer, J.: Local government and city states in Germany. In: The place and role of local government in federal systems, pp. 83–94. Konrad-Adenauer-Stiftung, Johannesburg (2005)
10. Briesen, D.: Gesellschafts-und Wirtschaftsgeschichte Rheinlands und Westfalens, vol. 9. Kohlhammer (1995)
11. Olsen, J.: The left party and the AfD: populist competitors in Eastern Germany. Ger. Polit. Soc. **36**(1), 70–83 (2018)
12. Goodhart, D.: If only we could be more like Bavaria. https://unherd.com/2019/11/if-only-we-could-be-more-like-bavaria/. visited on 11 November 2022
13. Reitzenstein, A., et al.: Structural change in coal regions as a process of economic and social-ecological transition: lessons learnt from structural change processes in Germany. German Environment Agency (2022)
14. Liesbrock, P.: The giant is lagging behind. How the German electronic ID fails to reap its potential. Master thesis. Stockholm University (2022). https://doi.org/10.13140/RG.2.2.32159.64169
15. Mergel, I.: Digital transformation of the German state. In: Kuhlmann, S., Proeller, I., Schimanke, D., Ziekow, J. (eds.) Public administration in Germany. GPM, pp. 331–355. Springer, Cham (2021). https://doi.org/10.1007/978-3-030-53697-8_19
16. Tsap, V., Pappel, I., Draheim, D.: Factors affecting e-ID public acceptance: a literature review. In: Kő, A., Francesconi, E., Anderst-Kotsis, G., Tjoa, A.M., Khalil, I. (eds.) EGOVIS 2019. LNCS, vol. 11709, pp. 176–188. Springer, Cham (2019). https://doi.org/10.1007/978-3-030-27523-5_13

Exploring the Influence of Technology Use on Teleworking Benefits: A Gender Multigroup Analysis

Arielle Ornela Ndassi Teutio[1] and Jean Robert Kala Kamdjoug[2(✉)]

[1] GRIAGES, Faculty of Social Sciences and Management, Catholic University of Central Africa, 11628 Yaoundé, Cameroon
[2] ESSCA School of Management, 1 Rue Joseph Lakanal, 49003 Angers, France
`jean-robert.kala-kamdjoug@essca.fr`

Abstract. The contribution of technology in the context of the COVID-19 crisis has had a reasonably disruptive effect on the organization of work, creating a new generation of employees willing to telework more. However, this trend is also a concern for gender inequality in the workplace that existed before the crisis. This study investigates the gender gap in the influence of technology use on job autonomy, job satisfaction, well-being, and intention to continue teleworking. A sample of 500 teleworkers, 250 men and 250 women, was tested using the PLS-SEM method. The results show a gender gap in the influence of technology use on job satisfaction, and in the effect of job autonomy on the intention to continue teleworking. Technology use had more influence on job satisfaction for women, while well-being had more influence on job satisfaction for men. However, technology use did not affect the intention to continue teleworking for both genders, as did job autonomy for men. Women show a higher intent to continue teleworking compared to men.

Keywords: Teleworking · gender · remote working · technology use · digitalization

1 Introduction

Since the beginning of the COVID-19 pandemic, and the implementation of the containment measures, we are witnessing a way of working that was not widespread enough and known by all [1]. Digital technologies have played a vital role in the COVID-19 response; the same goes for teleworking. Teleworking is now booming in organizations and tends to be the dominant way of working in the future. For its magnitude, teleworking has received attention in the recent literature, and the need to thoroughly study its effects shows there are future opportunities [2]. Although the issue of gender equality in telework has been addressed [3], very few studies have focused on the contribution of technology use to gender differences in telework [4]. The present study seeks to fill this gap and contribute to the existing literature by examining the gender difference in the influence of technology use on the intention to continue teleworking and the tele-working benefits, which in this study are job autonomy, job satisfaction, and well-being. Regarding gender, we are considering people who are biologically born (sex) male or female. The paper is divided as follows: Sect. 2 presents the conceptual framework of the research and the hypotheses development. Section 3 describes the methodology used. Section 4 presents the results. Section 5 highlights the discussions and implications, and finally, a conclusion is given in Sect. 6.

2 Conceptual Framework and Hypotheses Development

The model presented in Fig. 1 highlights three constructs validated by the literature as the significant benefits of telework for both female and male employees. These factors are determinants of teleworker behavior and work performance. The intention to continue teleworking is also highlighted to assess whether the con-structs studied influence the willingness to continue.

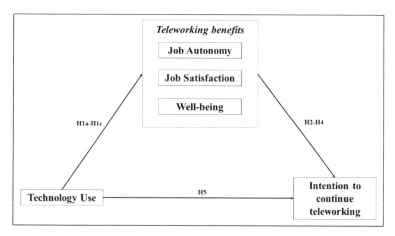

Fig. 1. Research model

Job autonomy in this study refers to the extent to which teleworking offers substantial freedom, independence, and discretion to the teleworker in planning work and determin-ing the procedures to perform it [5]. To this end, telework technologies offer improved tools to support teleworkers in accomplishing their tasks. Thus, they aim to help the employee feel personally responsible for their work's results and take the initiative and make their own decisions easily. The literature on work establishes that autonomy allows teleworkers more flexibility and a better ability to reorganize their work according to their preference. From this perspective, the higher the perceived degree of independence, the more willing employees are to continue teleworking [6]. Further, depending on their commitment to managing family life, women may sometimes have a lower level of autonomy than men [7]. We, therefore, put forward the following hypotheses:

H1a: There is a gender gap in the positive influence of technology use on job autonomy.

H2: There is a gender gap in the positive influence of job autonomy on the intention to continue teleworking.

Job satisfaction refers to the teleworker's pleasant emotional state that results from the evaluation of their work about the achievement of their work values [8]. This evaluation can be made through the teleworker's positive or negative feelings about their work. The literature shows that women are more satisfied than men regarding interpersonal relationships, working conditions, and fulfillment [9]. However, other studies show that the gender difference is insignificant [10]. In this study, we examine the role played by technology in providing teleworkers with a positive feeling about the expectations they have of their work and the performance they must achieve. Therefore, we hypothesize the following:

H1b: There is a gender gap in the positive influence of technology use on job satisfaction.

H3: There is a gender gap in the positive influence of job satisfaction on intention to continue teleworking.

Well-Being is a whole set of positive perceptions about the working conditions that allow the employee to flourish in the telework environment [11]. The literature identifies work-family conflict as the main barrier to teleworkers' well-being [12]. When stress levels are high, it is difficult for them to organize their work time. To this end, well-being measures the quality of life of teleworkers. It has been shown that women experience less well-being than men because they are the most affected by work-family conflicts [13]. However, multiple features of new technologies assist employees in increasing the impact on the well-being and organization of teleworkers. We, therefore, hypothesize the following:

H1c: There is a gender gap in the positive influence of technology use on well-being.

H4: There is a gender gap in the positive influence of well-being on intention to continue teleworking.

The advancement of technological functionalities has reshaped the practice of telework. The COVID-19 crisis has contributed to a more favorable acceptance of telework by managers and employees. The usual working methods and processes have been changed, making the use of technology unavoidable. Technology has taken a prominent place, whether for managing meetings, signing documents, managing teams, communicating, working with customers, managing the supply chain, etc. Research has shown that technology makes teleworkers more productive than office workers [14]. In this sense, technology can stimulate engagement in teleworking. We, therefore, hypothesize the following:

H5: There is a gender gap in the positive influence of technology use on intention to continue teleworking.

3   Methodology

Reviewing the literature on the gender issue at work allowed us to design our conceptual research model. The partial least squares structural equation modeling (PLS-SEM) method tests the model. We developed a questionnaire through a careful selection of measurement items proposed by the literature on the research model's five (05) constructs. Technology use was adapted from González-Sánchez et al. [3], Job autonomy was adapted from Weinert et al. [15], Job satisfaction was adapted from Lepold et al. [16], Well-being was adapted from Ab Wahab, Tatoglu [17] and Intention to continue teleworking was adapted from Weinert et al. [15]. A 7-point Likert scale ranging from "Strongly disagree" to "Strongly agree" was adapted to measure each item. The questionnaire was set up on the Google Forms platform to perform an online collection. We conducted a pretest to ensure that the items were understood and coherent. We asked six (06) master students in information systems management to read the questionnaire. Their task was to read the questionnaire thoroughly and to detect any incoherence. This pretest phase allowed us to confirm that the questionnaire was free of ambiguity. The minimum sample size (92 responses) was obtained using G*power 3.1.9.2 software [18]. The questionnaire was distributed through the professional platform "LinkedIn" during the COVID-19 pandemic containment period, to reach the maximum number of teleworkers. We informed them about the anonymity of the questionnaire and the use of the data for research purposes only. With the first fifty (50) responses, we conducted a pilot test to ensure the items were reliable. The results of the pilot test showed that all the measurement items presented an acceptable level of reliability (>0.7 [19]). We, therefore, continued the data collection, which took place from September to November 2020. From the data obtained, we isolated 500 responses, including 250 responses from men and 250 from women. These data were analyzed using SmartPLS 4 software [20]. We first assessed the measurement model to ensure the reliability and validity of the constructs. Then, we proceeded to the structural model assessment to analyze the causal relations between the constructs. Finally, the multi-group analysis was performed to establish the gender difference (Group 1 = Man; Group 2 = Woman).

4   Results

4.1   Sample Description

The study population comprises 500 teleworkers, of which 250 are men (50%) and 250 are women (50%). Most of the sample is between 26 and 35 years old (including 123 men and 93 women). The study did not focus on a particular geographical area. Teleworkers were eligible to complete the questionnaire regardless of their country of residence. For the entire study population, 45.8% reside in Africa, 44.2% in Europe, 9.4% in America, and 0.6% in Asia.

4.2   Model Assessment

The measurement model was assessed by first examining the outer loadings of each measurement item. Then, examine Cronbach's alpha, rho_A, and each construct's composite reliability (CR) for internal consistency reliability. Their values must be greater than or equal to 0.7 [19]. Convergent validity was examined through the average variance extracted (AVE), which must be greater than or equal to 0.50 [19]. We removed all outer loadings whose values were less than 0.7 and which negatively influenced the convergent validity. Table 1 shows that all criteria meet the recommended thresholds for the measurement model. From this, we conclude that the research model is stable. The discriminant validity is reviewed through the heterotrait-monotrait (HTMT) correlation ratio, whose values must respect the threshold of 0.90 [19, 21]. Table 2 shows that HTMT values are less than 0.90. So, the discriminant validity is not present.

Table 1. Constructs' reliability and validity/R^2.

Constructs	Cronbach's alpha	rho_A	CR	AVE	R^2 Overall sample	R^2 Man	R^2 Woman
INT	0.881	0.891	0.944	0.893	0.421	0.412	0.447
JA	0.834	0.857	0.900	0.751	0.146	0.187	0.133
JS	0.883	0.899	0.915	0.685	0.207	0.151	0.275
TU	0.627	0.804	0.768	0.531	–	–	–
WB	0.824	0.832	0.883	0.655	0.166	0.193	0.174

INT = Intention to continue teleworking; JA = Job Autonomy; JS = Job Satisfaction; TU = Technology Use; WB = Well-Being

Table 2. Discriminant validity- heterotrait-monotrait ratio (HTMT)

	INT	JA	JS	TU	WB
INT					
JA	0.320				
JS	0.643	0.433			
TU	0.333	0.484	0.472		
WB	0.669	0.546	0.669	0.492	

INT = Intention to continue teleworking; JA = Job Autonomy; JS = Job Satisfaction; TU = Technology Use; WB = Well-Being

On the other hand, the structural model was assessed by first examining the collinearity issues. To do this, we ensured that the internal VIFs were all less than 3 [19]. Since collinearity was not an issue, we continued the analysis by examining the dependent constructs' explanatory power (R^2). The results (Table 1) show that the Intention to continue teleworking is explained by 41.2% ($R^2 = 0.412$) for men and 44.7% ($R^2 = 0.447$) for women. Job Autonomy is explained at 18.7% ($R^2 = 0.187$) for men and 13.3% ($R^2 = 0.133$) for women. Job Satisfaction is explained by 15.1% ($R^2 = 0.151$) for men and 27.5% ($R^2 = 0.275$) for women. And finally, Well-Being at work is explained at 19.3% ($R^2 = 0.193$) for men and 17.4% ($R^2 = 0.174$) for women.

We conducted the multigroup analysis (MGA) [22] to analyze the research model's causal relationships and assess the significant differences between men and women. To do this, we first ensured that each group (Group 1 = Man; Group 2 = Woman)

had sufficient statistical power, i.e., that the sample size for each group exceeded the minimum sample size needed to test the model (n = 92). We then generated the groups and evaluated the measurement invariance (MICOM).

First, we validated the configural invariance (Step 1) by ensuring that (1) the indicators are identical, reliable, and valid for each group, (2) that there are no missing values, and (3) that the settings of the PLS algorithm are similar. Then, we ran the permutation analysis (1000 permutations and a significance level of 0.05). The results (Table 3) show that the original correlation is greater than or equal to the 5% quantile and that the permutation P values are greater than 0.05 [22]. Thus, the compositional invariance (Step 2) is established. The results (Table 4) also show that the mean and variance values of the original difference fall between the lower (2.5%) and upper (97.5%) boundaries [22], except for the WB construct concerning the mean original difference (Step 3a). So, we conclude the partial measurement invariance. We then run the multigroup bootstrap analysis.

Table 3. MICOM-Step 2

Constructs	Original correlation	Correlation permutation mean	5.0%	Permutation p value (>0.05)	Interpretation
INT	1.000	1.000	1.000	0.425	Good
JA	0.996	0.999	0.995	0.054	Good
JS	1.000	0.999	0.998	0.909	Good
TU	0.965	0.989	0.963	0.058	Good
WB	0.998	0.999	0.997	0.185	Good

Table 4. MICOM-Step 3

Constructs	Original difference	Permutation mean difference	2.5%	97.5%	Permutation p value (>0.05)	Interpretation
Step 3a-Mean						
INT	−0.004	0.004	−0.166	0.176	0.952	Good
JA	−0.124	0.001	−0.193	0.174	0.190	Good
JS	−0.040	0.002	−0.164	0.167	0.674	Good
TU	0.061	0.001	−0.164	0.174	0.488	Good
WB	−0.236	0.003	−0.163	0.172	0.009	Bad

(continued)

Table 4. (*continued*)

Constructs	Original difference	Permutation mean difference	2.5%	97.5%	Permutation p value (>0.05)	Interpretation
Step 3b-Variance						
INT	−0.118	−0.003	−0.163	0.157	0.173	Good
JA	−0.047	−0.001	−0.215	0.220	0.681	Good
JS	−0.131	−0.007	−0.222	0.205	0.215	Good
TU	−0.107	−0.002	−0.261	0.254	0.453	Good
WB	0.084	−0.002	−0.238	0.241	0.513	Good

The MGA results (Table 5) show that a significant difference exists in the relationship between Technology Use and Job Satisfaction (P < 0.1) and the relationship between Job Autonomy and Intention to continue teleworking (P < 0.1). The other relationships are not significant.

Table 5. Hypotheses testing

Bootstrap results						
Hypotheses		P value (Overall sample)	P value (Man)	Significant Influence	P value (Woman)	Significant Influence
H1a	TU - > JA	0.000****	0.000****	Yes	0.000****	Yes
H1b	TU - > JS	0.000****	0.000****	Yes	0.000****	Yes
H1c	TU - > WB	0.000****	0.000****	Yes	0.000****	Yes
H2	JA - > INT	0.443 n.s	0.443 n.s	No	0.039**	Yes
H3	JS - > INT	0.000****	0.000****	Yes	0.000****	Yes
H4	WB - > INT	0.000****	0.000****	Yes	0.000****	Yes
H5	TU - > INT	0.796 n.s	0.508 n.s	No	0.453 n.s	No

(*continued*)

Table 5. (*continued*)

Bootstrap MGA- Man vs Woman

Hypotheses		Difference (Man - Woman)	1-tailed (Man vs Woman) p-value	2-tailed (Man vs Woman) p-value	Sign. Level	Conclusions
H1a	TU - > JA	0.068	0.222	0.443	n.s	Rejected
H1b	TU - > JS	-0.135	0.953	0.093	*	Supported
H1c	TU - > WB	0.022	0.393	0.786	n.s	Rejected
H2	JA - > INT	0.167	0.035	0.070	*	Supported
H3	JS - > INT	-0.055	0.707	0.587	n.s	Rejected
H4	WB - > INT	0.019	0.416	0.833	n.s	Rejected
H5	TU - > INT	-0.090	0.841	0.318	n.s	Rejected

*** * $P < 0.001$; *** $P < 0.01$; ** $P < 0.05$; * $P < 0.1$; n.s. not significant; INT = Intention to continue teleworking; JA = Job Autonomy; JS = Job Satisfaction; TU = Technology Use; WB = Well-Being

We analyzed the mediating effects (specific indirect effect) of JA, WB, and JS on the relationship between TU and INT. Table 6 reveals the full mediation of these constructs for women. On the other hand, for men, JA has no mediating effect on the relationship between TU and INT.

Table 6. Mediation analysis

Mediation relationship	P values (Overall sample)	P values (Man)	Interpretation	P values (Woman)	Interpretation
TU - > JA - > INT	0.459	0.453	No mediation	0.060	Full mediation

(*continued*)

Table 6. (*continued*)

Mediation relationship	P values (Overall sample)	P values (Man)	Interpretation	P values (Woman)	Interpretation
TU - > WB - > INT	0.000	0.000	Full mediation	0.000	Full mediation
TU - > JS - > INT	0.000	0.001	Full mediation	0.000	Full mediation

*** * P < 0.001; *** P < 0.01; ** P < 0.05; * P < 0.1; n.s. not significant; INT = Intention to continue teleworking; JA = Job Autonomy; JS = Job Satisfaction; TU = Technology Use; WB = Well-Being

5 Discussions and Implications

The study shows a gender difference regarding the positive influence of TU on JS and the positive influence of JA on INT. Indeed, TU has more impact on JS for women (R^2 = 27.5%) compared to men (R^2 = 15.1%). At the same time, JA influences INT only for women.

The study also shows that TU contributed positively to teleworking. We find that TU significantly affects JA (R^2 = 18.7%) and WB (R^2 = 19.3%) for men, although there is no significant gender difference for these two constructs. On the other hand, the results show that TU significantly influences for both men and women, the teleworking benefits highlight in this study.

JS and WB influence INT for both groups: men and women. However, JA does not influence INT for men. Also, TU does not affect INT for either men or women. Finally, INT is more explained among women (R^2 = 44.7%) than men (R^2 = 41.2%).

These results make us understand that, although technologies contribute to women's satisfaction in fulfilling their tasks at work, they have little effect on women's well-being and autonomy at work. On the other hand, from the men's point of view, technologies contribute most to their well-being and job autonomy but less to their job satisfaction. However, the study does show that job autonomy is a problematic factor that hinders the intention to continue teleworking. The fact that most employees experienced telework for the first time during the COVID-19 period may explain this situation. Forced changes in work methods and conditions disrupted workers' habits in managing their work. In addition, the fact that TU does not significantly influence INT shows that improvements are needed for the practice of telework in the future.

Regarding implications, we suggest that technology designers develop assistive technology for women in telework situations to manage work-life balance better. In stressful teleworking, women's physical and mental health is also essential to their well-being. Technology can significantly contribute to the development and implementation of mental health programs for teleworkers in general. In the case of men, to bring them more autonomy and satisfaction, managers and technology designers could work more on the contribution of digital tools in achieving the teleworker's objectives according to his field or management grade. In addition, managers must ensure that all teleworkers have the necessary digital capabilities and tools to do their job correctly and are fully equipped to do so. However, it is essential to ensure that the proliferation of digital work and communication tools does not amplify the risk of hyper-connectivity and degrade the health of teleworkers over time. The improvement of technological contributions to the practice of telework should also consider reducing competition between office-based and remote work. Through technology, teleworkers will need to be perceived as having the same level of commitment as those who will choose to work on-site. In this perspective, human resource managers must ensure that tomorrow's hybrid working model will not discriminate against women who choose to telework and men who will decide to continue working in the office. Indeed, managers must treat career development, salary compensation, and promotions equally. This research has shown that job satisfaction and well-being indirectly influence the effect of technology use on the intention to continue teleworking. Thus, the more technology designers focus on enhancing teleworker's sense of satisfaction and well-being while emphasizing their job autonomy, the more willing they will be to continue teleworking. This study contributes to the teleworking literature by examining gender differences in the use of technology during teleworking. Indeed, existing studies have focused on the gender perspective of teleworking conditions and the subsequent consequences for equality and benefits for teleworkers. This study deepens the knowledge by showing that technologies will be crucial in sustaining the continued practice of telework for both men and women.

6 Conclusion

This research reflected the importance of technology for teleworking from both men's and women's perspectives. The study reveals a gender difference in the relationship between technology use and job satisfaction and the relationship between job autonomy and intention to continue teleworking. However, the study has some limitations. We did not take into account work sectors. Future studies could investigate whether gender differences persist among workers based on their sector of activity (health, trade, service provider, etc.). Future studies could also examine the contribution of technology for teleworkers with disabilities and investigate the gender gap. We did not consider the parental or non-parental status of respondents. Future research could examine whether technologies are more effective for non-parenting teleworkers and how these technologies can improve the productivity of men and women in single-parenting situations.

References

1. Çoban, S.: Gender and telework: Work and family experiences of teleworking professional, middle-class, married women with children during the Covid-19 pandemic in Turkey. Gend. Work. Organ. **29**(1), 241–255 (2022). https://doi.org/10.1111/gwao.12684
2. Dias, P., Lopes, S., Peixoto, R.: Mastering new technologies: does it relate to teleworkers' (in)voluntariness and well-being? J. Knowl. Manag. **26**(10), 2618–2633 (2022). https://doi.org/10.1108/JKM-01-2021-0003
3. González-Sánchez, G., Olmo-Sánchez, M.I., Maeso-González, E.: Challenges and strategies for post-COVID-19 gender equity and sustainable mobility. Sustainability **13**(5), 2510 (2021). https://doi.org/10.3390/su13052510

4. Rodríguez Pérez, R.E., Ramos, R.: How has teleworking highlighted gender differences in mexico in the face of COVID-19?. In: The Economics of Women and Work in the Global Economy, 1st edn., pp 13–39. Routledge (2022)
5. Hackman, J.R., Oldham, G.R.: Motivation through the design of work: test of a theory. Organ. Behav. Hum. Perform. **16**(2), 250–279 (1976). https://doi.org/10.1016/0030-5073(76)900 16-7
6. Diab-Bahman, R., Al-Enzi, A.: The impact of COVID-19 pandemic on conventional work settings. Int. J. Sociol. Soc. Policy **40**(9/10), 909–927 (2020). https://doi.org/10.1108/IJSSP-07-2020-0262
7. Halliday, C.S., Paustian-Underdahl, S.C., Ordóñez, Z., Rogelberg, S.G., Zhang, H.: Autonomy as a key resource for women in low gender egalitarian countries: a cross-cultural examination. Hum. Resour. Manage. **57**(2), 601–615 (2018). https://doi.org/10.1002/hrm.21874
8. House, R.J., Wigdor, L.A.: Herzberg's dual-factor theory of job satisfaction and motivation: a review of the evidence and a criticism. Pers. Psychol. **20**(4), 369–389 (1967)
9. Feng, Z., Savani, K.: Covid-19 created a gender gap in perceived work productivity and job satisfaction: implications for dual-career parents working from home. Gender Manag.: Int. J. **35**(7/8), 719–736 (2020). https://doi.org/10.1108/GM-07-2020-0202
10. Dartey-Baah, K., Quartey, S.H., Osafo, G.A.: Examining occupational stress, job satisfaction and gender difference among bank tellers: evidence from Ghana. Int. J. Product. Perform. Manag. **69**(7), 1437–1454 (2020). https://doi.org/10.1108/IJPPM-07-2019-0323
11. Donati, S., Viola, G., Toscano, F., Zappalà, S.: Not all remote workers are similar: technology acceptance, remote work beliefs, and wellbeing of remote workers during the second wave of the COVID-19 pandemic. Int. J. Environ. Res. Public Health **18**(22) (2021). https://doi.org/10.3390/ijerph182212095
12. Song, Y., Gao, J.: Does telework stress employees out? A study on working at home and subjective well-being for wage/salary workers. J. Happiness Stud. **21**(7), 2649–2668 (2020). https://doi.org/10.1007/s10902-019-00196-6
13. Beckel, J.L.O., Fisher, G.G.: Telework and worker health and well-being: a review and recommendations for research and practice. Int. J. Environ. Res. Public Health **19**(7) (2022). https://doi.org/10.3390/ijerph19073879
14. Davies, A.: COVID-19 and ICT-supported remote working: opportunities for rural economies. World **2**(1), 139–152 (2021). https://doi.org/10.3390/world2010010
15. Weinert, C., Maier, C., Laumer, S.: Why are teleworkers stressed? An empirical analysis of the causes of telework-enabled stress. In: Wirtschaftsinformatik Proceedings 2015, pp 1407–1421 (2015)
16. Lepold, A., Tanzer, N., Bregenzer, A., Jiménez, P.: The efficient measurement of job satisfaction: facet-items versus facet scales. Int. J. Environ. Res. Public Health **15**(7) (2018). https://doi.org/10.3390/ijerph15071362
17. Ab Wahab, M., Tatoglu, E.: Chasing productivity demands, worker well-being, and firm performance. Pers. Rev. **49**(9), 1823–1843 (2020). https://doi.org/10.1108/PR-01-2019-0026
18. Schoemann, A.M., Boulton, A.J., Short, S.D.: Determining power and sample size for simple and complex mediation models. Soc. Psychol. Pers. Sci. **8**(4), 379–386 (2017). https://doi.org/10.1177/1948550617715068
19. Hair, J.F., Risher, J.J., Sarstedt, M., Ringle, C.M.: When to use and how to report the results of PLS-SEM. Eur. Bus. Rev. **31**(1), 2–24 (2019). https://doi.org/10.1108/EBR-11-2018-0203
20. Ringle, C.M., Wende, S., Becker, J.-M.: SmartPLS 4. SmartPLS (2022)
21. Henseler, J., Ringle, C.M., Sarstedt, M.: A new criterion for assessing discriminant validity in variance-based structural equation modeling. J. Acad. Mark. Sci. **43**(1), 115–135 (2015). https://doi.org/10.1007/s11747-014-0403-8
22. Cheah, J.-H., Thurasamy, R., Memon, M.A., Chuah, F., Ting, H.: Multigroup analysis using smartPLS: step-by-step guidelines for business research. Asian J. Bus. Res. **10**(3), 1–19 (2020)

Developing a Foresight Model for Law Enforcement Organizations to Detect and Analyze Future Challenges and Threats

Mohamed SaadEldin[✉]

Foresight and Decision-Making Support Center, Dubai Police, Dubai, UAE
mmsaad@hotmail.com

Abstract. The Analytic Hierarchy Process (AHP) is a method for making decisions in situations where multiple criteria must be considered. It is particularly useful in crisis situations where there are a number of feasible choices and decision criteria with diverse priorities. The author has developed a new model using the Analytic Hierarchy Process (AHP) method to detect and analyze future risks and challenges in Dubai Police and provide early warning signs to decision makers. The AHP method allows decision makers to differentiate between the importance of each criterion and compare them according to their importance. This helps to avoid stimulating risks or crises by choosing the best alternative based on a systematic evaluation of the available options. A proof-of-concept has been implemented as part of the work done by the author. The model is designed to improve the future readiness of Dubai Police by identifying and analyzing potential risks and challenges. The model uses mathematical, operational research, and decision-making models and is intended to be more reliable and practical than current models used in the law-enforcement sector, which are often deterministic, simplified, or inconsistent in their application and assumptions.

Keywords: Foresight · Challenge · Threat · Priority · Decision Support · Criterion

1 Introduction

Foresight is defined as a systematic, participatory, future intelligence-gathering and medium-to-long-term vision-building process aimed at enabling present-day decisions and mobilizing joint actions [1]. The law-enforcement sector is facing a complex and evolving set of challenges and threats as it seeks to keep communities safe. From traditional crimes and emerging forms of cybercrime to the threat of domestic terrorism, there is a need for continued innovation and adaptability in order to effectively combat these threats. There has been a trend of increasing crime rates in various communities around the world. A report published by the United Nations Office on Drugs and Crime (UNODC) found that global crime rates have been steadily increasing since the 1990s. This trend is concerning as it can lead to a sense of fear and insecurity within communities and can have negative impacts on social cohesion and quality of life. It

© The Author(s), under exclusive license to Springer Nature Switzerland AG 2024
A. Rocha et al. (Eds.): WorldCIST 2023, LNNS 801, pp. 313–325, 2024.
https://doi.org/10.1007/978-3-031-45648-0_31

is important for governments and law enforcement agencies to address this issue and implement effective strategies to prevent and reduce crime [2].

The AHP method is a well-established decision-making tool that has been widely used in various fields, including law enforcement, to prioritize and evaluate complex issues [3]. The AHP is based on a hierarchical structure that allows decision makers to compare and evaluate alternative options or scenarios based on a set of predefined criteria [4]. The AHP has several advantages over other decision-making methods, including its ability to incorporate subjective and qualitative factors, its transparency and clarity, and its ability to handle complex and multi-criteria decision problems [3]. There have been several studies that have demonstrated the effectiveness of AHP in law enforcement. One such study, published in the Journal of Criminal Justice, found that AHP was able to accurately predict resource allocation decisions made by police managers in a simulated setting [5]. Another advantage of AHP is that it allows for the integration of both qualitative and quantitative data. This is particularly useful in law enforcement, where decision-making often involves both types of data [5, 6].

In their paper, Guozhong Zheng et al. (2012) used the fuzzy analytic hierarchy process (AHP) method to evaluate work safety in hot and humid environments. A safety evaluation framework containing three factors (work, environment, and workers) and ten sub-factors was established. The weights of the factors and sub-factors were calculated based on pair-wise comparisons. Decision makers evaluated these data according to their experience [7]. Moreover, Krasimira et al. (2004) presented in their paper a multi-criteria analysis decision support system called MultiChoice, which was designed to support decision makers in solving different multi-criteria analysis problems. In order to solve multi-criteria analysis problems, several criteria were simultaneously optimized in a feasible set of a finite number of explicitly given alternatives. The information that the decision maker gives, reflects his/her preferences with respect to the quality of the alternative sought [8, 9]. A study by Xiang (1993) employed a multi-objective linear programming technique. In many practical situations, it would be desirable to achieve a solution that is "best" with respect to multiple criteria rather than one criterion as in Chuvieco (1993) (i.e., maximizing labor productivity) [10, 11]. Based on the literature review, we have concluded that, the (AHP) technique has proved useful in situations that require the selection of the best alternative from a number of feasible choices in the presence of multiple decision criteria and diverse criterion priorities.

The goal of this paper is twofold. Firstly, it provides guidelines for professionals involved in implementing an effective model to measure future challenges and risks for law enforcement. Secondly, it presents a new approach to comparing the different criteria, evaluating their importance, and ranking them according to their critical effect on risks or threats. If one criterion gets a higher weight than others, this means that this criterion has a greater effect on stimulating the risk and hence the system will rank it first. Finally, the paper describes a case study as a proof-of-concept. The results of this research provide valuable insights into the nature and scope of law enforcement future challenges and suggest that a systematic, objective, and standardized approach to measuring these challenges is both feasible and effective. The proposed model can serve as a useful tool for law enforcement agencies to anticipate and respond to future challenges in a proactive manner, helping to ensure the continued effectiveness and efficiency of these agencies in the face of an increasingly complex and dynamic environment.

2   Using Analytic Hierarchy Process (AHP) in Risk Assessment

During the past three decades, many developed and developing countries in the world have experienced serious hazards in different fields. The Analytic Hierarchy Process (AHP) is a multiple-criteria decision-making methodology designed to help decision makers in prioritizing decisions. It involves both tangible and intangible criteria, and has been implemented extensively all over the world [12–14]. Moreover, AHP is easy to use, understand and it can effectively handle both qualitative and quantitative data [15]. Rather than prescribing a "correct" decision, the AHP helps decision makers find one that best suits their goals and their understanding of the problem and evaluating alternative solutions. The new methodology in this paper depends on ranking variables automatically and gives a weight to each variable according to its relative importance by comparing different variables and selecting the critical ones (the variables which have a greater effect on stimulating the risk). The main element in the weighting method is how to determine weights of the criteria, which reflects the decision maker's preferences to the highest degree. If we consider the set of criteria $C = \{c1, c2,..., cn,\}$ hence the weight of each value can be calculated through the position of the value of the criterion in the range and therefore the importance of any criterion is determined by identifying the total weight of all the values of this criterion. If the weight of any indicator is higher than the others, this means that the criterion has a higher importance in terms of stimulating the threat/risk. The new method proposed by the author is:

1. Inputting the time-series data for criterion (1)
2. Calculating the total number of values for the criterion (n)
3. Determining the lowest value (L).
4. Determining the highest value (H).
5. Subtracting the highest value from lowest value (D) = H - L.
6. Setting the Number of interval (I) (from 0.1 to 1) = 10
7. Calculating the increment = D/I
8. Determining the ranges.
9. Setting the weight according to the range and values.
10. Calculating the weights for all time-series values.
11. Repeating the above process for criterion (2)…to criterion (n)
12. Calculating the weights for all criteria (indicators).
13. Normalizing the weight values for all criteria by using the following formula:
    Weight of criterion 1/total weights (weight of criterion 1 + weight of criterion 2 +.weight of criterion n).
14. Ranking the criteria (1…n), the top criteria have higher weight and will affect stimulating the risk.
15. Building a prevention plan with priorities depending on the ranking table.

3   Case Study

The key objectives for the proof of concept are to demonstrate the theoretical concept and to show how the theoretical concept can be automated.

3.1   Detecting the Risks with the Highest Frequency

The system selected a recurring risk called "Traffic Accident", because this risk has the highest number of occurrences and after analyzing the data, the system found that the trend is increasing from January 2011 to December 2021 (see Table 1).

Table 1. Traffic Accident Occurrences (2011–2021)[1]

Year	2011	2012	2013	2014	2015	2016	2017	2018	2019	2020	2021
Occurrences	10	12	14	19	20	24	36	40	43	52	64

3.2   Determining the Indicators and Choosing the Critical Factors

Determining the criteria/indicators that best describe the risk: The system uses from 3 (minimum) to 10 (maximum) criteria for each risk/threat to work efficiently, the number of criteria varies from one risk/threat to another and is implemented by law enforcement experts such as:

- Criterion (A): Total number of traffic accidents (severe injury, fatality, etc.)
- Criterion (B): Total number of traffic-related deaths.
- Criterion (C): The percentage of road traffic injuries to the total population.
- Criterion (D): The percentage of fatality accidents to the total of accidents.

Consider the set of indicators C = {c1, c2., cn} The procedure can be applied as follows:

Indicator A

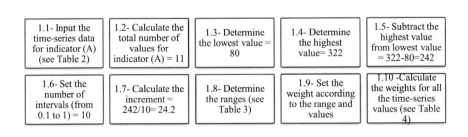

---

[1] Data represented in this paper is not a real data but used to test the model suggested by the author.

(Tables 2, 3 and 4)

Table 2. Indicator (A) Time-Series Data

Year	2011	2012	2013	2014	2015	2016	2017	2018	2019	2020	2021
Value	85	88	92	120	100	80	140	160	222	280	322

Table 3. Indicator (A) Range Values

Value + Increment	Range	Weight
$80 + 24.2 = 104.2$	≥ 80 to less than or equal 104.2	0.1
$104.2 + 24.2 = 128.4$	> 104.2 to less than or equal 128.4	0.2
$128.4 + 24.2 = 152.6$	> 128.4 to less than or equal 152.6	0.3
$152.6 + 24.2 = 176.8$	> 152.6 to less than or equal 176.8	0.4
$176.8 + 24.2 = 201$	> 176.8 to less than or equal 201	0.5
$201 + 24.2 = 225.2$	> 201 to less than or equal 225.2	0.6
$225.2 + 24.2 = 249.4$	> 225.2 to less than or equal 249.4	0.7
$249.4 + 24.2 = 273.6$	> 249.4 to less than or equal 273.6	0.8
$273.6 + 24.2 = 297.8$	> 273.6 to less than or equal 297.8	0.9
$297.8 + 24.2 = 322$	> 297.8 to less than or equal 322	1

Table 4. Indicator (A) Weights

Year	2011	2012	2013	2014	2015	2016	2017	2018	2019	2020	2021	Total Weight
Value	85	88	92	120	100	80	140	160	222	280	322	
Weight	0.1	0.1	0.1	0.2	0.1	0.1	0.3	0.4	0.6	0.9	1	3.9

Indicator B

1.1- Input the time-series data for indicator (B) (see Table 5)	1.2- Calculate the total number of values for indicator (B) = 11	1.3- Determine the lowest value = 127	1.4- Determine the highest value= 290	1.5- Subtract the highest value from lowest value = 290-127=163
1.6- Set the number of intervals (from 0.1 to 1) = 10	1.7- Calculate the increment = 163/10= 16.3	1.8- Determine the ranges (see Table 6)	1.9- Set the weight according to the range and values	1.10 - Calculate the weights for all time-series values (see Table 7)

(Tables 5, 6 and 7)

Table 5. Indicator (B) Time-Series Data

Year	2011	2012	2013	2014	2015	2016	2017	2018	2019	2020	2021
Value	132	136	139	140	127	143	143	150	186	210	290

Table 6. Indicator (B) Range Values

Value + Increment	Range	Weight
$127 + 16.3 = 143.3$	≥ 127 to less than or equal 143.3	0.1
$143.3 + 16.3 = 159.6$	>143.3 to less than or equal 159.6	0.2
$159.6 + 16.3 = 175.9$	>159.6 to less than or equal 175.9	0.3
$175.9 + 16.3 = 192.2$	>175.9 to less than or equal 192.2	0.4
$192.2 + 16.3 = 208.5$	>192.2 to less than or equal 208.5	0.5
$208.5 + 16.3 = 224.8$	>208.5 to less than or equal 224.8	0.6
$224.8 + 16.3 = 241.1$	>224.8 to less than or equal 241.1	0.7
$241.1 + 16.3 = 257.4$	>241.1 to less than or equal 257.4	0.8
$257.4 + 16.3 = 273.7$	>257.4 to less than or equal 273.7	0.9
$273.7 + 16.3 = 290$	>273.7 to less than or equal 290	1

Table 7. Indicator (B) Weights

Year	2011	2012	2013	2014	2015	2016	2017	2018	2019	2020	2021	Total Weight
Value	132	136	139	140	127	143	143	150	186	210	290	
Weight	0.1	0.1	0.1	0.1	0.1	0.1	0.1	0.2	0.4	0.6	1	2.9

Indicator C, D
(Tables 8, 9, 10, 11, 12 and 13)

Table 8. Indicator (C) Time-Series Data

Year	2011	2012	2013	2014	2015	2016	2017	2018	2019	2020	2021
Value	540	616	780	700	800	850	1050	1110	1206	1708	2000

Table 9. Indicator (C) Range Values

Value + Increment	Range	Weight
540 + 146 = 686	≥540 to less than or equal 686	0.1
686 + 146 = 832	>686 to less than or equal 832	0.2
832 + 146 = 978	>832 to less than or equal 978	0.3
978 + 146 = 1124	>978 to less than or equal 1124	0.4
1124 + 146 = 1270	>1124 to less than or equal 1270	0.5
1270 + 146 = 1416	>1270 to less than or equal 1416	0.6
1416 + 146 = 1562	>1416 to less than or equal 1562	0.7
1562 + 146 = 1708	>1562 to less than or equal 1708	0.8
1708 + 146 = 1854	>1708 to less than or equal 1854	0.9
1854 + 146 = 2000	> 1854 to less than or equal 2000	1

Table 10. Indicator (C) Weights

Year	2011	2012	2013	2014	2015	2016	2017	2018	2019	2020	2021	Total Weight
Value	540	616	780	700	800	850	1050	1110	1206	1708	2000	
Weight	0.1	0.1	0.2	0.2	0.2	0.3	0.4	0.4	0.5	0.8	1	4.2

Table 11. Indicator (D) Time-Series Data

Year	2011	2012	2013	2014	2015	2016	2017	2018	2019	2020	2021
Value	480	530	745	650	800	825	1000	1050	1200	1400	1650

Table 12. Indicator (D) Range Values

Value + Increment	Range	Weight
480 + 117 = 597	≥480 to less than or equal 597	0.1
597 + 117 = 714	>597 to less than or equal 714	0.2
714 + 117 = 831	>714 to less than or equal 831	0.3
831 + 117 = 948	>831 to less than or equal 948	0.4
948 + 117 = 1065	>948 to less than or equal 1065	0.5
1065 + 117 = 1182	>1065 to less than or equal 1182	0.6
1182 + 117 = 1299	>1182 to less than or equal 1299	0.7
1299 + 117 = 1416	>1299 to less than or equal 1416	0.8
1416 + 117 = 1533	>1416 to less than or equal 1533	0.9
1533 + 117 = 1650	>1533 to less than or equal 1650	1

Table 13. Indicator (D) Weights

Year	2011	2012	2013	2014	2015	2016	2017	2018	2019	2020	2021	Total Weight
Value	480	530	745	650	800	825	1000	1050	1200	1400	1650	
Weight	0.1	0.1	0.3	0.2	0.3	0.3	0.5	0.5	0.7	0.8	1	4.8

3.3 Normalizing the Factors Weight

1. Calculate the total weight = total weight for the indicator (A) + total weight for indicator (B) + total weight for indicator (C) + total weight for the indicator (D).
2. Calculate the optimized weights = Weight for Indicator (n)/Total Weight.
3. The next table (Table 14) shows that Indicator (D) needs more attention from decision and policy makers than other indicators in their future plans.
4. The system will carry out the above procedures for "Choosing the Critical Factors" and "Normalizing the Factors Weight" automatically as in (see Fig. 1).

Table 14. Indicators Ranking

Indicator	Weight	Optimized Weight	Rank
Indicator D	4.8	0.30	1
Indicator C	4.2	0.27	2
Indicator A	3.9	0.25	3
Indicator B	2.9	0.18	4
Total		1	

3.4 Designing a Forecasting Model for Each Indicator

1. The system will use four forecasting models to be applied (Linear, Quadratic, Exponential Growth and Cubic Models).
2. Linear equation $(y) = mx + b$ (where m and b constants, m is the slope of the line)
3. Quadratic equation $(y) = ax^2 + bx + c$ (where a, b, c are constants with a $\neq$ 0)
4. Exponential Growth equation $(y) = a + bx$ (where a, b are constants with a $\neq$ 0).
5. Cubic equation $(y) = ax^3 + bx^2 + cx + d$ (where a, b, c, d are constants with a $\neq$ 0).
6. MAPE (Mean Absolute Percentage Error, n = total number of actual values) =

$$\sum \frac{\left| \frac{Actual\ Value - Fitted\ Value}{Actual\ Value} \right|}{n} X\,100$$

R^2: The coefficient of determination[2].

$$r = \frac{n\left(\sum xy\right) - \left(\sum x\right)\left(\sum y\right)}{\left[n\left(\sum x^2\right) - \left(\sum x\right)^2\right]\left[n\left(\sum y^2\right) - \left(\sum y\right)^2\right]}$$

Indicator (A)

Table 15 shows the correlation of variables for each model. The system will automatically choose the cubic model because its correlation value is the largest in comparison to the other models (=0.964), and the value of (MAPE) is the smallest (=8.8), so the forecasting data for the cubic model will be more accurate and Table 16 shows the comparison between the forecasting data for the indicator (A).

Indicators (B), (C) and (D)

The system will repeat the previous processes for indicators B, C and D.

---

[2] The coefficient of determination R^2 is most often seen as a number between 0 and 1.0, used to describe how well a regression line fits a set of data. An R^2 near 1.0 indicates that a regression line fits the data well, while an R^2 closer to 0 indicates a regression line does not fit the data very well

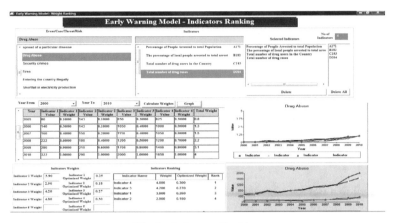

Fig. 1. Indicators Ranking

Table 15. Mathematical Measures for Indicator (A)

Model	Fitted Trend Equation	MAPE	R^2
Linear	$y = 19.2 + 22.3909\ x$	27.08	$R^2 = 0.772$
Quadratic	$y = 120.988 - 24.5881\ x + 3.91492\ x^2$	11.15	$R^2 = 0.956$
Exponential Growth	$y = 60.5547 * (1.14447^{\,x})$	17.86	$R^2 = 0.81$
Cubic	$y = 86.67 + 3.76\ x - 1.742\ x2 + 0.3143\ x3$	8.81	$R^2 = 0.964$

3.5 Generating Alerts (Sending Warning Messages)

a. Calculate the probability of each indicator by setting a range for each one. The probabilities will have values between 0.1 and 1 as in (Table 17).

b. Determine the level of danger (see Table 18), the level of danger equation is calculated as: [(The probability of indicator A + Probability of indicator B + Probability of indicator C + Probability of indicator D) / total Number of indicators] * 100.

c. Convert the result to the following color-coded scale (Warning Signals):

1. **Dangerous** (Red Color) = Probability at or above 90%.
2. **High** (Orange Color) = Probability at or above 70% but below 90%.
3. **Medium** (Yellow Color) = Probability above 55% but below 70%.
4. **Low** (Green Color) = Probability above 20%, but at or below 55%.
5. **Very Low** (Blue Color) = Probability at or below 20%.

d. The system found that in the year 2023 there is a dangerous level of the detected risk, the probability of a crisis and the time frame remaining before its emergence.

e. The system will generate alerts automatically based on the forecasting that was calculated. Once alerts are generated, they can be distributed through different channels to many parties who are exposed to hazards.

Table 16. Forecasting Models for Indicator (A) for the years (2011–2025)

Year	X	Actual Data (Indicator A)	Linear Model	Quadratic Model	Exponential Growth	Cubic Model
2011	1	85	41.59	100.32	69.30	89
2012	2	88	63.98	87.47	79.32	89.73
2013	3	92	86.37	82.46	90.77	90.76
2014	4	120	108.76	85.27	103.89	93.95
2015	5	100	131.16	95.92	118.89	101.21
2016	6	80	153.55	114.39	136.08	114.41
2017	7	140	175.94	140.70	155.74	135.44
2018	8	160	198.33	174.84	178.23	166.18
2019	9	222	220.72	216.80	203.98	208.53
2020	10	280	243.11	266.59	233.45	264.37
2021	11	322	265.50	324.22	267.18	335.58
2022	12		287.89	389.68	305.78	424.05
2023	13		310.28	462.96	349.95	531.67
2024	14		332.67	544.08	400.52	660.32
2025	15		355.06	633.02	458.38	811.88

Table 17. Probability Table

Indicator A		Indicator B		Indicator C		Indicator D	
Range	Probability	Range	Probability	Range	Probability	Range	Probability
≤80	0.1	≤132	0.1	≤540	0.1	≤480	0.1
81–128	0.2	133–165	0.2	541–832	0.2	481–715	0.2
129–177	0.3	166–198	0.3	833–1124	0.3	716–949	0.3
178–225	0.4	199–231	0.4	1125–1416	0.4	950–1183	0.4
226–274	0.5	232–264	0.5	1417–1708	0.5	1184–1417	0.5
275–322	0.6	265–297	0.6	1709–2000	0.6	1418–1651	0.6
323–370	0.7	298–330	0.7	2001–2292	0.7	1652–1885	0.7
371–419	0.8	331–363	0.8	2293–2584	0.8	1886–2119	0.8
420–467	0.9	364–396	0.9	2585–2876	0.9	2120–2353	0.9
>467	1	>396	1	>2876	1	>2353	1

Table 18. Final Results

Year	Indicator A	Probability	Indicator B	Probability	Indicator C	Probability	Indicator D	Probability	Level of Danger (%)	Warning Signals
2011	85	0.2	132	0.1	540	0.1	480	0.1	13%	Very Low
2012	88	0.2	136	0.2	616	0.2	530	0.2	20%	Very Low
2013	92	0.2	139	0.2	780	0.2	745	0.3	22.5%	Low
2014	120	0.2	140	0.2	700	0.2	650	0.2	20%	Very Low
2015	100	0.2	127	0.1	800	0.2	800	0.3	20%	Very Low
2016	80	0.1	143	0.2	850	0.3	825	0.3	22.5%	Low
2017	140	0.3	143	0.2	1050	0.3	1000	0.4	30%	Low
2018	160	0.3	150	0.2	1110	0.3	1050	0.4	30%	Low
2019	222	0.4	186	0.3	1206	0.4	1200	0.5	40%	Low
2020	280	0.6	210	0.4	1708	0.5	1400	0.5	50%	Low
2021	322	0.6	290	0.6	2000	0.6	1650	0.6	60%	Medium
2022	424	0.9	367	0.9	2539	0.8	1955	0.8	85%	High
2023	532	1	476	1	3196	1	2335	0.9	97.5%	Dangerous
2024	660	1	613	1	4007	1	2792	1	100%	Dangerous

4 Conclusion

In this paper, the author proposes a new approach using AHP method and presents the tool as a case study to prove the concept. Based on the literature review, the author concludes that the application of an AHP and multi-criteria approach is necessary when evaluating certain criteria (Indicators) and defining their relative importance in stimulating a threat. Using the integrated approach is beneficial as it helps policy makers decide which criteria or indicators are of greater importance and deserve more attention. Moreover, the ranking of the criteria helps in designing prevention plans that take into account certain priorities to prevent increasing future threats or risks. Furthermore, this approach uses the benefits of information systems to automatically calculate the relative importance of each value and set a weight for each of them without any interference from the users, in the aim of strengthening the ability of law enforcement organizations to prevent threats or risks before they occur.

References

1. EFP, European Foresight Platform: Foresight and Forward-Looking Activities – Exploring New European Perspectives, Kick-off Conference of the EFP, Vienna 14/15 June 2010 (2010). http://www.foresight-platform.eu/562/featured/efp-kickoff-conference
2. United Nations Office on Drugs and Crime (UNODC). Global Report on Crime and Justice (n.d.). https://www.unodc.org/unodc/en/data-and-analysis/crime/global-report-on-crime-and-justice.html
3. Saaty, T.S.: The Analytic Hierarchy Process. McGraw-Hill, New York (1980)
4. Saaty, T.S.: How to make a decision: the analytic hierarchy process. Eur. J. Oper. Res. **48**(1), 9–26 (1990). ISSN 0377-2217

5. Chen, J., Chen, H.: AHP applications in criminal justice and public safety. Eur. J. Oper. Res. **187**(1), 215–228 (2008)
6. Smith, M.R., Chen, J.: AHP applications in criminal justice and public safety: a review and assessment. J. Crim. Just. **42**(6), 435–443 (2014)
7. Guozhong, Z., Neng, Z., Zhe, T., Ying, C., Binhui, S.: Application of a trapezoidal fuzzy AHP method for work safety evaluation and early warning rating of hot and humid environments. Saf. Sci. **50**(2), 228–239 (2012)
8. Krasimira, G., Vassil, V., Filip, A., Mariyana, V., Silvia, K.:A multicriteria analysis decision support system. In: International Conference on Computer Systems and Technologies – CompSysTech 2004 (2004)
9. Korol, T.: Multi-criteria early warning system against enterprise bankruptcy risk. Int. Res. J. Financ. Econ. (61) (2011)
10. Xiang, W.: A GIS/MMP-based coordination model and its application to distributed environmental planning. Environ. Plann. B **20**(2), 195–220 (1993)
11. Chuvieco, E.: Integration of linear programming and gis for land-use modeling. Int. J. Geogr. Inf. Syst. **7**(1), 71–83 (1993)
12. Bagranoff, N.A.: Using an analytic hierarchy approach to design internal control system. J. Account. EDP **4**(4), 37–41 (1989)
13. Arbel, A.Y., Orgler, E.: An application of the AHP to bank strategic planning: the mergers and acquisitions process. Eur. J. Oper. Res. **48**(1), 27–37 (1990)
14. Moutinho, L.: The use of the analytic hierarchy process (AHP) in goal assessment: the case of professional services companies. J. Prof. Serv. Mark. **8**(2), 97–114 (1993)
15. Cengiz, K., Ufuk, C., Ziya, U.: Multi-criteria vendor selection using fuzzy AHP. Logist. Inf. Manag. **16**, 382–394 (2003)

Data Science Maturity Model: From Raw Data to Pearl's Causality Hierarchy

Luís Cavique[1]($\boxtimes$) (ID), Paulo Pinheiro[2] (ID), and Armando Mendes[3] (ID)

[1] Universidade Aberta and Lasige-FCUL, Lisbon, Portugal
luis.cavique@uab.pt
[2] Universidade Aberta and Cedis, Lisbon, Portugal
ppinheiro@cedis.pt
[3] Universidade Açores and LIACC, Ponta Delgada, Portugal
armando.b.mendes@uac.pt

Abstract. Data maturity models are an important and current topic since they allow organizations to plan their medium and long-term goals. However, most maturity models do not follow what is done in digital technologies regarding experimentation. Data Science appears in the literature related to Business Intelligence (BI) and Business Analytics (BA). This work presents a new data science maturity model that combines previous ones with the emerging Business Experimentation (BE) and causality concepts. In this work, each level is identified with a specific function. For each level, the techniques are introduced and associated with meaningful wh-questions. We demonstrate the maturity model by presenting two case studies.

Keywords: data science · maturity models · business experimentation · wh-questions · causality

1 Introduction

Data maturity models are a valuable and current topic since they allow organizations to plan their medium and long-term goals (Carvalho et al. 2019). Maturity models are an essential business management tool (Davenport 2018), allowing organizations to improve the planning of actions that should lead to the desired results. This problem is even more relevant as new concepts, keywords, and products are launched yearly in the information technology market, whose impact is rarely known.

In the 2010s, academic journals and companies began recognizing Data Science as an emerging discipline (Chiarello et al. 2021). The emergence of Data Science goes beyond Business Intelligence (BI) and Business Analytics (BA), abbreviated by BI&A. Data Science covers a broad spectrum of data and its derivatives, from initial data engineering (extraction, integration, transforming), exploration (aggregation, visualization), and modeling. Data Engineering (DE) arose in this decade as a synonym of Data Wangling, Feature Engineering, and Data Pre-processing, partially replacing the well-known

ETL (Extraction, transformation, loading). Data science overlaps multiple data-analytic disciplines, such as databases, statistics, operations research, and machine learning.

Most maturity models do not follow what is done in the technological sector, particularly in the GAFAM (Google, Amazon, Facebook, Apple, Microsoft), regarding Business Experimentation (BE). Experimentation is a simple method to test new ideas systematically. Experimentation can test a theory or hypothesis, help evaluate an existing product, and be valuable beyond the tech sector, making organizations brighter. A new period is rising in companies, the experimental revolution (Luca, Bazerman 2020).

In the requirements elicitation, the right questions lead to good application design. Moreover, questions illustrate the artifact's purpose, like in the well-known Gartner Analytic Ascendancy Model (GAAM) (Gartner 2012).

This work aims to present a comprehensible data science maturity model that includes the well-known business intelligence and analytics areas, the new practices in business experimentation (Thomke 2020), and Pearl's causality hierarchy (Pearl 2019). The proposed pipeline can be scratched as DE $\rightarrow$ BI $\rightarrow$ BA $\rightarrow$ BE. The proposed maturity model named _IABE is the Intelligence, Analytics, and Business Experimentation acronym.

In this paper, the keywords maturity models, hierarchy, and ascendancy models have the same meaning.

Two different contributions to the data science hierarchy are present. The levels of the model are given on the first contribution. Moreover, the techniques and the associated wh-questions are reported in the second contribution. A comprehensive approach to the typical causal questions is also provided.

The remaining paper is organized as follows. Section 2 presents the proposed maturity model _IABE with the levels and wh-questions. Two case studies demonstrate how the proposed model can be applied in Sect. 3. Finally, in Sect. 4, conclusions are drawn.

2 Proposed Maturity Model _IABE

The proposed Data Science Maturity Model comprises four stages: a previous level regarding Data Engineering (DE), followed by Business Intelligence (BI), Business Analytics (BA) at the second level, and finally, Business Experimentation (BE) at the upper level. The model pipeline is DE $\rightarrow$ BI $\rightarrow$ BA $\rightarrow$ BE.

Table 1 shows the rubric framework of the Data Science Maturity Model with four levels and four criteria. The criteria/dimensions include the business information system, the approach to planning, the guidance of the level, and finally, a synthesis function where Tc means the algorithmic running time complexity, represented by the big O notation, and T denotes the treatment of the experiment.

We associate the functions $f(X)$, $g(X)$, and $h(X, T)$ with BI, BA, and BE. Functions $f(X)$ and $g(X)$ use the same argument, X, where X denotes the set of attributes of the system. On the other hand, function $h(X, T)$ has two arguments, where T represents the treatment.

Since $O(N^2)$ is the time complexity between easy and hard problems, we create the threshold of $O(N^2)$ for the time complexity to distinguish between BI and BA, where N is the number of lines of the dataset. The function $f(X)$ can be exemplified by the sum of an attribute in an OLAP system, showing a running time complexity lower than $O(N^2)$,

$Tc(N) \leq O(N^2)$. Moreover, function $g(x)$ can be exemplified as a classification algorithm of a predictive model, with running time complexity usually upper than $O(N^2)$, $Tc(N) > O(N^2)$.

The second criterion in Table 1 is closely associated with how information system planning is developed. The BI behavior is reactive since it only cares about past events. Given the more complex models of BA, it is possible to elaborate on recommendations and take a proactive role in the company. Finally, in the BE stage, the ability to interact with customers by performing controlled trials moves the company to a new level in organizational learning with interactive planning.

Table 1. Data Science Hierarchy

		level 0	level 1	level 2	level 3
criteria	Information system	Data Engineering	Business Intelligence	Business Analytics	Business Experimentation
	approach to planning	inactive	reactive	proactive	interactive
	guidance	data pre-processing	data-driven	model-driven	experiment-driven
	function	n.a	$y = f(X)$, $Tc(N) \leq O(N^2)$	$y = g(X)$, $Tc(N) > O(N^2)$	$y = h(X, T)$

The previous level, named level 0, comprises Data Engineering tasks, where questions or answers are not presented, and there is no planning. This level includes pre-processing data features as an ETL process (Extraction, transformation, and loading) (Cavique et al. 2019).

Business Intelligence (BI) comprehends tools to support data-driven decisions, emphasizing reporting and data visualization. Data warehouse design is essential to provide multidimensional data tables that can be analyzed using OLAP (online analytical processing) systems (Cavique et al. 2020). Based on KPI (key performance indicators), alerts can be triggered in management by exception environments. The approach to planning is reactive based on the current information.

Business Analytics (BA) merges the areas of Data Mining/Machine Learning with Decision-Making tools. In Data Mining, two sub-areas should be mentioned: descriptive and predictive. The descriptive approach looks for relevant patterns in the data (Cavique 2007; Tiple et al. 2016; Cavique et al. 2018b), and the predictive models use supervised algorithms with labeled data to anticipate future events (Cavique et al. 2018a). Decision-making models include the techniques also referred to as Deductive modeling and studied in Operations Research, like decision analysis, simulation, and optimization (Cavique et al. 1999; Santos et al. 2013). These techniques aim to find the best solutions for each decision problem. BA comprehends tools to support model-driven decisions where the approach to planning is proactive.

In Business Experimentation (BE), experimentation interacts with individuals (customers, patients, or users), generating more data and feeding back to the system. A low-cost business experiment can change the way organizations design decision-making. BE comprehends tools to support experiment-driven decisions in interactive planning. Pearl's causality hierarchy refers to association, intervention, and counterfactuals, where the association is closely related to the traditional data mining approach. In the BE level, we add a new sub-level, the explanatory one, reflecting the title of the book 'The book of why' (Pearl, Mackenzie 2018).

level 3, BE	wh-questions	answers / techniques
	Why does treatment T cause this outcome?	Explanatory
	What if they received other treatment?	Counterfactual
	What if they received treatment T?	Intervention

level 2, BA	wh-questions	answers / techniques
	What is the best option?	Decision making
	What will happen?	Predictive models
	What are the interesting patterns?	Descriptive models

level 1, BI	wh-questions	answers / techniques
	What is happening now?	Alerts
	What is exactly the problem?	OLAP
	What happened?	Data Warehouse

Fig. 1. Sub-levels of the Data Science hierarchy with wh-questions

Figure 1 shows the three stages associated with the standard techniques and the related comprehensive questions that can be asked. BI&A is highly effective at answering questions of 'what'. On the other hand, BE answers questions of 'what if' and 'why', which implies causal relationships.

The sub-levels of BI, in addition to data warehouses, OLAP systems, and Alerts supported by KPI, are included. The BA sub-levels comprehend the descriptive, predictive, and decision-making models. As sub-levels of BE, we included, in addition to the randomized controlled trial, the last rungs of Pearl's causality ladder, the counterfactual, and explanatory reasoning.

3 Case Studies

Given the pipeline of levels, DE → BI → BA → BE, the two different case studies focus on the transitions, BI → BA and BA → BE, are presented. For each level, business wh-questions illustrate the potentialities of the reported techniques.

As stated before, we associate a function to each level: BI with f(x), BA with g(x), and BE with h(x, T). BI and BA use the same data, but different wh-questions can be pointed out. In BE, the concept of intervention/treatment T draws closer to the problem of causality identification.

3.1 From Business Intelligence to Business Analytics

In the first case study, we consider a hospital's information system that includes patient data, medications, surgeries, healthcare professionals, and billing.

Table 2 shows a data extract from the hospital's information system, where the 'complication' attribute indicates the patients who had postoperative difficulties (Dhar 2013).

Table 2. Extract the hospital's information system

patient	age	#medications	payments	complication
1	52	7	3,121 €	Yes
2	57	9	7,113 €	Yes
3	43	6	3,475 €	Yes
4	33	6	520 €	No
5	35	8	789 €	No
6	49	8	4,177 €	Yes
7	58	4	239 €	No
8	62	3	678 €	No
9	48	0	97 €	No
10	37	6	1,690 €	Yes

We present four relevant wh-questions at the BA level. The first three wh-questions are answered by the data mining area (classification, clustering, association), and the last one belongs to the area of management science (inventory management), as follows:

BA.1 – what are the rules that explain postoperative complications? (classification);

BA.2 – which patient groups can be found? (clustering);

BA.3 – which drug D is used with drug C? (association);

BA.4 – how to mitigate breaks in the medication stock? (inventory management).

The first two wh-questions use the data presented in Table 2. The other two wh-questions use the available data from the hospital's information system.

Classification algorithms are supervised methods that make predictions based on a discriminant class. In Table 2, the discriminant class corresponds to the 'complication' attribute. The pattern retrieved by BA.1 can predict whether a new patient 50 years old and not taking medication will have any complications in the postoperative period.

Descriptive methods do not use discriminating attributes; they are also unsupervised since any attribute with unique characteristics does not guide them. Wh-questions BA.2 and BA.3 exemplify this set of methods. For wh-question BA.2, the clustering algorithm divides patients into K groups by age, the number of medications, and payment that best characterizes the dataset. In Table 2, regarding payment, four patients are identified as 'major users' of the hospital, having spent more than the average.

An association rules algorithm is used to answer wh-question BA.3, which finds the most frequent sets in the same attribute. For example, people who use paracetamol also use a nasal decongestant, creating a rule as follows: paracetamol $=>$ nasal decongestant.

For wh-question BA.4, different inventory management policies can be applied to determine when and how much to order; deterministic or stochastic models can be applied. The most common determinist models use continuous or periodic reviews.

Given the BA wh-question, in BI, similar wh-questions can be asked. In BI, the wh-questions are more straightforward, as follows:

BI.1 – which patients have postoperative complications?

BI.2 – which patients paid more than the average?

BI.3 – what are the two most used medications?

BI.4 – which medications are in short supply?

Wh-questions BI.1 and BA.1 are similar, considering patients with postoperative complications. However, the first has an easy answer using an SQL query, while the second lacks a classification algorithm.

Likewise, wh-questions BI.2 and BA.2 refer to the amounts paid by patients in the hospital, in which the first wh-question is answered with SQL, and the second to determine the groups of patients is to use a clustering algorithm.

Wh-questions BI.3 and BA.3 also refer to similar issues, considering medications. However, for the first wh-question, it is enough to order the medication consumption, and the second one lacks an association rule algorithm.

Finally, BI.4 and BA.4 refer to medication inventory, but the first is answered using a database query, and the second involves more complex policies and algorithms, like the other BA wh-questions.

A particularity should be highlighted in the case of wh-questions BI.1 and BA.1.

Time complexity distinguishes between BI and BA. BI.1 is solved by a SQL query running time complexity of $O(N)$ in the worst case. On the other hand, BA.1 requires a classification algorithm with a time complexity larger than $O(N^2)$.

In wh-question BI.1, we intend to find the patients who had complications, with patients 1, 2, 3, 6, and 10 retrieved. The SQL query is as follows: Select patient, age, #medications From Table 2 Where complication $=$ 'yes'.

On the other hand, in wh-question BA.1, the data are provided, intended to extract patterns. Thus, we intend to know the attributes that cause postoperative complications.

With a classification algorithm, the following rule is found: (Age $\geq$ 35 and #medications $\geq$ 4) => Complications = 'Yes'. Those 35 years or older who take four or more medications have medical complications.

The two approaches treat the same data from information systems in different ways. Although the wh-questions are similar, BI.1 presents a pattern (e.g., SQL query), and data are retrieved. On the other hand, in BA.1, the data is provided, and the patterns are extracted.

3.2   From Business Analytics to Business Experimentation

The transition from BA to BE is exemplified in the second case study, using the Telco Customer Churn (2018) public dataset that contains information on eighteen covariates potentially related to both the outcomes of interest (churn or no-churn). Telco's churn is around 26%, revealing the importance of customer retention interventions that require concrete and personalized actions.

Traditional data mining studies focused primarily on predictive mining, where the cause-and-effect scenario is described. However, this information alone is insufficient as it does not benefit the final user. What becomes more exciting and critical to organizations is to mine patterns to create actionable knowledge (Cao 2007). Some attributes cannot influence or be changed, such as the attribute 'age' or 'gender', denominated by non-actionable attributes. On the other hand, the attributes that allow operational changes are called actionable attributes. Actionable attributes operationalize actionable knowledge.

Cao (Cao 2007) (Cao 2010) presents a new approach, which opposes data-driven to domain-driven. Data-driven corresponds to traditional data mining, while domain-driven is related to the business domain or business area. The domain-driven data mining, D3M, closes the gap between researchers and practitioners by generating actionable knowledge for real user needs. D3M approach moves away from BA and goes towards BE. Beyond the usual data, treatment T is introduced in the function y = h(x, T).

Each row represents a customer in the Telco dataset, and each column contains the customer's attributes. Those attributes can be grouped in customer demographic information, like gender, age range, and if they have partners and dependents. The second group of attributes describes customers' account information, like how long they have been a customer, contract, payment method, paperless billing, monthly charges, and total charges. Moreover, the third group of attributes presents each customer's services like phone, multiple lines, internet, online security, online backup, device protection, tech support, and streaming TV and movies. There is also an attribute, churn, which indicates whether or not the customer has abandoned services in the last month.

The best actionable attribute is the type of contract since customers with annual or biannual contracts tend to be more loyal than customers with monthly contracts. As shown in Fig. 2, to avoid churn (Y), the actionable attribute 'contract' (T) was chosen from the attributes of the decision tree. We aim to find any actionable attributes in which we can intervene to avoid dropouts and measure their causal effects and impact on the business (Pinheiro, Cavique 2022).

In order to exemplify the transition of information from BA to BE, two wh-questions are formulated. BA.u is a prediction question about the churn variable, and BE.u measures the effect of an intervention T regarding the churn variable, as follows:

BA.u – what are the rules that explain the churn Y?

BE.u – what is the impact of treatment T on the reduction of the churn Y?

As in the previous sub-section, the two approaches treat the same dataset using different information systems.

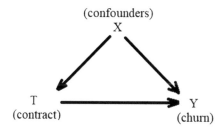

Fig. 2. DAG for Telco customer churn

4 Conclusions

Maturity models (MM) allow organizations to plan their actions to achieve the desired results. The maturity model requires a multi-stage planning tool to identify and control which advances should be made. However, most maturity models do not follow what is done in the digital sector concerning experimentation.

Several maturity models study Business Intelligence (BI) and Analytics (BA) domains (Carvalho et al. 2019). Our first goal aims to find a maturity model in the Data Science domain, including the recent Business Experimentation (BE), approaches (Thomke 2020), and new causality hierarchy (Pearl 2019). Our second objective is to find clear and meaningful maturity levels using a function-based approach and illustrated with wh-questions.

This study proposed the maturity model named _IABE, the acronym for Intelligence, Analytics, and Business Experimentation. Each level can be summarized by a function and a set of techniques associated with meaningful wh-questions.

Two case studies illustrated with wh-questions were presented for the transitions BI-BA and BA-BE. Transition BI-BA showed different questions set side-by-side with the techniques based on a hospital database. The transition from Business Analytics to Experimentation, BA-BE, used the Telco Customer Churn dataset.

This work clarified with comprehensive wh-questions the current levels of a Data Science Hierarchy, where each level was associated with a specific function.

References

Cao, L.: Domain-driven, actionable knowledge discovery. In: IEEE Intelligent Systems, pp. 78–79. IEEE Computer Society, Sydney (2007)

Cao, L.: Domain-driven data mining: challenges and prospects. IEEE Trans. Knowl. Data Eng. **22**(6), 755–769 (2010). https://doi.org/10.1109/TKDE.2010.32

Carvalho, J.V., Rocha, A., Vasconcelos, J., Abreu, A.: A health data analytics maturity model for hospitals information systems. Int. J. Inf. Manage. **46**, 278–285 (2019). https://doi.org/10.1016/j.ijinfomgt.2018.07.001

Cavique, L., Mendes, A.B., Martiniano, H.F.M.C., Correia, L.: A bi-objective feature selection algorithm for large omics datasets. Expert Syst. e12301 (2018a). https://doi.org/10.1111/exsy.12301

Cavique, L.: A scalable algorithm for the market basket analysis. J. Retail. Consum. Serv. Spec. Issue Data Min. Retail. Consum. Serv. **14**(6), 400–407 (2007)

Cavique, L., Rego, C., Themido, I.: Subgraph ejection chains and tabu search for the crew scheduling problem. JORS J. Oper. Res. Soc. **50**(6), 608–616 (1999)

Cavique, L., Cavique, M., Gonçalves, A.: Extraction of fact tables from a relational database: an effort to establish rules in denormalization. In: Rocha, Á., Adeli, H., Reis, L.P., Costanzo, S. (eds.) WorldCIST'19 2019. AISC, vol. 930, pp. 936–945. Springer, Cham (2019). https://doi.org/10.1007/978-3-030-16181-1_88

Cavique, L., Cavique, M., Santos, J.: Supply-demand matrix: a process-oriented approach for data warehouses with constellation schemas. In: Rocha, Á., Adeli, H., Reis, L., Costanzo, S., Orovic, I., Moreira, F. (eds.) WorldCIST 2020. AISC, vol. 1159, pp. 324–332. Springer, Cham (2020). https://doi.org/10.1007/978-3-030-45688-7_33

Cavique, L., Marques, N.C., Gonçalves, A.: A data reduction approach using hypergraphs to visualize communities and brokers in social networks. Soc. Netw. Anal. Min. **8**, 60 (2018b). https://doi.org/10.1007/s13278-018-0538-6

Chiarello, F., Belingheri, P., Fantoni, G.: Data science for engineering design: State of the art and future directions. Comput. Ind. **129**, 103447 (2021). https://doi.org/10.1016/j.compind.2021.103447. ISSN 0166-3615

Davenport, T.H.: DELTA plus model & five stages of analytics maturity: a primer, international institute for analytics (2018)

Dhar, V.: Data science and prediction. Commun. ACM **56**(12), 64–73 (2013)

Gartner. Gartner analytic ascendancy model. Gartner.com (2012)

Luca, M., Bazerman, M.H.: The Power of Experiments: Decision Making in a Data-Driven World. MIT Press (2020). ISBN 978-0262043878

Pearl, J.: Causality: Models, Reasoning, and Inference. Cambridge University Press, Cambridge (2000)

Pearl, J.: The seven tools of causal inference, with reflections on machine learning. Commun. ACM **62**(3), 54–60 (2019)

Pearl, J., Mackenzie, D.: The Book of Why: The New Science of Cause and Effect. Basic Books, New York (2018). ISBN: 978-0-465-09760-9

Pearl, J., Glymour, M.: Causal Inference in Statistics: A Primer. Wiley (2016). ISBN 978-1-119-18684-7

Pfeffer, J., Sutton, R.I.: Knowing 'what' to do is not enough: turning knowledge into action. Calif. Manage. Rev. **42**, 83–108 (1999)

Pinheiro, P., Cavique, L.: Uplift modeling using the transformed outcome approach. In: Marreiros, G., Martins, B., Paiva, A., Ribeiro, B., Sardinha, A. (eds.) EPIA 2022. LNCS, vol. 13566, pp. 623–635. Springer, Cham (2022). https://doi.org/10.1007/978-3-031-16474-3_51

Santos, J., Negas, E.R., Santos, L.C.: Introduction to data envelopment analysis. In: Mendes, A., L. D. G. Soares da Silva, E., Azevedo Santos, J. (eds.) Efficiency Measures in the Agricultural Sector, pp. 37–50. Springer, Dordrecht (2013). https://doi.org/10.1007/978-94-007-5739-4_3. ISBN 978-94-007-5738-7

Telco Customer Churn. Dataset (2018). https://www.kaggle.com/blastchar/telco-customer-churn. Accessed 01 Nov 2021

Thomke, S.H.: Experimentation Works: The Surprising Power of Business Experiments. Harvard Business Review Press (2020) ISBN 978-1633697102

Tiple P., Cavique, L., Marques, N.C.: Ramex-forum: a tool for displaying and analyzing complex sequential patterns of financial products. Expert Syst. 1–16 (2016). https://doi.org/10.1111/exsy.12174

Exploring Potential Drivers of Citizen's Acceptance of Artificial Intelligence Use in e-Government

Joaria Moreira and Mijail Naranjo-Zolotov[✉] [iD]

NOVA Information Management School (NOVA IMS), Campus de Campolide, 1070-312
Lisbon, Portugal
`{m20200527,mijail.naranjo}@novaims.unl.pt`

Abstract. The current advancement of information technologies has created the
conditions to introduce and popularize e-government, bringing citizens closer
to public administration. Yet, e-government faces challenges such as the digital divide, civic data overload, lack of trust in government institutions and their
online services. Artificial intelligence has the potential to address many of those
challenges but also raises ethical, privacy, and security concerns. Which requires
that before successfully adopting such a disruptive technology, it is imperative
to delve into the drivers leading to citizens' acceptance first. Consequently, this
study proposes an empirical model to explore and better understand the citizens'
acceptance towards the use of AI in e-government. We used an online survey to
collect data (N = 208). The results reveal that the perceived usefulness and trust
of AI and social influence significantly contribute to the acceptance of AI use in
e-government. Despite the majority being aware of AI and e-government, some
are not or are not aware of how AI can be used in e-government. The findings of
this study can help local and national governments assess the acceptance of the
adoption of AI-based technologies in e-government and define tailored strategies
to respond to citizens' concerns and highlight benefits to society.

Keywords: Artificial Intelligence · e-Government · Acceptance of AI use ·
Features of the technology · Characteristics of individuals

1 Introduction

Currently, Artificial Intelligence (AI) empowers a constellation of mainstream technologies [1] that may reshape how we behave, work, cooperate, and even make decisions. The
impact of societal and economic innovations created by these technologies is already
noticeable in several industries, including financial services, healthcare, telecommunications, transportation, among others [2]. However, compared to the private sector, the
public sector is lagging in AI adoption [3]. Several governments in Europe are trying
to catch up and close the current gap. As uncertainty increases and demands shift, governments have realised that it is crucial to innovate and incorporate new technologies to
deliver better services to citizens [4]. According to a study commissioned by Microsoft

and conducted by EY, 65% of surveyed European public organisations have recognised the value of AI, seeing it as a priority, and 67% have adopted at least one AI application [5].

AI-based technologies can enhance the quality of public services by improving their efficiency and responsiveness. For instance, guiding decision-making by summarising vast amounts of data and increasing citizen engagement by helping them be more informed about key policy issues [4]. On the other hand, these technologies can also be used with malicious intentions by facilitating control over information and communication, spreading misinformation, reinforcing filter bubbles, and manipulating citizens [6, 7].

Although research on the adoption of AI in the public sector is gradually increasing [8], its implementation is still in an embryonic stage. To successfully adopt AI technologies in the public sector, it is vital to first explore and understand the drivers leading to citizens' acceptance. Therefore, the objective of this study is to explore the extent to which some of the features of the technology itself and personal characteristics can influence the acceptance of AI use in e-government. We propose a research model and then collect data using an electronic questionnaire to evaluate the model.

2   e-Government and its challenges

E-government is defined as governments' use of information and communication technologies (ICTs) in its structures and procedures, combined with organisational change [9]. However, e-government alone does not guarantee better government responsiveness, citizen satisfaction, or increased citizen engagement and participation [10]. Presently, it faces several challenges that must be addressed:

Digital Divide: Low levels of digital literacy and inequality of access to ICTs may prevent some citizens from benefiting from the efficiency and diligence of digital government [11]. ***Civic data overload*:** Despite decision-makers encouraging mass citizen participation and desiring to use their inputs and suggestions to make decisions, governments still lack the necessary tools and resources to process and analyse them effectively [12]. In addition, citizens exhibit difficulty accessing and comprehending government information available online [13]. These limitations and difficulties decrease the prospect of a meaningful exchange of ideas and lead to a decline in the overall quality of mass participation [14]. ***Trust in government institutions and their online services:*** Citizens' adoption and use of e-government primarily depend on their trust in government institutions and their online services. Unfortunately, many citizens complain about the lack of quality and effectiveness of public online services or, due to personal beliefs, still prefer traditional means of reaching public entities [11, 15]. ***Privacy and security concerns*:** As a result of the increase of data leakage and misuse, systems intrusion and cyber incidents, citizens' concerns about privacy and security have also increased [11].

3   Acceptance of AI Use in e-Government: Hypotheses Building

Adopting AI-based technologies in e-government has far-reaching economic, legal, political, and regulatory implications [16]. However, to successfully adopt these technologies, it is paramount to first understand the factors leading to citizen acceptance. Previous studies have identified (i) *features of the technology itself* (usage and outcome-of-usage characteristics) and (ii) *characteristics of individual users* (demographic and psychographic characteristics) as fundamental for comprehending the attitudes, intentions, and behaviours that influence the acceptance and adoption of technology [17].

AI awareness, i.e. familiarity and knowledge of AI and its developments, plays a significant role in how citizens perceive AI. Citizens aware of the practical value of these technologies, due to past interactions and experiences, tend to trust them and recognize them as useful, that is, capable of improving their performance [18–21].

H1. *AI Awareness positively influences perceived usefulness of AI.*

H2. *AI Awareness positively influences trust in AI.*

One of the major concerns that arise among citizens when considering the adoption of AI, regardless of the context, is related to privacy [8, 22]. In e-government, these concerns are decisive because public services deal with sensitive and confidential information, such as personal data [23]. According to the Center for Strategic and International Studies (CSIS), the number of cyber incidents against government agencies has grown significantly in recent years, and citizens are increasingly concerned about how their data is collected, stored and used.

H3. *Privacy concerns negatively influence perceived usefulness of AI.*

H4. *Privacy concerns negatively influence trust in AI.*

Previous studies reveal that online self-efficacy, i.e. belief in the self's ability to protect their privacy online, impacts perceived usefulness and trust in AI [24]. As e-government encompasses *"the use of ICTs, in particular the internet, to achieve better government"* [9], it is expected that citizens who believe they can protect their privacy online tend to trust AI and consider it useful because of its benefits.

H5. *Online self-efficacy positively influences perceived usefulness of AI.*

H6. *Online self-efficacy positively influences trust in AI.*

Acceptance of AI flourishes when citizens benefit from its application, i.e. when it is considered useful, in some way, for society [20]. When AI is used, citizens can benefit from rational decisions that do not account for emotions [25, 26]. However, citizens who prefer to interact with humans due to AI's lack of empathy and technological maturity may find it hard to identify and recognize its benefits [21].

H7. *Perceived usefulness of AI positively influences the acceptance of AI use in e-government.*

In the context of the adoption of technologies such as AI, trust, defined "*as a degree of trustworthy of being able to fulfil a user's purpose of usage*" [27], is one of the main drivers of acceptance. Therefore, we propose the following hypothesis:

H8. *Trust in AI positively influences the acceptance of AI use in e-government.*

Social influence plays an important role when citizens face disruptive change with little or no information [28]. Citizens may be influenced by interpersonal sources, e.g. opinions of peers and superiors, or external sources of information, e.g. reports and news disseminated by the media [18, 29]:

H9. *Social influence positively influences the acceptance of AI use e-government.*

According to Starke et al. (2020), the greater the political interest of citizens, the less their satisfaction and trust in the government. This finding suggests that citizens interested in political affairs and familiarised with the decisions made by the government have low expectations about the government's ability to produce favourable outcomes [30]. Thus, political interest is expected to negatively impact the acceptance of AI use in e-government. We propose the following hypothesis:

H10. *Political interest negatively influences the acceptance of AI use in e-government.*

4 Research methodology

The study uses a convenience sample to assess the research model. The convenience sample was collected through an online questionnaire, developed on the Qualtrics platform, and distributed using a snowball technique through social networks. Before being distributed, the questionnaire was reviewed and approved by NOVA IMS Ethics Committee. The data collected does not contain any information that puts in risk the privacy of the respondents and was used exclusively for research purposes, being treated in a completely anonymous and confidential manner.

The eight constructs of the research model were operationalized into twenty-three measurable items: *AI awareness* was measured using four items about self-reported knowledge of AI, and interest in its applications and development (adapted from [18, 31]). *Privacy concerns* were measured with three items adapted from previous research [32]. *Online self-efficacy* – people's belief in their own ability to protect their privacy and personal information online – was measured using four items (adapted from [33, 34]). *Perceived usefulness of AI* was measured with three items adapted from earlier research on user acceptance of information technology [19, 35]. *Trust in AI* was measured using three items (adapted from [36]. *Social influence* was measured using three items about interpersonal and external influence (adapted from [18]. *Political interest* was measured with two items (adapted from [37, 38]. The main outcome measure – *acceptance of AI use in e-government* - was measured with a single item adapted from earlier research [21]: "*In the next 12 months, if available, I would use e-government powered by AI*". A seven-point Likert scale ranging from strongly disagree (1) to strongly agree (7) was used to assess each item.

5   Results

5.1   Sample Characterization

For this study, 208 responses were collected from Portuguese citizens. The sociodemographic characterization of the sample is represented in Table 1.

Table 1. Sociodemographic characterization (N = 208).

	N	%
Age		
Between 18 and 24 years old	114	54.8
Between 25 and 31 years old	55	26.4
Between 32 and 38 years old	16	7.7
Between 39 and 45 years old	7	3.4
Between 46 and 52 years old	6	2.9
53 years or older	10	4.8
Gender		
Female	131	63.0
Male	76	36.5
Other / Prefer not to answer	1	0.5
Education Level		
Less than secondary (high) school	2	1.0
Secondary (high) school or equivalent	36	17.3
Bachelor's degree	116	55.8
Master's degree	53	25.5
PhD	1	0.5

5.2   Measurement model

Indicator Reliability. Indicator reliability assesses the adequacy of the items (i.e., indicators) operationalized to measure the respective constructs – all standardised item loadings must be greater than 0.7 [39].

Internal Consistency Reliability. To ensure that the items operationalized to measure the same construct are consistent and mutually associated, we verified Cronbach's alpha and composite reliability [40] values. These measures consider the same thresholds: values above 0.7 are considered satisfactory and indicate high levels of reliability [39]. The results in Table 2 indicate that all the constructs of the study have high levels of internal consistency reliability.

Table 2. Cronbach's Alpha and Composite Reliability

Constructs	Cronbach's Alpha	Composite Reliability	Average Variance Extracted (AVE)
AIA	0.879	0.917	0.735
AIUEG	1.000	1.000	1.000
OSE	0.813	0.876	0.641
PC	0.744	0.853	0.661
PI	0.865	0.936	0.879
PUAI	0.882	0.927	0.809
SI	0.700	0.833	0.626
TAI	0.887	0.930	0.816

Note: AIA – AI awareness; AIUEG – Acceptance of AI use in e-Government; OSE – Online self-efficacy; PC – Privacy concerns; PI – Political interest; PUAI – Perceived usefulness of AI; SI – Social influence; TAI – Trust in IA.

Convergent Validity. Convergent validity measures the correlation between items used to assess the same construct. The metric used to evaluate it is the average variance extracted (AVE) that must be equal to or greater than 0.5, which indicates that the items positively correlate with their respective constructs [39]. AVE values of the constructs of this study ranged between 0.626 and 1.000, above the minimum requirement of 0.5 (Table 3).

Discriminant Validity. Discriminant validity indicates the extent to which a construct is empirically distinct from other constructs in the model is assessed based on the Heterotrait-Monotrait Ratio of correlations (HTMT) [39] which must be less than 0.9 to establish discriminant validity between constructs. Table 3 indicates that the HTMT criterion is fulfilled.

5.3 Structural Model

After confirming the reliability and validity of the measurement model, it is essential to assess the structural model [39]. Its assessment, made through a Bootstrapping procedure, focuses on analysing the significance of the structural model's relationships (path coefficients) and explaining the variance of its dependent variables (R^2) – perceived usefulness of AI, trust in AI and acceptance of AI use in e-government. The levels of significance dictate the rejection or acceptance of the proposed hypotheses. A result is significant if the p-value is smaller than 0.05 or greater than 0.95. The structural model and achieved results are shown in Fig. 1.

Table 3. Heterotrait-Monotrait Ratio (HTMT).

	AIA	AIUEG	OSE	PC	PI	PUAI	SI	TAI
AIA								
AIUEG	0.337							
OSE	0.264	0.393						
PC	0.097	0.331	0.174					
PI	0.260	0.046	0.140	0.065				
PUAI	0.335	0.795	0.350	0.295	0.036			
SI	0.333	0.752	0.296	0.243	0.118	0.717		
TAI	0.457	0.660	0.426	0.334	0.084	0.656	0.506	

Note: AIA – AI awareness; AIUEG – Acceptance of AI use in e-Government; OSE – Online self-efficacy; PC – Privacy concerns; PI – Political interest; PUAI – Perceived usefulness of AI; SI – Social influence; TAI – Trust in IA.

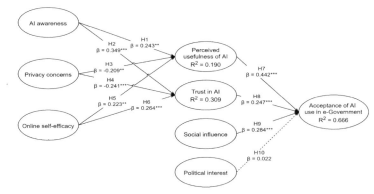

Fig. 1. Structural model. **Note:** Non-significant paths are in dashed lines. Significant at *p < 0.01; **p < 0.005; ***p < 0.001.

6   Discussion

Results of the survey with a sample of the Portuguese population show that AI awareness, privacy concerns and online self-efficacy explain 19% of the variation in the perceived usefulness of AI and 31% of the variation in AI trust. AI awareness, i.e. familiarity and knowledge of AI and its developments, positively influences perceived usefulness and trust in AI. These findings, corroborated by previous studies [18, 20, 21, 24], suggest that citizens more aware of AI applications and developments tend to trust it and perceive it as useful since they have more knowledge about the practical value of these technologies and have a solid personal predisposition about the targeted behaviour - perceive AI as useful and trustworthy [18, 41]. Privacy concerns negatively impact perceived usefulness and trust in AI, which is in line with findings from previous studies [24]. If AI-based technologies fail, people are unlikely to trust and use them again. Therefore, it is essential to ensure that these technologies can protect the data collected and analysed. This finding is also validated by the fact that citizens who consider themselves capable of protecting their privacy online - online self-efficacy - tend to view AI as useful and trustworthy.

Results also indicate that perceived usefulness and trust in AI and social influence explain 67% of the variation in acceptance of AI use in e-government. Perceived usefulness of AI, followed by social influence, are the main drivers of acceptance of AI use in e-government, supporting previous studies [18]. Acceptance of AI thrives when citizens consider it useful for society [20]. Furthermore, both the opinions of peers and superiors (interpersonal influence) and the news disseminated by the media (external influence) impact the acceptance of the use of AI in the context of e-government. Social influence may be of particular interest to citizens facing an innovation with limited information [28]. In the lack of their own experience or well-formed beliefs, individuals are influenced by interpersonal and external sources of information [18, 29]. Political interest was expected to negatively influence the acceptance of AI in e-government; however, our results show that political interest is actually not statistically significant for the acceptance of AI in e-government. This finding is inconsistent with studies as Chen & Wen (2021) and Starke et al. (2020), that show evidence that citizens who distrust the government tend to have an unfavourable perception of AI [29, 30].

7 Conclusions

The adoption of technology as powerful and disruptive as AI could be the answer to solving some of the current problems of e-government. However, for it to be successfully adopted, it is essential to understand the factors that influence its acceptance in this context. This study indicates that the perceived usefulness of AI, followed by social influence and trust in AI, are strong predictors of acceptance of the use of AI in e-government. Though, to trust AI and recognize the benefits of using it, it is essential to be familiar with it and its applications (AI awareness). These findings suggest that governments that already use e-government and intend to incorporate AI-based technologies into it should invest in citizens' understanding that AI is already present in their daily lives and address its advantages and disadvantages. However, governments must also do it for e-government by clarifying its meaning and applications and solving its current technical problems. These investments could also result in increased trust in the government. This study can help local and national governments assess the acceptance of the adoption of AI-based technologies in e-government and define tailored strategies to respond to citizens' concerns and highlight benefits to society.

References

1. Stone, P., et al.: One hundred year study on artificial intelligence (2015)
2. Twentyman, J.: Intelligent economies: AI's transformation of industries and societies (2018)
3. Council of Europe: Ad hoc Committee on Artificial Intelligence (CAHAI) Feasibility Study (2020)
4. Berryhill, J., Heang, K.K., Clogher, R., McBride, K.: Hello, world: artificial intelligence and its use in the public sector. OECD Observatory of Public Sector Innovation (OPSI)., pp.1–148 (2019). https://doi.org/10.1787/726fd39d-en

5. EY, Microsoft: Artificial Intelligence in the Public Sector: European Outlook for 2020 and Beyond. Microsoft (2020)
6. Savaget, P., Chiarini, T., Evans, S.: Empowering political participation through artificial intelligence. Sci. Public Policy **46**, 369–380 (2019). https://doi.org/10.1093/scipol/scy064
7. König, P.D., Wenzelburger, G.: Opportunity for renewal or disruptive force? How artificial intelligence alters democratic politics. Govern. Inf. Q. **37** (2020). https://doi.org/10.1016/j.giq.2020.101489
8. Fast, E., Horvitz, E.: Long-term trends in the public perception of artificial intelligence. In: 31st AAAI Conference on Artificial Intelligence, AAAI 2017, pp. 963–969 (2017)
9. OECD: The Case for E-Government: Excerpts from the OECD Report The E-Government Imperative. OECD J. Budget. **3**, 1987–1996 (2003)
10. Mishra, S.S., Geleta, A.T.: Can an E-government system ensure citizens' satisfaction without service delivery? Int. J. Public Adm. **43**, 242–252 (2020). https://doi.org/10.1080/01900692.2019.1628053
11. le Blanc, D.: E-participation: a quick overview of recent qualitative trends. DESA Working Paper. Though the goal of realising citizen centricity ha (2020)
12. Chen, K., Aitamurto, T.: Barriers for crowd's impact in crowdsourced policymaking: civic data overload and filter hierarchy. Int. Public Manag. J. **22**, 99–126 (2019). https://doi.org/10.1080/10967494.2018.1488780
13. Toots, M.: Why E-participation systems fail: the case of Estonia's Osale.ee. Govern. Inf. Q. **36**, 546–559 (2019). https://doi.org/10.1016/j.giq.2019.02.002
14. Arana-Catania, M., et al.: Citizen participation and machine learning for a better democracy. Digit. Govern.: Res. Pract. **2**, 1–22 (2021). https://doi.org/10.1145/3452118
15. Al-Mushayt, O.S.: Automating E-government services with artificial intelligence. IEEE Access **7**, 146821–146829 (2019). https://doi.org/10.1109/ACCESS.2019.2946204
16. Zuiderwijk, A., Chen, Y.C., Salem, F.: Implications of the use of artificial intelligence in public governance: a systematic literature review and a research agenda. Govern. Inf. Q. 101577 (2021). https://doi.org/10.1016/j.giq.2021.101577
17. van Ittersum, K., Rogers, W., Capar, M.: Understanding technology acceptance: phase 1– literature review and qualitative model development. Technology Report …. 0170, 1–123 (2006)
18. Belanche, D., Casaló, L.V., Flavián, C.: Artificial Intelligence in FinTech: understanding robo-advisors adoption among customers. Ind. Manage. Data Syst. **119**, 1411–1430 (2019). https://doi.org/10.1108/IMDS-08-2018-0368
19. Davis, F.: Perceived usefulness, perceived ease of use, and user acceptance of information technology. MIS Q. **13**, 319–340 (1989)
20. Albarrán Lozano, I., Molina, J.M., Gijón, C.: Perception of artificial intelligence in Spain. Telematics Inform. **63**, 101672 (2021). https://doi.org/10.1016/j.tele.2021.101672
21. Nadarzynski, T., Miles, O., Cowie, A., Ridge, D.: Acceptability of artificial intelligence (AI)-led chatbot services in healthcare: a mixed-methods study. Digit. Health **5** (2019). https://doi.org/10.1177/2055207619871808
22. Kelley, P.G., et al.: Exciting, useful, worrying, futuristic: public perception of artificial intelligence in 8 countries. In: AIES 2021 - Proceedings of the 2021 AAAI/ACM Conference on AI, Ethics, and Society, pp. 627–637. Virtual Event (2021). https://doi.org/10.1145/3461702.3462605
23. Cho, S.H., Oh, S.Y., Rou, H.G., Gim, G.Y.: A study on the factors affecting the continuous use of e-government services - focused on privacy and security concerns -. In: Proceedings - 20th IEEE/ACIS International Conference on Software Engineering, Artificial Intelligence, Networking and Parallel/Distributed Computing, SNPD 2019, pp. 351–361. IEEE (2019). https://doi.org/10.1109/SNPD.2019.8935693

24. Araujo, T., et al.: In AI we trust? Perceptions about automated decision-making by artificial intelligence. **35**, 611–623 (2020). https://doi.org/10.1007/s00146-019-00931-w

25. Gesk, T.S., Leyer, M.: Artificial intelligence in public services: when and why citizens accept its usage. Gov. Inf. Q. **39**, 101704 (2022). https://doi.org/10.1016/j.giq.2022.101704

26. Lichtenthaler, U.: Extremes of acceptance: employee attitudes toward artificial intelligence. J. Bus. Strateg. **41**, 39–45 (2019). https://doi.org/10.1108/JBS-12-2018-0204

27. Bitkina, O.V., Jeong, H., Lee, B.C., Park, J., Park, J., Kim, H.K.: Perceived trust in artificial intelligence technologies: a preliminary study. Hum. Factors Ergon. Manuf. **30**, 282–290 (2020). https://doi.org/10.1002/hfm.20839

28. Taylor, S., Todd, P.A.: Understanding information technology usage: a test of competing models. Inf. Syst. Res. **6**, 144–176 (1995). https://doi.org/10.1287/isre.6.2.144

29. Chen, Y.N.K., Wen, C.H.R.: Impacts of attitudes toward government and corporations on public trust in artificial intelligence. Commun. Stud. **72**, 115–131 (2021). https://doi.org/10.1080/10510974.2020.1807380

30. Starke, C., Marcinkowski, F., Wintterlin, F.: Social networking sites, personalization, and trust in government: empirical evidence for a mediation model (2020). https://doi.org/10.1177/2056305120913885

31. Gefen, D.: E-commerce: the role of familiarity and trust. Omega (Westport) **28**, 725–737 (2000). https://doi.org/10.1016/S0305-0483(00)00021-9

32. Baek, T., Morimoto, M.: Stay away from me. J. Advert. **41**, 59–76 (2012). https://doi.org/10.2753/JOA0091-3367410105

33. Boerman, S.C., Kruikemeier, S., Zuiderveen Borgesius, F.J.: Exploring motivations for online privacy protection behavior: insights from panel data. Communic Res. **48**, 953–977 (2021). https://doi.org/10.1177/0093650218800915

34. LaRose, R., Rifon, N.J.: Promoting i-safety: effects of privacy warnings and privacy seals on risk assessment and online privacy behavior. J. Consum. Aff. **41**, 127–149 (2007). https://doi.org/10.1111/j.1745-6606.2006.00071.x

35. Bhattacherjee, A.: Acceptance of e-commerce services: the case of electronic brokerages. IEEE Trans. Syst. Man Cybern. - Part A: Syst. Hum. **30**, 411–420 (2000). https://doi.org/10.1109/3468.852435

36. Pechar, E., Bernauer, T., Mayer, F.: Beyond political ideology: the impact of attitudes towards government and corporations on trust in science. Sci. Commun. **40**, 291–313 (2018). https://doi.org/10.1177/1075547018763970

37. Marcinkowski, F., Starke, C.: Trust in government: what's news media got to do with it? Stud. Commun. Sci. **18**, 87–102 (2018). https://doi.org/10.24434/j.scoms.2018.01.006

38. Starke, C., Lünich, M.: Artificial intelligence for political decision-making in the European union: effects on citizens' perceptions of input, throughput, and output legitimacy. Data Policy **2** (2020). https://doi.org/10.1017/dap.2020.19

39. Hair, J.F., Ringle, C.M., Sarstedt, M.: Partial least squares structural equation modeling: rigorous applications, better results and higher acceptance. Long Range Plann. **46**, 1–12 (2013). https://doi.org/10.1016/j.lrp.2013.01.001

40. Jöreskog, K.G.: Simultaneous factor analysis in several populations. Psychometrika **36**, 409–426 (1971). https://doi.org/10.1007/BF02291366

41. Castañeda, J.A., Muñoz-Leiva, F., Luque, T.: Web Acceptance model (WAM): moderating effects of user experience. Inf. Manage. **44**, 384–396 (2007). https://doi.org/10.1016/j.im.2007.02.003

Assessing the Technological and Commercial Readiness for Innovative Projects

Stanislav Cseminschi, Elena Cojocari, Andreea Ionica$^{(\boxtimes)}$, Monica Leba ,
and Vlad Florea

University of Petrosani, Petrosani, Romania
andreeaionica@upet.ro

Abstract. We live in a world governed by innovation that imprints a rapid pace of change. Innovative start-up projects are the engine of these changes. The low success rates of these projects are well known and somehow assumed in the idea of fail, fail again, fail better. We learn from failure, but it's best to prepare for success. Thus, the paper proposes the use of the CloverLeaf methodology in the evaluation of innovative higher education projects. The identification of the technological, marketing, commercial and management readiness for the projects within the University of Petrosani represents the main results of the present research. These findings will be helpful in directing decisions regarding the following steps to be taken regarding the efforts to be invested based on the readiness dimension that needs to be improved, the crowdfunding campaigns (reward and equity) to be launched, and/or business accelerators that can raise the project's chances of success.

Keywords: Assessment · Innovation · Readiness · Technology

1 Introduction

In the current world, it is crucial to have the knowledge required to decide when the concept of a proposed project has developed enough to move on to the following stage. An evaluation enables project workers to comprehend how much innovation a certain technology needs before being put to use. A rating might be useful in gauging a project's advancement. An examination of a project's competence for a significant change is called a readiness assessment.

Robert Cooper, a process specialist, claims that just one new product out of every seven is successful and about four new items out of every seven go through the development phase [1]. The great powers are part of a global race in the field of technological innovation. This competition is capable not only of generating conflicts and economic problems for some companies or states caught in the middle [2], but at the same time, it can bring major benefits for people as well as technologies to simplify processes and help economy growth. A trend can be observed through the international policies applied in favor of innovation, with the big global players recognizing its importance in a post-pandemic world and with increasingly important geopolitical context. We also see a competition between the great powers to be the first to reach a higher level of technological innovation, a kind of arms war that no longer uses nuclear weapons, but advanced technologies capable of sustaining economic growth [3].

Despite the fact that innovation has been extensively studied, most of the work has been conceptual. In other words, researchers have looked at innovation and have attempted to categorize, explain, and find techniques to best succeed in it. As a result, there is surprisingly little quantitative research on the subject, maybe because innovation is frequently challenging to measure. Many technological advancements have not yet been fully adopted in the real world and have not been commercialized. The simplest form of the technology commercialization process begins with a researcher developing the technology, which then reaches the market. Recent researches [4, 5] have suggested some obstacles related to innovations that can be seen in Fig. 1. Inability to assess the innovation's readiness is one of the biggest barriers.

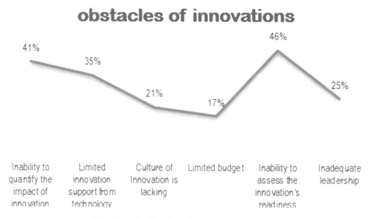

Fig. 1. The barriers for the innovations

The technology by itself cannot produce wealth; rather, wealth is produced through the applications of technological inventions or innovations in the form of goods and services. So, it is important to evaluate a technology's "readiness for commercialization" at an early stage.

Problem Statement - The lack of reliable prediction techniques determines the low success rate in technology transfer from universities.

The **main objective** of the paper is the assessment of readiness of the selected projects. From the fulfillment of this objective derives the importance of research for the development of future projects. Thus, the results will be useful in evaluating the next steps, namely analyzing the opportunity to use crowdfunding platforms and/or participating in business accelerators and/or try to improve the poor dimensions of the innovative projects so they can be ready for the other steps, such as one from the first two declared above.

The paper is divided into 5 parts, begins with an introduction and ends with conclusions. After the introduction, the next part shows the features of the readiness assessments and their descriptions. The third part shows the insides of the innovative projects of the University of Petrosani and leads to the part of applying the Cloverleaf Model on them and to the main conclusions.

2   Materials and Methods

2.1   Theoretical Background - Readiness Assessments

Researchers begin to assess their inventions from the first step until the finished product is ready for use in the marketplace in order to prevent failures. The NASA-developed scale with Technology Readiness Levels (TRLs) [6] is one of the most extensively used tool for evaluating technology maturity.

In order for a technology to go from conception to research, development, and deployment, the TRLs must be methodically addressed [7]. Universities concentrate on TRLs 1–4, whereas businesses concentrate on TRLs 7–9.

Even if the TRL is frequently used for knowing the readiness of a product, it mostly shows the technological part of an innovation. So, for showing a more commercial part of an innovation was created the CRL.

The Commercial Readiness Level (CRL) methodology [8] evaluates a number of variables that affect market and commercial conditions in addition to technology maturity. This makes it possible to overcome important obstacles to support the commercialization of a technology. Similar to TRL, CRL rates [9] the commercial viability of a technology on a scale of 1–9.

Frequently, these methodologies such as TRL and CRL are used separately by researchers. The scholars, Heslop, McGregor, and Griffith, combined a number of success dimensions into what they call the Cloverleaf framework [10]. The model, which has already been validated, incorporates all of the major categories of criteria mentioned in the literature and offers a balance between the many areas so that no one area predominates in the total evaluation. All these contributed to the choice of the model for use in the research. This model claims that four major dimensions are required for a science and technology venture success. These are technology readiness, market readiness, commercial readiness, and finally, management readiness dimensions. Technology readiness is, of course, a starting point. Technological readiness refers to the propensity of people to accept and use new technologies to achieve goals in personal life and at work [11]. Design can be seen as a general state of mind resulting from a gestalt of psychic activators and inhibitors that together determine a person's predisposition to use new technologies. The principal aspects of the technological readiness are presented in Fig. 2.

Market readiness is the process of ensuring that a company or organization has a product ready to launch on the market [12]. This process often involves interviewing potential users to determine whether the product meets the needs of specific users or whether the product will help them complete their tasks more efficiently. Investing in marketing to push more business into an inefficient or ineffective pipeline can be a recipe for disaster. Lack of proper support structure can jeopardize a company's growth and stability. Marketing Readiness dimension helps identify key areas that need attention before marketing and sales activities begin. It not only reveals strengths, but also gaps in how well the company articulates who it is, what it does, how it does it, and what makes it different.

Fig. 2. The principal aspects of the technological readiness

Market maturity is very important and corresponds to technologies that offer identifiable and quantifiable advantages [13], with clear advantages over competing products resulting in significant market growth. In Fig. 3 are presented the quantitative and qualitative aspects of market readiness.

Fig. 3. Quantitative and qualitative aspects of market readiness

A third related dimension is called commercial readiness. It reflects the development stage of the commercialization business model. One of the most interesting and crucial steps in invention is the commercial readiness. Customer satisfaction is ensured by commercial readiness, which is a crucial prerequisite for bridging the gap to commercial growth in mainstream markets [14]. Prior to the launch of a new product for sale, commercial readiness must be rigorously validated.

350   S. Cseminschi et al.

This dimension summarizes the result of the market assessment, taking into account the price and consumer qualities of competitors' products introduced to the market and developed. It can be used even to determine the market readiness of the technology.

Commercial readiness fixes the organization of bilateral exchange of information with potential clients in order to receive feedback on interest and clarify the characteristics of the development object. Also, it fixes successive approximations of the pricing model and the corresponding adjustments in production technologies, taking into account the constraint on the price of products.

The commercial aspect defines innovation as an economic necessity realized through the needs of the market [15]. The "materialization" of inventions, and developments into new, technologically sophisticated types of industrial products, means and objects of labor, technologies, and production organization, as well as "commercialization", which makes them into a source of revenue, are two points that should be noted.

After technology readiness, market readiness, and commercial readiness, we come to the fourth dimension called the management readiness. Management readiness assesses the readiness of the management team that is responsible for the technology. It addresses issues like the inventor's capacity to lead the innovation as a team player, whether the inventor's expectations for success are reasonable [16], whether the inventor is well-known and regarded in the industry, whether commercialization skills like sales and marketing capabilities are available, whether management capabilities are available, and whether the inventor is the owner of patents for innovations developed in government laboratories.

In Fig. 4 can be seen that only 4% of executives did not designate innovation as a strategic goal and do not have any intentions to do so in the future. The information is according to the McKinsey Global Survey conducted in 2010 [17].

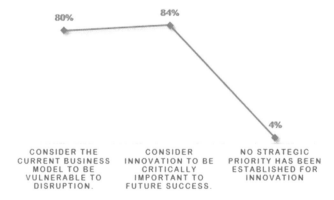

Fig. 4. How executives take innovation into consideration

Ideally, the management team, will have the experience, the expertise, the energy and excitement to commercialize the invented technology. The link between people, processes, and performance assessment is indicated by a management readiness. Without coordination and synchronization, no implementation will be successful. In order to coordinate operations and convey changes, the organization needs to have systems and personnel in the right place. Learning and adapting to change are ongoing processes. The objective is to transform the company and alter people's perspectives. A creative team needs to conduct thorough research on the technology in order to succeed [18]: its capacity, readiness, and fundamental market qualities, as well as its freedom or ability to run a firm, its marketing strategy, and how well it fits in with the team as a whole. The degree to which each requirement is met was determined by quantifying the judgements made for each criterion of the four Cloverleaf framework regions. The technology is prepared for commercialization if all of the requirements listed for the four "leaves" determining readiness are met [16].

2.2 Research Approach

The problem statement is the outcome of an exploratory research, which is followed by a formal research design that uses techniques from descriptive and causal research to gather the necessary data, analyze, and draw conclusions (see Fig. 5).

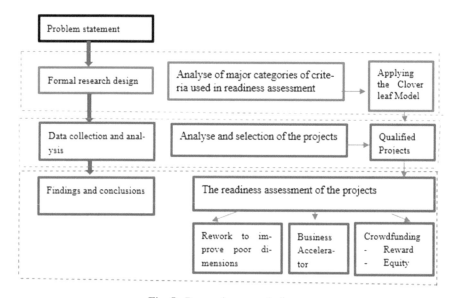

Fig. 5. Research approach diagram

After examining the main categories of readiness evaluation criteria, the Clover Leaf model was selected since it is a validated model and it covers the majority of the criteria described in the literature. The methodology was used for the qualified projects that emerged from a qualitative study conducted at a university-based business accelerator. The projects' readiness assessments are helpful for future enhancements, such as rework to improve one or more of the dimensions, participate in dedicated accelerators or create a crowdfunding campaign, either reward or equity type.

3   Insides of the University of Petrosani Innovational Projects

Project activities carried out as part of the educational process at the University of Petrosani are mainly aimed at innovational projects. These initiatives have financial costs and are geared towards producing a beneficial outcome.

In Table 1 are presented three of the ongoing innovational projects related to the University of Petrosani. These were selected based on an accelerator organized in the university, completed with a demo day where pitches and Business Model Canvas were presented for 14 student projects.

Table 1. Selected innovational projects

Name, Logo; TRL	Description
TRL 4: Small Scale Prototype Prototype in testing: Techsylvania2022, ADRVest Accel 2022	A bracelet that lights up based on a color code to gently signal a stress state. This is part of a wearable device for alarming the burnout state using an application that we develop, that estimates this state based on a personal profile, taking into account the physiological and the organizational climate parameters.
S M R T TRL 5: Large Scale Prototype Prototype tested: Innovation Labs 2022, Techsylvania 2022, ADRVest Accel 2022	Small electric car with two seats side-by-side, usable on various types of terrain, like parks or urban centers, to reduce pollution by using alternative sources such as regenerative braking and solar energy panels to power the motor.
ware TRL 4: Small Scale Prototype Prototype in testing: Techsylvania2022, ADRVest Accel 2022	Customized wearable devices for quality-of-life improvement for people with low mobility of the upper limb. The problem that we address is related to medical conditions due to upper limb impairments with main cause strokes, but also accidents or genetic disorders.

4   Applying the Cloverleaf Model to the Innovational Projects of the University of Petrosani

The four dimensions of the Cloverleaf Model were used to analyze the projects and identify the best traits for each innovation that can be seen in the Table 2.

By using the Cloverleaf Technology Readiness Assessment Tool [10] we quantified the best traits found in each of the four dimensions for the innovative projects of the university. Using the Cloverleaf Model, the three above mentioned technologies of the University of Petrosani were evaluated, using a quantitative research based on questionnaires having 112 validated responses and a qualitative research based on interviews with business mentors and entrepreneurs that were part of the ADRVest Accel 2022. This accelerator brought the best start-ups from Arad, Timiș, Caraș-Severin and Hunedoara counties face to face with potential investors. The entrepreneurs, including the ones from the University of Petrosani had the opportunity during the program to present their business ideas to the audience.

The interviews were made with the 26 entrepreneurs that were selected from the 200 registered, within the business acceleration program and with over a hundred mentors and investors that were present. The results provided information on the differences between the projects and a possibility to reflect on the perception of technology commercialization readiness.

Table 2. Best traits in the four dimensions of the innovative projects

	Technology readiness	Market readiness	Commercial readiness	Management readiness
BCalm	A recently new technology No other dominant patents exist Resources for research and development are available for this innovation	It clearly offers advantages over other products This innovation has definable market potential Market is large and is growing Manufacturing is viable	Customer experience is superior Inventor has industry contacts Expected positive return on investment Low marketing costs	Communication is efficient and effective Inventors are good players as a team
XWare	Fixes an issue Has clearly visible advantages Offers unique benefits to customers or end users Products and services are highly usable	It has significant long-term potential No fierce or aggressive competition is present	Availability of raw materials Potential partners with similar technology Aligned to the values of society	Proactive approach to problem solving Inventors are realistic to success high expectations
SmartBuggy	A quality product may result from it The capacity to drive industry change The prototype has been successfully shown	There is need for this technology Users are prepared to pay a fair price Low dissonance applies to the technology	Good reputation Lack of debt Positive impact on people and planet Products align to customer needs	Management skills are available Satisfied employees Willingness to take calculated risk

These technologies with their respective grading are shown in Table 3.

Table 3. Evaluation of the three projects from University of Petrosani

	BCalm	XWare	SmartBuggy
Technology Readiness (*Max 90 points*)	33	18	26
Market Readiness (*Max 45 points*)	30	15	35
Commercial Readiness (*Max 63 points*)	32	30	31
Management Readiness (*Max 45 points*)	31	31	29
Total Score (*Max 243 points*)	126	94	121

The BCalm project has the biggest score, but it is below 2/3 of the maximum of 243 points, meaning that in this moment none of the projects is ready for commercialization.

By analyzing the total scores, Bcalm and SmartBuggy fall in the interval 1/3 and 2/3 of maximum, meaning that these are suitable for participating in dedicated business accelerators, in order to benefit from a guided improvement of weak dimensions.

The project XWare needs more rework on technology readiness in order to be accepted and fully benefit from a dedicated accelerator.

Neither of these projects is ready for a crowdfunding campaign.

The evaluation criteria utilized have enhanced the quality of decisions regarding the selection of technology for commercialization, even though this assessment could yet be improved.

5   Conclusions

From presenting innovative ideas to debating some of the most important issues that arise when business meets technology, the assessment of the readiness can be the key to the movement to the next stage.

The significance of successful technological commercialization can emerge from the study. Many times, the company behind an idea or discovery did not find the right time to develop, manufacture, and sell the particular technology or item. But when compared to 100 years ago, most technological advancements can be attributed to the ongoing inventiveness of a big population.

The process of constant progress can continue for a very long time, at least until the market is saturated or a newer innovation emerges that has a competitive advantage, it is simpler to use, more dependable, or accomplishes the job better.

The paper highlights the usefulness of finding out the technological commercialization readiness of an innovation. By using the Cloverleaf Model, we made the four assessments making it possible to differentiate, at least for now, between normal and good projects. All these results guide us in exploring the most effective opportunities for the evolution of the Higher Education innovative projects carried out by teams of students from the University of Petroşani.

The critical analysis allowed the identification of the weak points and the best suited future enhancement directions for each project.

We believe that in order for these projects to go from invention to innovation, the present research takes an important step at this moment in the new product development life cycle. Nowadays, technology is not just a hobby. For some, it may be a profession, but for some, it is the meaning of life.

References

1. Cooper, R., Kleinschmidt, E.J.: An investigation into the new product process: steps, deficiencies, and impact. J. Prod. Innov. Manage. **3**(2), 71–85 (1986)
2. Vogels, A.E., Rainie, L., Anderson, J.: Power dynamics play a key role in problems and innovation. Pew Research Center (2020)
3. Chin, W.: Technology, war and the state: past, present and future. Int. Afairs **95**(4), 765–783 (2019)
4. Am, J.B., Furstenthal, L., Jorge F., Roth, E.: Innovation in a crisis: why it is more critical than ever. In: McKinsey & Company, Strategy & Corporate Finance Practice (2020)
5. Arza, V., López, E.: Obstacles affecting innovation in small and medium enterprises: quantitative analysis of the Argentinean manufacturing sector. Res. Policy **50**(9), 104324 (2021)
6. Mankins, J.: Technology readiness level – a white paper, advanced concepts office office of space access and technology NASA (1995)
7. Dunbar, B.: Technology readiness level. NASA (2022)
8. Animah, I., Shafiee, M.: A framework for assessment of Technological Readiness Level (TRL) and Commercial Readiness Index (CRI) of asset end-of-life strategies. In: European Safety and Reliability Conference (ESREL) (2018)
9. Gerdsri, N., Manotungvorapun, N.: Readiness assessment for IDE startups: a pathway toward sustainable growth. Sustainability **13**, 13687 (2021)
10. Heslop, L.A., McGregor, E., Grifith, M.: Development of a technology readiness assessment measure: the cloverleaf model of technology transfer. J. Technol. Transf. **26**, 369–84 (2001)
11. Persons, T.M., Mackin, M.: Technology readiness assessment guide: best practices for evaluating the readiness of technology for use in acquisition programs and projects. US Government Accountability Office, Washington, DC, USA (2020)
12. Hjorth, S.S., Brem, A.M.: How to assess market readiness for an innovative solution: the case of heat recovery technologies for SMEs. Sustainability **8**(11), 1152 (2016)
13. Verma, S.: New product newness and benefits, A study of software products from the firms' perspective. Malardalen University Press Dissertations No. 81 (2010)
14. Pakurár, M., Haddad, H., Nagy, J., Popp, J., Oláh, J.: The service quality dimensions that affect customer satisfaction in the Jordanian banking sector. Sustainability **11**, 1113 (2019)
15. Huang, Z., Farrukh, C., Shi, Y.: Commercialisation journey in business ecosystem: from academy to market, entrepreneurial, innovative and sustainable ecosystems. Appl. Qual. Life Res. (2018)
16. Oosthuizen, R., Buys, A.J.: The development and evaluation of an improved cloverleaf model for the assessment of technology readiness for commercialization. South Afr. J. Ind. Eng. **14** (2012). https://doi.org/10.7166/14-1-302
17. Capozzi, M., Gregg, B., Howe, A.: Innovation and commercialization. McKinsey Global Survey results (2010)
18. Haghbin, A.: Readiness assessment for technology management system implementation within a conglomerate. In: IEEE International Conference on Engineering, Technology and Innovation (2019)

Progressive Web Apps: An Optimal Solution for Rural Communities in Developing Countries?

Carlos Cuenca-Enrique[1] , Laura del-Río-Carazo[1(✉)] ,
Santiago Iglesias-Pradas[1] , Emiliano Acquila-Natale[1] , Julián Chaparro-Peláez[1] ,
Iván Armuelles Voinov[2] , José Gabriel Martín Fernández[3],
and Cristina Ruiz Martínez[3]

[1] ETSI de Telecomunicación, Universidad Politécnica de Madrid, Madrid, Spain
{carlos.cuenca,laura.delrio,s.iglesias,emiliano.acquila,
julian.chaparro}@upm.es
[2] Universidad de Panamá, Panamá, Panamá
ivan.armuelles@up.ac.pa
[3] Fundación acciona.org, Alcobendas, Spain
{josegabriel.martin.fernandez,
cristina.ruiz.martinez}@acciona.com

Abstract. ICTs can play a key role in addressing the challenge of energy access in rural areas of developing countries. It is essential to choose the right technology to suit the context, as these areas have characteristics that must be considered for the correct appropriation of solutions, such as the instability of the network or the age of the devices. The aim of this study is to analyze which devices and which development technologies are the most suitable for the development of digital solutions in rural areas of developing countries. To this end, on the one hand a comparison is made between mobile and fixed terminals, with the result that mobile devices are the most suitable due to penetration figures, availability of mobile networks and affordability. On the other hand, mobile development solutions such as, Native Apps, Hybrid Apps, Mobile Web Apps and Progressive Web Apps (PWAs) are analyzed. After analyzing the technologies from the user's point of view, considering aspects such as adoption, affordability and usage, it is concluded that the optimal solution is PWA technology, as it combines the favorable aspects of web and application development. Finally, a successful case of the development of an energy access management tool developed with PWA technology for a rural electrification project in Ngäbe Buglé, Panama, is presented.

Keywords: Developing Countries · ICT · Energy access · Management · Mobile · Progressive Web Apps

1 Introduction

Today, 6.96 billion people live in a developing country, which is 87.10% of the world's population [1]. A total of 137 countries are considered developing countries, including all Central and South America, all of Africa, almost all Asian countries and many island states [1]. One of the main challenges that people are facing in these countries is energy poverty [2]. Despite the overall improvement in energy access in recent years, from 83% in 2010 to 90.2% in 2021, rural areas in developing countries still show levels of access that are far from the targets set in the 2030 Agenda. In these cases, the main challenge remains to address the "last mile" problem, i.e. reaching the most vulnerable and isolated communities: 759 million people do not have access to energy yet, 85% of whom live in rural areas [3]. Communities in the last mile are often isolated and with very complex access, so operation and maintenance tasks are often very costly for projects.

Solar home systems (SHS) are traditionally used to provide last mile energy access, rather than conventional solutions such as grid extension, as they provide energy to dispersed and remote points, being individual systems, and at a much lower cost. In addition, these systems significantly improve their impact and sustainability when used with energy-as-a-service (EaaS) business models [4]. This model allows users to pay regular electrification fees instead of having to make a significant upfront investment to purchase the equipment. In addition, maintenance is included in the periodic fees so that they do not have to pay large amounts of money when the SHS needs to be repaired. Both SHS and the EaaS model bring remarkable benefits for the inhabitants of the last mile, however, the operation and maintenance management they require is quite complex for the providers [5]. ICT access is a key point in the last mile energy access challenge, as it will enable universal access to be achieved faster and in a sustainable way. But it is not a solution in itself; it must be properly designed to ensure the impact of the solution [4, 6].

In this study, we will focus on the design of digital solutions adapted to rural environments in developing countries. The aim of the research is to identify which technologies have the most suitable characteristics to implement appropriate ICT. The study is divided into the following sections. Section 2, reviews existing technologies for the design of solutions and compares them from the point of view of the needs of developing countries. Section 3 describes a practical case. Section 4 presents the conclusions and future directions of the study.

2   Analysis and Discussion

The disparity in access to technology between developed and developing countries makes it clear that the same solution will not be valid for both areas of the world. When designing technology solutions for rural dwellers in developing countries, attention must be paid to multiple factors that will influence all decisions, from the choice of technology and the design of the solution to the type of maintenance it will require [7, 8]. It is necessary to identify the critical factors that will influence the correct technological implementation and thus ensure the impact and sustainability of the solution. Technical, economic and social issues are the main areas that hinder the introduction of ICTs in developing countries [9–11]. The most prominent barriers in each area are:

- **Technical**: aspects related to the availability and quality of networks and devices, such as a) the technologies to which they have access are old and often obsolete for existing developments; b) technologies are complex, which complicates their maintenance and use; c) the lack of access to connectivity networks makes the implementation of many solutions difficult; d) when they do have connectivity, they have poor signal reception, low bandwidth and/or an unstable connection, so solutions that work offline are required, and e) the lack of electricity supply prevents the implementation of digital solutions in many cases.

- *Economic:* refers to the financial aspects of ICT, such as a) the lack of investment in appropriate solutions, b) the low profitability of projects due to specificity and c) the high costs of ICT, including equipment, development and maintenance costs.
- *Social*: these factors include social barriers and human capacities to implement ICT. They include factors such as a) high resistance to change, which makes it difficult to implement solutions that are too innovative, b) lack of technological skills, which makes it necessary to prioritize user experience in these solutions, c) shortage of qualified staff to maintain and adapt solutions, and d) high levels of illiteracy, which makes it necessary to adapt solutions.

In this context, it would be necessary to analyze which solutions are best suited to address the development of ICT solutions in developing countries. It is necessary to bear in mind that the selection of a technology will not have an impact on all barriers, since, as mentioned in the previous section, technology alone cannot solve a problem. From the users' perspective the analysis will be divided into two parts, first we will analyze the devices on which to design solutions and then we will analyze which development technologies to choose.

2.1 Appropriate Devices for ICT Solutions in Developing Countries

To assess which devices would most easily overcome the barriers to ICT deployment, the digital divide between rural areas in developing countries and the rest of the world is contextualized.

Eighty-eight percent of the rural population in developing countries is covered by a 3G or more advanced mobile network, which is considered adequate connectivity for the use of online applications [12]. However, the figures are significantly lower for fixed broadband and fixed telephony in developing countries, at 13% and 8% respectively. Although no specific data are available for rural areas, it can be assumed from the overall data that the gap between mobile telephony and the rest will be even wider in rural areas. In terms of available devices, mobile phones and computers have a large gap in their numbers, 78% versus 35%, which is likely to be even higher in rural areas, as it can be assumed that most of the complex devices will be located in urban areas in these countries [13]. Moreover, in terms of affordability, fixed broadband tariffs in developing countries exceed the UN price ceiling of 2% of monthly gross national income (GNI) per capita. However, the average price of mobile broadband in developing countries does not exceed this figure, making mobile phones the most affordable option for rural dwellers [14]. Finally, in terms of power supply, the energy required to recharge a mobile phone is much lower than for other devices [15, 16].

From all these aspects, it can be stated that, in technical and economic terms, mobile devices are the most suitable for rural areas in developing countries, although the technology used to develop the solution will be key, as we will see in the next section.

2.2 Appropriate Technologies for Mobile Development in Developing Countries

The requirements for mobile application development are becoming increasingly complex. There is a need to develop solutions that are cross-platform, compatible with multiple devices and provide an optimal user experience [17]. Below, we outline the different options for developing mobile solutions: native applications, mobile web applications, hybrid mobile applications and progressive web apps (PWA).

Native Applications
A native application is a software developed specifically to run on a particular operating system (Android, iOS, Windows Phone…). The advantage of this technology is that the development of these apps is done using a language and tools specific to each operating system [18]. This allows these apps to be optimized for these devices, making them faster and more efficient, as well as being able to access specific system resources [17]. In addition, because they are installed on the device, they can be accessed offline. On the negative side, when developing an app, it is necessary to make a different development for each operating system, which implies high costs [19]. In addition, as they are solutions installed on the devices, updates must be developed to keep the solution operational. Another disadvantage of this type of app is that they usually require a distribution channel (Play Store on Android and App Store on iOS) so that the end user can obtain the app on their device; this requires each version of the application to be approved by the organizations that control it (Google, Apple…), which means extra time and little control over the process [20].

Hybrid Applications
A hybrid application is software developed in a specific language or with specific tools that allow it to be transformed into one or more native applications [21]. They have the advantage that they use specific system resources and are therefore more efficient, just like native applications. In addition, unlike native applications, they require shorter development times as they do not require specific developments for each system and device. Regarding the negative points, although they improve the development aspects compare to native apps, they maintain the disadvantage of requiring updates to remain operative, implying that the devices on which they are installed cannot be very old since in that case there will be no updates available [20].

Mobile Web Applications
A web application is a software developed to be executed in a web browser [19]. The main advantages are that they allow developments to be agnostic to the operating system that runs them, so it is not necessary to make specific developments for each operating system. On the negative side, because these applications run on top of the browser layer, they are less efficient and cannot access all the available resources of the device. In addition, since they run on the browser, they require an Internet connection to work, unlike native and hybrid applications [21].

Progressive Web Applications
A PWA is a website that looks and behaves as a mobile app. The main advantage is that

it is a software developed to be executed in a web browser, avoiding the inconveniences of obsolescence and updates and allowing it to be executed offline and installed on any operating system that supports it just like native and hybrid applications. This allows the creation of applications that can be installed on any device without the need for specific development for each operating system [22]. In addition, they are much easier to develop than the other options, require shorter development times and do not rely on third parties. They do not require intermediaries, and therefore the download, installation and update tasks are fully controlled by the developers, which means shorter development times. In short, they have similar characteristics to web apps, but with all the positive aspects of native apps, good performance in unstable network conditions and reduced loading times [23].

Table 1. Developing countries ICT Barriers – Technology Analysis

Barrier	Native App	Web Mobile App	Hybrid App	PWA
Technical				
Obsolescence	X	✓	≈	✓
Complexity	X	✓	≈	✓
Lack of access to connectivity	–	–	–	–
Quality connectivity	✓	X	✓	✓
Lack of power supply	–	–	–	–
Economic				
Lack of investment	X	✓	✓	✓
Low profitability	X	✓	✓	✓
High costs	X	✓	✓	✓
Social and educational				
High resistance to change	–	✓	–	✓
Lack of TIC skills	≈	X	≈	✓
Shortage of skilled personnel	X	✓	≈	✓
High levels of illiteracy	–	–	–	–

"✓": Positive Impact; "≈": Medium Impact; "X": Negative Impact; "–": None Impact

Having analyzed the different technologies for the development of mobile solutions, an assessment has been made of which option is the most suitable to overcome the barriers to ICT development in developing countries (Table 1).

On a technical level, it is observed that Native and Hybrid App solutions, thanks to being installed on the devices, allow adapting to the quality of connectivity, however, they present problems of technological complexity and obsolescence, since they require updates that must be compatible with the models of mobile devices. In this area, PWAs manage to overcome all barriers, offering offline solutions and running on web browsers, without the need to adapt to each device and operating system.

At an economic level, almost all the technologies analyzed overcome the barriers at the same level; only native apps have a negative impact in this aspect, since they require much more resources and the developments are not reusable, as they must be developed from scratch for each device and operating system. In addition, the technical profile required to develop native applications is much more specialized and therefore will involve higher costs.

Finally, at the social level, cell phones present facilities for illiterate users, given the ease of use of the microphone and speaker, and, given the high penetration rate of these devices, resistance to change will be significantly lower. In terms of technologies, the only one that adapts to all barriers is the PWA, because it is much more common to find personnel specialized in web than app, and, since it does not need to be installed or updated, it is easier to handle for people with lower technological skills. Finally, the user experience is more similar to the solutions to which they are already accustomed.

In conclusion, the mobile solution technology best suited to the barriers of ICT deployment in developing countries is PWA, as it combines the positive aspects of both worlds (web and app), and manages to create an ideal solution to address the barriers identified in the literature. This does not mean that there are not areas for improvement in the development of mobile solutions, but PWA is currently the technology best suited to this environment.

3   Practical Case: Client Portal App

This chapter describes a pilot project in Ngäbe Buglé, an indigenous area in Panama where a PWA has been developed to facilitate the management of a rural electrification project. The project uses Solar Home Systems with an Energy as a Service model.

3.1   Context

The Client Portal App has been developed in collaboration with the acciona.org Foundation to meet the digitalization needs of the rural electrification project in Ngäbe Buglé, Panama. The project began in 2018 under the leadership of *Acciona Microenergía Panama*, a Panamanian non-profit association, and its partner, the acciona.org Foundation. Currently, 2,500 households have access to electricity thanks to the supply of home photovoltaic systems. The Ngäbe Buglé region is the largest indigenous region in Panama, and only 4% of its population has access to electricity. The Ngäbe Buglé region has lower access to communication services and devices than the rest of Panama's indigenous peoples, and mobile networks and mobile devices continue to outnumber other options. In Panamá, 2G and 3G networks cover 88% of the population, while 4G networks cover 12%. In the region, 0.9% of families have residential fixed telephony service, compared to 26.9% who have access to mobile telephony service. Prior to 2020, almost no Ngäbe Buglé families had home internet access, with the mobile phone serving as the only available system for accessing Internet Services. In terms of affordability, the global figures remain unchanged, with the cost of mobile subscriptions amounting to 2.1% of GNI, while fixed-broadband networks account for 5.7%. It seems clear that mobile solutions are the most appropriate for this Panamanian region, for reasons of availability and affordability [24].

3.2   Problematic and Requirements

The Ngäbe Buglé communities in which the foundation operates are located in scattered parts of the region and it is difficult and time-consuming to travel to them. To solve this problem, operation centers were created with the support of local agents in the region, where users can go to pay their electrification fees and report equipment breakdowns. Despite the fact that owing to the existence of these centers it has been possible to provide maintenance and assistance services to users, distances are still too long for users to have constant communication with the Foundation, and many of the trips to these centers could be avoided, since in many cases breakdowns are resolved with very simple instructions that the users themselves can carry out at home (e.g.: remove dust and dirt from the solar panel or disconnect and connect the connection cables of the light bulbs). In addition to the maintenance aspects, users must visit the centers monthly to pay their 5.00 USD fee. In many cases, getting to and from their homes costs as much or more than the monthly electrification fee.

As seen in the previous section, the region has connectivity and access figures that make mobile solutions the most viable solutions to improve the quality of life of its inhabitants. To design a viable solution, the requirements of the area were identified. Despite having access to the Internet network, access is unstable and with reduced bandwidth. The project users have mobile devices although they do not have the latest versions of their operating systems. Finally, the cost needs to be low and the platform must have a user experience similar to that of other tools that users are already used to, in order to facilitate adoption and use.

Bearing in mind the requirements of the area and the operational problems of the project, and taking advantage of the researchers' experience in the design and implementation of ICT solutions for projects in developing countries, the architecture of a mobile solution using PWA technology was designed.

3.3   Description of the Solution

The Client Portal App platform has been developed with the aim of facilitating the management of electricity supply to rural populations in developing countries that do not have a conventional electricity service but rely on isolated photovoltaic systems to have energy access. Figure 1 describes the basic components of the platform. The functionalities available on the platform are a) access to their personal and utilities information, b) creation and consultation of failures, c) possibility of mobile payment for electrification fees and d) access to resources such as user manuals, frequently asked questions and catalogue of electrical appliances.

A screenshot of the tools main page is shown in Fig. 2a. The appearance is that of an App, with a menu in the lower area, no navigation bar and occupying the entire screen. In addition, when opening the tool, a notification appears to offer the installation of the tool on the device, simulating the installation process of an App and placing a shortcut on the main screen of the phone (Fig. 2b). The application works offline, and therefore customers can access all their information without continually relying on the network, which facilitates its management since in the area they have unstable connectivity.

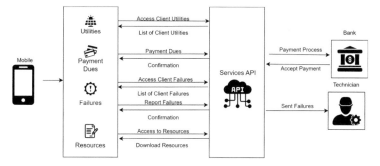

Fig. 1. Software components and architecture Client Portal App

(a)   Initial screen   (b)   Install Notification

Fig. 2. Client Portal App Screens

3.4 Results

After designing and developing the first version of the application, which took three months of work, the development met all the required functionalities and was suitable for the application context. After validating the first version, during the following two months, user experience improvements were made and the suitability of the solution was analyzed with early adopters in the region and with the Foundation's team working on the ground in the region (20 users), obtaining a 100% success rate in terms of installation and adaptation to the connectivity context. We are currently working to launch the first pilot of the application and expect to have 100 active users during 2023. The pilot study will analyze user behavior in two main areas: adoption and use.

4   Conclusions

ICTs can play a key role in helping developing countries to meet the challenge of last-mile energy access. However, for technological solutions to have a real impact, it is necessary to overcome existing barriers to ICT deployment in developing countries. Twelve barriers have been identified, classified into three different areas: technical, economic and social.

Mobile solutions are best suited to overcome ICT implementation barriers in developing countries. However, it is necessary to carefully select the technology to develop these mobile solutions, as not all of them will achieve the same results. Of the four technologies studied (native applications, mobile web applications, hybrid applications and PWA), only PWA overcomes all the barriers.

As a practical case study, a tool developed to improve the management of energy access in the Ngäbe Buglé region in Panama has been presented. The tool has been developed with PWA technology and it manages to fulfill the requirements, by combining the positive aspects of native applications and mobile web applications. As future lines of research, the user experience with the tool will be evaluated by means of a pilot and it will be assessed whether it is necessary to make design modifications in the application to improve the implementation of the solution and its impact on the electrification project.

References

1. United Nations. Revision of World Population Prospects 2022. (2022). https://population.un.org/wpp/. Accessed 14 Nov 2022
2. United Nations Conference on Trade and Development UNCTAD. Four key challenges facing least developed countries. (2022). https://unctad.org/news/four-key-challenges-facing-least-developed-countries. Accessed 31 Oct 2022
3. World Bank. Universal Access to Sustainable Energy Will Remain Elusive Without Addressing Inequalities. (2021). shorturl.at/kqLO6. Accessed 13 Nov 2022
4. SEGIB, ARIAE, and MAUE. ODS 7 en Iberoamérica. Alcanzar la última milla, p. 148 (2021)
5. Del-Río-Carazo, L., Acquila-Natale, E., Iglesias-Pradas, S., Hernández-García, Á.: Sustainable Rural Electrification Project Management : An Analysis of Three Case Studies. Energies **15**, 1203 (2022). https://doi.org/10.3390/en15031203
6. Labelle, R., Pramanik, J., Sarkar, B., Kandar, S.: Impact of ICT in rural development: perspective of developing countries ICTs for e-environment guidelines for developing count ries, wit h a focus on climat e change I… impact of ICT in rural development: perspective of developing countries. Am. J. Rural Dev. **5**(4), 117–120 (2017). https://doi.org/10.12691/ajrd-5-4-5
7. Del-Río-Carazo, L., Iglesias-Pradas, S., Acquila-Natale, E., Martín-Fernández, J.G.: Appropriate technology for access to universal basic services: a case study on basic electricity service provision to remote communities in the Napo River Basin. Sustainability (2022). https://doi.org/10.3390/su14010132
8. Cuenca-Enrique, C., Del-Río-Carazo, L., Acquila-Natale, E., Macarrón, C.: Semi-automatic document processing and text-recognition for digital transformation of low-resources organizations. Iber. Conf. Inf. Syst. Technol. Cist. **2022**, 22–25 (2022). https://doi.org/10.23919/CISTI54924.2022.9820135
9. Touray, A., Salminen, A., Mursu, A.: ICT barriers and critical success factors in developing countries. Electron. J. Inf. Syst. Dev. Ctries. **56**(1), 1–17 (2013). https://doi.org/10.1002/j.1681-4835.2013.tb00401.x
10. Shahadat, M., Khan, H., Hasan, M., Kum, C., Prof, C.: Barriers to the introduction of ict into education in developing countries: the example of Bangladesh. Int. J. Instr. July **5**(2), 1694–1609 (2012)
11. Kapurubandara, M., Lawson, R.: Barriers to adopting ICT and e-commerce with SMEs in Developing Countries: an Exploratory study in Sri Lanka. Framework, pp. 1–12 (2001). http://www.collecter.org/archives/2006_December/07.pdf

12. Lamba, A., Yadav, J., Devi, G.U.: Analysis of technologies in 3G and 3.5G mobile networks. In: Proceedings of the International Conference on Communication System Network Technology (CSNT 2012), pp. 330–333 (2012). https://doi.org/10.1109/CSNT.2012.79
13. Pew Research Center. Smartphone Ownership Is Growing Rapidly Around the World, but Not Always Equally. (2019). https://www.pewresearch.org/global/2019/02/05/smartphone-ownership-is-growing-rapidly-around-the-world-but-not-always-equally/. Accessed 31 Oct 2022
14. Union International Telecommunication ITU. 2021 Measuring Digital Development: Facts and Figures (2021)
15. Tawalbeh, M., Eardley, A., Tawalbeh, L.: Studying the energy consumption in mobile devices. Procedia Comput. Sci. **94**(MobiSPC), 183–189 (2016). https://doi.org/10.1016/j.procs.2016.08.028
16. Mayo, R.N., Ranganathan, P.: Energy consumption in mobile devices: why future systems need requirements-aware energy scale-down. Lect. Notes Comput. Sci. (including Subser. Lect. Notes Artif. Intell. Lect. Notes Bioinform.) **3164**, 26–40 (2004). https://doi.org/10.1007/978-3-540-28641-7_3
17. Majchrzak, T.A., Biørn-Hansen, A., Grønli, T.M.: Progressive web apps: the definite approach to cross-platform development? Proc. Annu. Hawaii Int. Conf. Syst. Sci. **2018**, 5735–5744 (2018). https://doi.org/10.24251/hicss.2018.718
18. de Andrade Cardieri, G., Zaina, L.M.: Analyzing user experience in mobile web, native and progressive web applications: a user and HCI specialist perspectives. ACM Int. Conf. Proc. Ser. (2018). https://doi.org/10.1145/3274192.3274201
19. Jobe, W.: Native Apps Vs. Mobile Web Apps. Int. J. Interact. Mob. Technol. 7(4), 27 (2013). https://doi.org/10.3991/ijim.v7i4.3226
20. R, N.: Beyond Web/Native/Hybrid: A New Taxonomy for Mobile App Development (2018)
21. Malavolta, I.: Beyond native apps: web technologies to the rescue! (keynote). Proceedings of the 1st International Working Mobile Development Co-located with SPLASH 2016 (Mobile! 2016), pp. 1–2 (2016). https://doi.org/10.1145/3001854.3001863
22. Russell, A., Berriman, F.: Progressive web apps: escaping tabs without losing our soul. Ingrequently Noted (2015). https://infrequently.org/2015/06/progressive-apps-escaping-tabs-without-losing-our-soul/. Accessed 01 Nov 2022
23. Tandel, S.J.A.: Impact of progressive web apps on web app development. Int. J. Innov. Res. Sci. Eng. Technol. 7(9), 9439–9444 (2018). https://doi.org/10.15680/IJIRSET.2018.0709021
24. Union International Telecommunication ITU. Digital Development Dasboard Panama. (2021). shorturl.at/gKO69. Accessed 09 Nov 2022

Aspects for Implementations of Decentralized Autonomous Organizations (DAO) in Switzerland

Moë Caviezel[1], Florian Spychiger[2(✉)], and Valerio Stallone[2]

[1] Hotelplan Group, Opfikon, Switzerland
[2] Zurich University of Applied Sciences, Winterthur, Switzerland
florian.spychiger@zhaw.ch

Abstract. The so-called decentralized autonomous organization (DAO) is a new concept that has emerged in recent years through the development of blockchain technology. The DAO represents a new form of organization in which the rules and regulations are embedded on the blockchain in the form of smart contracts and can operate without central control or third-party intervention. However, DAOs face many challenges such as security and legal issues due to their novelty. This paper describes the DAO concept and analyzes various aspects from business, social, technical, security and legal perspectives. Furthermore, the advantages and disadvantages of the concept have been elaborated. A qualitative approach was adopted by conducting literature research and three semi-structured expert interviews. From the results, aspects and guiding questions were developed to help initiators with the implementation of DAOs. It can be concluded that the concept DAO is a means to an end. With the blockchain technology, the DAO has advantages such as transparency and traceability. The ideal that a use case can be controlled in a completely decentralized manner is unrealistic. Therefore, a combination of centralized and decentralized components should be used in the setup.

Keywords: decentralized · autonomous · organization · DAO · blockchain · smart contracts

1 Introduction

With the rise of Web3.0, the concept of Decentralized Autonomous Organizations (DAOs) emerged, which is a new form of decentralized organization based on blockchain technology [1]. DAOs are entities, such as companies, institutions, or objects, that operate without a manager or a central authority [2]. DAOs are considered autonomous because the rules of this organization are based on so-called smart contracts on the blockchain that execute the decisions of the organization autonomously and its governance emerges within the community [3, 4]. Every decision must go through a voting process [5]. Once the conditions of the rules are met, the agreement is automatically executed by the majority consensus of the blockchain network [6, 7]. Additionally, blockchain technology is open source providing characteristics such as decentralization, transparency, and independence that can be seen as advantages [4]. Due to these advantages in comparison to traditional organizations, a DAO is useful and worth to be implemented. However, the first DAO projects launched in recent years were rather moderately successful [8]. Since implementing the concept of DAO still faces some challenges, this paper addresses the research question:

What aspects need to be considered to establish a DAO in Switzerland?

The focus of this study is the analysis of the individual aspects of a DAO in business, legal, security, technical and social perspectives in relation to Switzerland.

2   Basic Aspects of DAO Applications

Although DAO is already an established term in the blockchain environment, the concept is not yet clearly understood [9, 10]. In the paper by Chohan [10], the DAO concept is discussed in relation to governance processes, legal frameworks and liability, DAO construction and voting rights. Hassan & Filippi [9] examine the levels of decentralization and the autonomy aspect. The authors [9, 10] conclude that some theoretical as well as practical solutions need to be found. In recent years several application categories of DAOs have been established [11]. Depending on their mode of operation, structure and technology, DAOs can be divided into different forms such as protocol, social, collector, service, investment, grant, entertainment, media, and research DAOs [12–15]. In the following, we depict the five different perspectives of DAOs being the business-related, social, technical, legal, and security-related of DAO applications.

With the implementation of the blockchain-based DAO concept, *business* models, roles, and processes can be transformed [16, 17]. With a DAO, a new organizational structure emerges that requires new components and makes the traditional hierarchy model disappear [18]. Furthermore, the degree to which centralized and decentralized elements are used for decision-making should be decided based on the use case and the available resources [19]. Moreover, the knowledge from the operational and strategic level as well as supporting and primary processes must be translated into smart contracts [20].

The *social* driver of a DAO is the community, which brings suggestions, votes on them, and interacts together [3]. To communicate about votes, chat rooms and other social channels should be provided to the community [21, 22]. Additionally, to create incentives, many DAOs rely on reputation systems, reward programs, or development initiatives that allow the community to earn tokens or prizes for their engagement in social media [18, 22, 23].

The *technical aspects* include, first, the basic technology of the blockchain, as well as all other technical components, such as control by rule sets such as smart contracts, cryptography to ensure the integrity of the data, events that occur over the network, creation of new blocks, arrival of new peers, system management, internet, wallets, and tokens [3, 18, 24]. In addition, since there are different types of blockchains in terms of access, a permissioned DAO needs an authorization concept [25].

Due to the various definitions and types of DAOs and their smart contracts, the assignment of the *legal* situation and therefore its liability is particularly difficult [26].

If a DAO has a corporate law structure, it can be assigned to one of the corporate forms regulated under Swiss law [27]. A distinction can be made as to whether it is an association, foundation, simple partnership, cooperative or does not belong to any legal form at all.

One of the characteristics of a DAO is transparency, which means that the source code can be viewed by anyone [28, 29]. This benefits hackers as they could analyze the code and discover potential *security* vulnerabilities [29]. Anonymity is another aspect of the security perspective, as due to anonymity, it is tempting to make illegal transactions [28]. Another point mentioned is the immutability of transactions in a blockchain [29]. Further, if the consensus mechanism is adjusted or a hacker attacks the DAO, then the DAO is usually left with only a soft fork or hard fork as a measure [8].

Based on the articles presented above, we deduced several advantages and disadvantages of the application of DAOs (see Table 1).

Table 1. Advantages and disadvantages of a DAO

Advantages	Disadvantages
• Democratic decision making by community [3] • Cannot be shut down by third parties such as the state or police [9] • Trustworthy through consensus mechanism and does not need to rely on a central entity such as managers or a bank for trust [30] • Opensource and is therefore completely transparent [3] • Transaction data is encrypted, and data is distributed across all servers in the network, thereby generating synchronization of databases [24] • Data cannot be manipulated because it is immutable [24] • Boundless collaboration in the community [3] • Motivation through incentives and reputation [23] • Avoidance of single point of failure [31]	• Possible hacking attacks due to open source [28] • No trade secrets on the blockchain (R&D) [32] • Data cannot be manipulated because it is immutable and cannot be deleted [24] • Laws, liability, and responsibility are unclear; at best, the community would have to be jointly liable for it [27] • Computing power high for few transactions per second (tps) [24] • Anonymity and KYC [33] • Longer voting [34]

3   Methodology

In this study, a qualitative research approach is applied [35] using MAXQDA. First, the different areas of a DAO were identified and the various concepts of DAOs were examined based on the literature review. For the literature review, the keywords DAO, blockchain, and smart contracts were used and the relevant papers in relation to Switzerland were selected. Within the scope of this research, 70 literature sources were analyzed and the findings of the authors [1–3, 5, 9, 19, 27] made the most significant contribution to establish a knowledge basis from a legal, economic, social, security and technical perspective. Furthermore, the advantages and opportunities as well as the disadvantages and risks were analyzed.

Subsequently, three semi-structured interviews with experts (A, B, and C) were conducted [36]. The experts were selected according to their knowledge base and involvement with DAOs. In particular, they needed to have experience with implementing DAOs. The concept of DAOs is still in its initial phase, which is why the literature on DAOs is limited and expert knowledge is essential for deriving the different aspects. Experts bring great value to this study due to their experiences and interpretations from a practical perspective. The open structure of the interview makes it possible to obtain important additional information that cannot be found in the literature. Since there are few experts in this field in Switzerland, only three experts could be persuaded for an interview. The experts are already involved in DAO projects (B), are reconsidering the concept (C) or are available for consulting on blockchain and DAO topics (A).

We used an interview guide with open questions and asked follow-up questions if necessary [36]. In the interview guide, questions were asked about the various perspectives, which were slightly adapted accordingly on the expert's project or company. The interviews took place online via Microsoft Teams or in person. All interviews were recorded, and 45 to 60 min were scheduled for the interviews.

After conducting, the interviews were transcribed and analyzed. For the manual transcription in German the program MAXQDA was used, because it offers some advantageous features like the automatic speaker change and the speed settings especially for interview transcription. The evaluation follows the qualitative content analysis according to Mayring [37]. In this process, the texts in the transcripts were categorized and structured also using the MAXQDA program according to a coding guide [37]. The categorization derived from the different perspectives. The experts' answers were then compared, and the core statements of the experts were summarized in the results in a structured way [37]. From the literature review and the expert interviews, the artifact (aspects) was developed as a scientific output, which serves as an assistance for start-ups, companies and other entities that want to establish a DAO and facilitates the entry into the decentralized form of organization.

4   Results

4.1   Interviews Evaluation

From the coding process, especially in the *business-related perspective*, the use case and the purpose are an important subject for the respondents. It can thus be considered an essential aspect and is a relevant influencing factor. The governance should therefore be based on the purpose and, starting from the use case, an ecosystem should be built, and the individual components determined and estimated, whether they should be controlled decentralized or centralized. All respondents are unanimous that especially at the beginning of the project, most of the components are centralized and can be gradually decentralized. The members are equal, the hierarchy and mediating instances disappear, but nevertheless roles remain. For example, operators, developers, investors, and consumers can be found in DAOs. The interviewees agree that there is no completely decentralized business case (yet). According to interviewee A it is unrealistic that there is no need for centralized entities in a DAO. Interviewee a sees clean documentation and transparency as important factors. interviewee B also mentions transparency and the choice between centralization and decentralization as an important economic point. Interviewee C says that you have to be aware of why you want to decentralize the organization. Moreover, decision-making in decentralized organizations is usually less efficient than in hierarchical structures.

Regarding the *social perspective*, interviewees A and B emphasize incentives, attractiveness, and onboarding. Interviewee C sees the questions of whether the members can help, shape and form and then how they can do it as crucial. All interviewees are sure that the community represents the minimal structural element of a DAO and that it cannot function without it. Furthermore, Interviewee A recommends that channels should be created where members can exchange information with each other. Advertising for the use case is also an option. Interviewee B would use incentive systems in the form of tokens for the community. Interviewees B and C believe that it also depends on permissions and the weighting of votes. Interviewee B brings up the example that when controlling physical objects (e.g., a house) with a DAO, that members who are in proximity and thus more likely to be affected when the physical object is changed, should have a greater weighting in co-determination. Interviewee C says, starting from the beginning of the project, a DAO is gradually decentralized and therefore it needs a certain group that does preliminary work, pre-structures and co-organizes. According to interviewee C, these members who have done the initial groundwork should be given veto rights.

When it comes to *technical perspective*, all interviewees brought up other points to consider. Interviewee C primarily sees no major challenges in the technical aspects, as there are many tools such as Aragon or Colony that simplify the technical setup of a DAO. Sound, technical know-how is not mandatory for setting up a DAO, but it can help. Interviewee A particularly addresses the functions of a DAO and their usability as well as user acceptance. The handling of the functions should be simple and well-integrated for the user. Another point mentioned by interviewee A is that the functions of a DAO should be designed so that a user does not notice that it is running on a blockchain. The technical components should be customized according to the use case. Since setting up a smart contract can be complex, it is worth to choose a blockchain with native on-chain functions. Depending on the use case, custody solutions, multisig functions and KYC and AML can be used. Interviewee B looks at it from the developer perspective and sees general coding difficulties, as blockchain technology and its programming languages and software packages are still relatively new and therefore not much documentation exists yet. Furthermore, not only the blockchain components should be considered, but also the central components such as hardware parts, depending on the case.

Regarding *the legal perspective*, in Switzerland, there is still no legal framework specifically for DAOs. Interviewee A does not expect a DAO to be recognized as a legal form in the Swiss legal system in the future. Rather, a framework is to be created that specifies what has to be done. In this regard, Swiss laws should formulate technology independently such that it can be applied to different organizational forms. Since anyone can participate, it is not the operator who assumes liability, but the creator of the software.

Interview partner B has a similar opinion and sees the liability with the creator in the event of damage. Interviewee B sees it as a gray area and tries to circumvent the legal regulations. Interviewee C would welcome the recognition of the DAO as a legal form in Switzerland.

Regarding the *security perspective*, interviewees A and B agree that problems can occur at different levels, which is why putting measures in place is necessary. If a DAO is hacked, interviewee A does not see a problem because the DAO can be reset to the status before the hacking due to its traceability. In addition, attributes can be defined for wallets, such as that a wallet is locked. Based on the use case, the three most important questions about the measures according to interviewee A depend on what is to be protected, what the requirements are and what the worst-case scenario looks like. Interviewee B finds testing particularly relevant. Not only on the blockchain and in smart contracts but also in oracles and communications between the physical components and the blockchain.

4.2   Implementation Aspects

Based on the answers of the experts and the literature, we derived aspects for implanting a DAO and corresponding guiding questions. Table 2 lists the aspects to consider and the guiding questions for setting up a DAO.

Generally, it can be said that the following aspects should be considered when setting up a DAO. In the *business-related perspective*, the use case must be considered precisely in terms of its purpose and goal. Accordingly, the individual components should be identified from the use case and classified in the ecosystem, and the degree of decentralization should be determined. In the *technical perspective*, the choice of blockchain or platform and its functions coordinated according to the use case is of great importance. There are different ways to map these technical components in the DAO. Functions for submitting proposals and votes must be included. Which user has which rights in a DAO must be considered in an authorization concept. In the *security perspective*, it is important to consider what must be protected and what risks are involved. Measures should be defined so that the DAO can act immediately in the event of a risk. In the *legal perspective*, attention must be paid to how the DAO fits into the Swiss legal system. Depending on the application, the DAO can take different forms and can be declared as an association, foundation, cooperative or simple partnership. This aspect is important since the question of liability follows on the basis of it. In the *social perspective*, the community is the driver of a DAO, since governance is created through it and therefore the promotion and preservation of the community is essential. Consequently, various measures such as incentive and reward systems in the form of reputation tokens should be issued or attractiveness of the use case should be communicated. In addition, communication channels should be created in which the community can exchange information about votes.

Table 2. Aspects and guiding questions for implementing DAOs

	Aspects	Guiding Questions
Business	Purpose of a DAO	• What is the common goal and what is the purpose? • Does a DAO help achieve the goal?
	Use Case and Business Case	• What requirements must be met? • What is the business model? • What are the constraints (time and cost and scope) of the project?
	Knowledge management	• What know-how and resources are needed for the business case? • Which know-how is needed for which processes? • Which knowledge is needed centralized or decentralized? • Which decisions can be determined centralized or decentralized?
	Ecosystem	• Which stakeholders are involved? • Which physical or digital components are needed? • How are the components controlled? Are they physical or digital components and to what degree are they centralized or decentralized?
Social	Community	• Onboarding, education, and training community about benefits • How to reward the community and create incentive? E.g., reputation tokens • Is a communication channels for sharing needed? • What marketing strategy should be used? social media?
	Governance	• To what extent can the community influence governance? • Should there be restrictions from central units? • How decentralized can the DAO be controlled?
	Anonymity	• Should users be anonymous? • Which users have which rights and accesses? (cf. authorization)
Technical	Blockchain	• Which blockchain or platform should be used? • Which functions already exist on-chain? • Which ones would have to be programmed individually? • How is the consensus formed? PoW or PoS?

(*continued*)

Table 2. (*continued*)

	Aspects	Guiding Questions
	Authorization concept	• Who has access to the DAO and who gives permission? • What rights (read-only, voting rights, etc.) are granted?
	Rules	• How to put know-how and documentation on blockchain? • How to bring operational & management rules to Blockchain? e.g., in Smart Contracts or in native functions of Blockchain
	Wallet and Tokens	• Are wallet and tokens, required to make transactions, created?
	Voting	• What entitlements and weightings of votes should be assigned? • What voting mechanism should be used?
	Hardware	• Are hardware components needed?
	Usability	• How to design simple functions, UI, and usability?
Legal	Legal Form	• What is the purpose of the DAO? • Who is liable in the event of a claim?
	No legal form	• Are there no members and no company? • Is the DAO creator liable if it has no legal form?
Security	Security Risks	• Is the DAO protected from hackers? What needs to be protected? • What kind of amounts/risks are involved?
	Measures	• What attributes needs to be defined? e.g., Wallet locked • What to do when hacked? forking measures/regenerate tokens • Which conventional measures should be used? Testing, code review, audits • How to protect Data? E.g., Encryption, multisig. secure passwords, KYC, and AML (for Exchange)

5   Conclusion

To facilitate the start of a DAO, an attempt was made to derive the most important aspects for the implementation of a DAO and thus answers the research question: *What aspects need to be considered to establish a DAO in Switzerland?* The answer to the research question is based on the different forms a DAO can take: It depends on the use case. From the analysis of the interviews, it becomes clear that the literature differs sometimes from the opinions of the experts. The ideal conception of a DAO that can operate completely decentral, as described by some authors [1, 9] does not exist. The experts and authors [19] agree that it should be determined to which degree the elements in a DAO should be decentralized. A DAO is a flexible concept and therefore a means to a purpose and should not be regarded as a completely decentralized entity. The derived aspects provide a general basis for the different perspectives of a DAO to guide the initiators of a DAO through their use case.

For future research, the different application areas of DAOs can be studied in the business area. In addition, this paper did not consider any aspects of the cost of building a DAO. Especially from a legal point of view, it is necessary to examine whether a new legal framework for DAOs or DAO-like organizations is needed in Switzerland. Furthermore, it also remains an open question to which extent our results can be generalized to DAOs in the European Union or the USA. Nevertheless, since DAOs act globally, we believe that certain aspects derived in this study can benefit the implementation of DAOs beyond Switzerland.

References

1. Wang, S., Ding, W., Li, J., Yuan, Y., Ouyang, L., Wang, F.-Y.: Decentralized autonomous organizations: concept, model, and applications. IEEE Trans. Comput. Soc. Syst. **6**(5), 870–878 (2019). https://doi.org/10.1109/TCSS.2019.2938190
2. Faqir-Rhazoui, Y., Arroyo, J., Hassan, S.: A comparative analysis of the platforms for decentralized autonomous organizations in the Ethereum blockchain. J. Internet Serv. Appl. **12**, 9 (2021). https://doi.org/10.1186/s13174-021-00139-6
3. Hexlant, J.C.: DAOs: empowering the community to build trust in the digital age. Stanford J. Blockchain Law Policy (2022). https://stanford-jblp.pubpub.org/pub/dao/release/1
4. Samuel, R.: Decentralized Autonomous Organizations (DAO): The Professional and In Depth Guide to Mastering DAO (2020). https://de.1lib.ch/book/7115143/631651
5. Jentzsch, C.: Decentralized Autonomous Organization to Automate Governance (2016). https://archive.org/details/DecentralizedAutonomousOrganizations/WhitePaper
6. Hellwig, D., Karlic, G., Huchzermeier, A.: Smart contracts. In: Entwickeln Sie Ihre eigene Blockchain, pp. 81–105. Springer, Heidelberg (2021). https://doi.org/10.1007/978-3-662-629 66-6_4
7. Voshmgir, S.: Token economy: how the Web3 reinvents the internet. Token Kitchen. 2nd edn (2020). https://books.google.ch/books?id=YWaZzQEACAAJ
8. Shier, C., et al.: Understanding a revolutionary and flawed grand experiment in blockchain: the DAO attack. SSRN (2017). https://doi.org/10.2139/ssrn.3014782
9. Hassan, S., De Filippi, P.: Decentralized autonomous organization. Internet Policy Rev. **10**(2) (2021). https://doi.org/10.14763/2021.2.1556
10. Chohan, U.W.: The decentralized autonomous organization and governance issues. SSRN (2017). https://doi.org/10.2139/ssrn.3082055
11. Samman, G., Freuden, D.: A Decentralized Governance Layer for the Internet of Value (2020). https://www.monsterplay.com.au/blockchain/dao-a-decentralized-governance-layer-for-the-internet-of-value/
12. Cointelegraph. Types of DAOs and How to Create a Decentralized Autonomous Organization. https://cointelegraph.com/decentralized-automated-organizations-daos-guide-for-beginners/types-of-daos-and-how-to-create-a-decentralized-autonomous-organization
13. Ledger. Your DAO Guide—The Most Important DAO Categories Defining the Space (2021). https://www.ledger.com/academy/your-dao-guide
14. Hennekes, B.: The 8 Most Important Types of DAOs You Need to Know (2022). https://alchemy.com/blog/types-of-daos

15. Hamburg, S.: A Guide to DeSci, the Latest Web3 Movement. Future (2022). https://future. a16z.com/what-is-decentralized-science-aka-desci/
16. Adams, R., Parry, G., Godsiff, P., Ward, P.: The future of money and further applications of the blockchain. Strateg. Chang. 26(5), 417–422 (2017). https://doi.org/10.1002/jsc.2141
17. Li, J., Greenwood, D., Kassem, M.: Blockchain in the built environment and construction industry: a systematic review, conceptual models and practical use cases. Autom. Constr. 102, 288–307 (2019). https://doi.org/10.1016/j.autcon.2019.02.005
18. Hou, J., Ding, W., Liang, X., Zhu, F., Yuan, Y. Wang, F.: A study on decentralized autonomous organizations based intelligent transportation system enabled by blockchain and smart contract. In: 2021 China Automation Congress (CAC), pp. 967–971 (2021). https://doi.org/10.1109/CAC53003.2021.9727429
19. Beck, R., Müller-Bloch, C., King, J.L.: Governance in the blockchain economy: a framework and research agenda. J. Assoc. Inf. Syst. 19(10), 1020–1034 (2018). https://doi.org/10.17705/1jais.00518
20. Sreckovic, M., Windsperger, J.: Decentralized autonomous organizations and network design in AEC: a conceptual framework. SSRN (2019). https://doi.org/10.2139/ssrn.3576474
21. Calderon, H.: Why community is key to blockchain and crypto project success. Forkast.News. (2021). https://forkast.news/why-community-key-blockchain-crypto-project-success/
22. Ziolkowski, R., Miscione, G., Schwabe, G.: Exploring Decentralized Autonomous Organizations: Towards Shared Interests and 'Code is Constitution' (2020). https://doi.org/10.5167/UZH-193663
23. Esber, J., Kominers, S.: A novel framework for reputation-based systems. Future (2021). https://future.a16z.com/reputation-based-systems/
24. Zheng, X., Zhu, Y., Si, X.: A survey on challenges and progresses in block-chain technologies: a performance and security perspective. Appl. Sci. 9(22), 4731 (2019). https://doi.org/10.3390/app9224731
25. Rikken, O., Janssen, M., Kwee, Z.: Governance challenges of blockchain and decentralized autonomous organizations. Information Polity 24(4), 397–417 (2019). https://doi.org/10.3233/IP-190154
26. Pranata, A.R., Tehrani, P.: MThe Legality of Smart Contracts in a Decentralized Autonomous Organization (DAO). In: Tehrani, P.M. (eds.) Advances in Computational Intelligence and Robotics, pp. 112–131. IGI Global (2022). https://doi.org/10.4018/978-1-7998-7927-5.ch006
27. Gyr, E.: Dezentrale Autonome Organisation DAO. Jusletter (2017). https://jusletter.weblaw.ch/juslissues/2017/917/dezentrale-autonome-_ad06f55b2f.html
28. Lin, I.-C., Liao, T.-C.: A survey of blockchain security issues and challenges. Int. J. Netw. Secur. 19(5), 653–659 (2017). https://doi.org/10.6633/IJNS.201709.19(5).01
29. Madnick, S.E.: Blockchain isn't as unbreakable as you think. SSRN (2019). https://doi.org/10.2139/ssrn.3542542
30. Meinel, C., Gayvoronskaya, T.: Blockchain: Hype Oder Innovation. Springer, Heidelberg (2020). https://doi.org/10.1007/978-3-662-61916-2
31. Isler, M.: Datenschutz auf der Blockchain. Jusletter (2017). https://jusletter.weblaw.ch/juslissues/2017/917/datenschutz-auf-der-_fbecc2b55b.html
32. Pesch, P.J.: Blockchain, smart contracts und datenschutz. In: Fries, M., Paal, B.P. (eds.) Smart Contracts. Mohr Siebeck (2019). https://doi.org/10.1628/978-3-16-156911-1
33. Goldbarsht, D., de Koker, L.: Financial Technology and the Law: Combating Financial Crime, vol. 47, Springer, Cham (2022). https://doi.org/10.1007/978-3-030-88036-1
34. Ruane, J., McAfee, A.: What a DAO Can—And Can't—Do. Harvard Business Review (2022). https://hbr.org/2022/05/what-a-dao-can-and-cant-do
35. Jonker, J., Pennink, B.: The Essence of Research Methodology: A Concise Guide for Master and PhD Students in Management Science. Springer, Heidelberg (2009). https://doi.org/10.1007/978-3-540-71659-4

36. Bryman, A., Bell, E.: Business Research Methods, 3rd edn. Oxford University Press (2011). https://www.uwcentre.ac.cn/haut/wp-content/uploads/2018/11/Alan_Bryman_Emma_Bell_Business_Research_Methodsb-ok.cc.pdf
37. Mayring, P., Fenzl, T.: Qualitative inhaltsanalyse. In: Baur, N., Blasius, J. (eds.) Handbuch Methoden der Empirischen Sozialforschung, pp. 633–648. Springer, Fachmedien Wiesbaden (2019). https://doi.org/10.1007/978-3-658-21308-4_42

Modeling Blockchain System for Fashion Industry

Aleksandra Labus(✉) , Dušan Barać , Petar Lukovac , Vukašin Despotović ,
and Milica Simić

Faculty of Organizational Sciences, University of Belgrade, Jove Ilića 154, Belgrade, Serbia
aleksandra@elab.rs

Abstract. The digitalization process affected the process of conducting supply chain management in the fashion industry. Innovative information technologies enabled improved communication between supply chain stakeholders, a more efficient and transparent way of conducting business transactions and protecting intellectual property rights. Blockchain technology has the potential to improve security and trust within fashion industry supply chains, reduce the placing of counterfeit products on the market, and provide customers insight into product authenticity. This paper focuses on the possibilities of using blockchain technologies in the supply chain of the fashion industry. The main goal of the paper is to propose a model of a blockchain system for the fashion industry. The proposed system should provide insight into the supply chain flow in the fashion industry based on using blockchain technology. Furthermore, the proposed model enables tracking all business transactions among stakeholders in the fashion industry in a transparent and immutable way, and for the customers enables insight into the product origin, authenticity and all relevant product data.

Keywords: Blockchain · Supply chain · Fashion industry

1 Introduction

The development of the digital economy, from digitization, digitalization, to digital transformation, affected the development of new business models based on innovative technologies such as cloud computing, big data, blockchain, augmented reality, the Internet of Things, and artificial intelligence [1, 2]. In modern business ecosystems, for stakeholders these technologies enable secure, private, real-time communication, exchange, storing and analysis of data and business transactions. Furthermore, for customers, these technologies enable new experiences during the purchase process and create new values for the products they are buying. These are very important issues in the fashion industry, having in mind the existence of strong competition in the fashion market.

The fashion industry surrounds us every day and plays an important role in everyone's life. The development of the technological production system, both in many industries and in fashion, has caused significant changes in terms of industrial automation. Constant investment in automation ensures not only survival in the market but also the competitiveness of the products themselves. Designing products, manufacturing, marketing, sales, and shipping are the main steps in the apparel supply chain. Digital technologies have the full potential to improve business processes within the supply chain. Within this industry, there are certain threats and challenges that need special attention. Threats in terms of brand preservation, protection from counterfeiting fashion apparel, security and privacy, traceability, and transparency of all business transactions within the supply chain in the fashion industry. In the fashion industry blockchain technologies can solve these problems.

This paper focuses on the possibilities of using blockchain technologies in the fashion industry. The main goal is to model a blockchain system for the fashion industry. The proposed system should provide insight into the supply chain flow in the fashion industry based on using blockchain. Furthermore, the proposed model shows the conducting of business transactions among stakeholders and enables the customers to track product origin and all relevant product data.

2 Blockchain

Blockchain can be defined as a complex mathematical algorithm aimed at maximizing the security of financial transactions using cryptographic methods [3]. It is based on a distributed database, which consists of encrypted data that cannot be changed or copied [4]. This is very important in conducting business transactions in order to reduce fraud. Blockchain provides a distributed, unchangeable, transparent and secure ledger in which all transactions of money are reliably recorded. All business transactions are validated and "chained" into a blockchain [5]. Each conducted transaction became block in the scope of the blockchain ledger.

In order to conduct transactions in the blockchain some basic criteria must be fulfilled [6–8]. Each block must contain a block header and block data. The Block header contains the block number, the hash value of the previous block header, the timestamp, the nonce value, and the size of the block. The hash value of the previous block header is stored in the block header of the current block (so-called chaining of blocks). This is so-called auditability. In a blockchain network must be a consensus among different members. This means that each member of the network has the right to verificate and validate transactions, record data and sign smart contracts. Smart contract enables finishing transactions between interest parties. It contains contract clauses written in computer programs and they are automatically executed when predefined conditions are met [3]. All transactions in the blockchain network are transparent to all network members. A copy of transactions is stored in the distributed ledger of each node in the network. This is so-called reliability. For conducting transactions it is not necessary to involve a central authority because of the decentralization rules. Once the transaction is added and recorded it cannot be changed. This is so-called immutability.

3 Blockchain in the Fashion Industry

The fashion and textile industry are one of the fastest growing industries, involving a complex supply chain both locally and globally, for the procurement of raw materials and the delivery of finished products. This complexity of the industry requires a system that is transparent, distributed, and can protect intellectual property rights. The supply chain of the fashion ecosystem comprises sourcing raw materials (fiber, yarn, fabric), manufacturing apparel in-house or through outside vendors, distributing apparel through distributors and wholesalers, and selling apparel products in retail stores to customers. In order to meet consumer demand on time, the supply chains need to provide a quick accurate and real-time response [9].

One of the biggest problems facing the fashion industry is counterfeit products. Counterfeit products are consumer goods that are not genuine but are designed to look identical to authentic branded products. Such products have a negative impact on the image and value of the brand itself. Blockchain has the ability to establish authenticity in the fashion industry by using immutable business records and creating secure digital identities for all supply chain participants [10].

3.1 Blockchain in Fashion Supply Chain

A supply chain generally involves five roles, namely supplier, manufacturer, distributor, retailer, and consumer [11]. They are linked through logistics activities. In the supply chain of the fashion industry, Supply Chain Management enables tracking of all processes from the supply of raw materials, to designing, manufacturing, storage, distribution and selling products to customers. These processes should be tracked between involved stakeholders such as suppliers of raw materials, fashion designers, manufacturers, distributors, sellers and customers [10–12].

Blockchain in the supply chain enables [9, 13–15]:

- Tracking of all business transactions among different stakeholders. Each transaction presents a decentralized immutable record that cannot be misused. For all transactions financial flow within the whole supply chain is transparent.
- Insight into each participant within the supply chain. Only legitimate participants of the system are allowed the full authority to access the same information, while no other participant can have any access to the system. Participants are enabled to interact with each other without fear of the possibility of personal data being compromised.
- Reduction of problems of counterfeiting by enabling tracking product origin furthermore, reducing trading on the gray market. Blockchain-based tags and chips can be used to ensure the product's authenticity.
- Use of smart contracts among participants during the conducting transactions among stakeholders with no intermediaries.
- Measurement of performance indicators for environmental, economic, and social sustainability.
- Intellectual property protection for fashion designers and brand creators.

To customers, blockchain technology provides information about the clothes that they are buying. It provides the customer with information about the raw materials used to make the product, the life cycle of the product and how the finished product is made. For some consumers, the most important thing is the quality, material from which the product is made, and way of maintenance, while for others, the most important information is about the product's place of origin, who is the designer that created it. Blockchain can also be useful for creating loyalty programs. In this way, customers can have insight into all their transactions with a fashion company and collect points that can be used for discounts during future purchases. Information from the blockchain can add some additional value to the product during the purchase decision-making process, i.e. by giving insight into the celebrity that wore clothes on some special occasion.

4   Modeling Blockchain System for the Fashion Industry

A blockchain system for the fashion industry is proposed in Fig. 1.

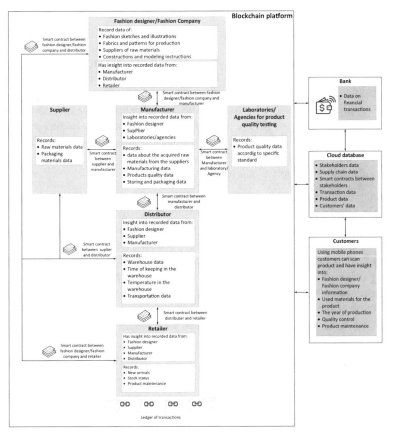

Fig. 1. Blockchain system for the fashion industry

The main stakeholders in the proposed blockchain system for the fashion industry are:

1. **Fashion designer/Fashion Company**. This stakeholder record data of fashion sketches and illustrations of fashion clothes, fabrics and patterns for the production, chosen suppliers from which raw materials will be procured, constructions and modeling instructions for the manufacturing process. Furthermore, they have insight into the recorded data from the manufacturer, distributor and retailer. All transactions between fashion designer/fashion company and manufacturer are conducted via smart contracts. These smart contracts have specific requirements from both sides. For fashion designer/company it is important that clothes be manufactured according to specific designs using constructions and modeling instructions, that specific fabrics be used, that the final product is of adequate quality and not distribute this information to those who can misuse them. For manufacturer it is important to receive all mentioned information related to clothes manufacturing, all funds needed for the procurement of raw materials, manufacturing services, additional costs for the testing of clothes quality and distribution process.

2. **Supplier**. This stakeholder records raw materials data and packaging materials data. The supplier sells materials to a manufacturer and transactions between them are regulated using smart contracts.

3. **Manufacturer**. Collects all necessary information from the fashion designer/company, procures raw materials for the supplier, records manufacturing data, products quality data received from laboratory/agency for product quality testing, storing and packaging data.

4. **Laboratories/Agencies for product quality testing**. Test product quality data according to specific standards. These also can include testing raw materials. The manufacturer and laboratory/agency sign the smart contract related to this process.

5. **Distributor**. The role of the distributor is to record warehouse data, time of keeping products in the warehouse, temperature levels in the warehouse, and transportation data between the fashion designer/company and manufacturer, between supplier and manufacturer. Distributor signs smart contracts with fashion designer/company, manufacturer, and supplier.

6. **Retailer**. When the products (clothes) arrive at the store, the retailer records new arrivals, stock status and product maintenance. Furthermore, he has insight into recorded data from fashion designer/company, supplier, manufacturer and distributor. The retailer procures clothes from fashion designer/company and receives them for distributor. All transactions are conducted using smart contracts between these participants in the supply chain.

7. **Customers**. During the purchase process in the retail store or in the web shop, customers have insight into the product's origin, values, used fabrics, production process, date of production, information about the testing quality, and products' maintenance. To receive these data customers can scan QR codes related to the products.

In the fashion industry, there is a growing need to ensure traceability throughout the whole production chain. It is difficult for customers to access information about the product, the origin of the material and the authenticity of the product itself. It is necessary to ensure visibility and meet consumer demands in terms of transparency and quality assurance. All stakeholders in the supply chain have the challenge of sharing key information with their ecosystem due to the risk of manipulation and exploitation of confidential data. It would be important for each participant in the supply chain to receive all the necessary information from their associates transparently and to constantly have insight into all processes within the supply chain. Smart contracts are formed between participants in the supply chain, which are transparent to each participant, and each participant in the contract states its requirements and conditions that must be respected. With the help of these contracts, the level of trust increases and there is no room for fraud and abuse between the participants. All financial transactions among all participants in the blockchain system are conducted using smart contracts, digital wallets and e-banking services. All transaction data are stored in the blockchain ledger of transactions in the scope of the chosen blockchain platform. The following data are stored in the cloud database: Stakeholders data, Supply chain data, Smart contracts between stakeholders, Transaction data, Product data, Customers' data.

5   Conclusion

This paper presents a blockchain system for the fashion industry that should enable the transparency of all data in the supply chain, increase customer confidence when purchasing and create security for all stakeholders. Trust can be built using a distributed database to store and authenticate all transactions in the supply chain. All authorized participants have the opportunity to monitor their supply network and create a sustainable and transparent supply chain.

The fashion industry is one of the largest industries in the market today and many problems such as counterfeiting of goods cause insecurity and mistrust in the purchasing process. For blockchain technology to be relevant for consumption, it is necessary that they remove the fear of online shopping and become aware that the role of blockchain is to keep all data safe and to make all transactions transparent. Consumers will have insight into all the necessary information about the product quality, and that the purchase is safe.

Future directions of research will refer to the examination of consumer readiness for the application of the blockchain in the fashion industry and the implementation of the proposed system.

References

1. Verhoef, P.C., et al.: Digital transformation: a multidisciplinary reflection and research agenda. J. Bus. Res. **122**, 889–901 (2021). https://doi.org/10.1016/j.jbusres.2019.09.022
2. Reis, J., Amorim, M., Melao, N., Matos, P.: Digital transformation: a literature review and guidelines for future research. In: Proceedings of the10th European Conference on Information System Management Acadademy Conference Publication Limited, vol. 1, pp. 20–28 (2016). https://doi.org/10.1007/978-3-319-77703-0
3. Zorica Bogdanović, T.N., Radenković, B., Despotović-Zrakić, M., Barać, D., Labus, A.: Blockchain technologies: current state and perspectives. In: Second International Scientific Conference on Digital Economy (DIEC 2019), pp. 1–12 (2019). http://ipi-akademija.ba/file/zbornik-diec-2019/44

4. Sanka, A.I., Irfan, M., Huang, I., Cheung, R.C.C.: A survey of breakthrough in blockchain technology: adoptions, applications, challenges and future research. Comput. Commun. **169**, 179–201 (2021), https://doi.org/10.1016/j.comcom.2020.12.028
5. Nofer, M., Gomber, P., Hinz, O., Schiereck, D.: Blockchain. Bus. Inf. Syst. Eng. **59**(3), 183–187 (2017). https://doi.org/10.1007/s12599-017-0467-3
6. Agrawal, T.K., Kumar, V., Pal, R., Wang, L., Chen, Y.: Blockchain-based framework for supply chain traceability: a case example of textile and clothing industry. Comput. Ind. Eng. **154**, 107130 (2021). https://doi.org/10.1016/j.cie.2021.107130
7. Wang, H., Zheng, Z., Xie, S., Dai, H.N., Chen, X.: Blockchain challenges and opportunities: a survey. Int. J. Web Grid Serv. **14**(4), 352 (2018). https://doi.org/10.1504/ijwgs.2018.100 16848
8. Fu, B., Shu, Z., Liu, X.: Blockchain enhanced emission trading framework in fashion apparel manufacturing industry. Sustain. **10**(4), 1–19 (2018). https://doi.org/10.3390/su10041105
9. Tripathi, G., Tripathi Nautiyal, V., Ahad, M.A., Noushaba, F.: Blockchain technology and fashion industry-opportunities and challenges. In: Blockchain Technology : Applications and Challenges, p. 300 (2021)
10. Mukherjee, P., Pradhan, C.: Blockchain 1.0 to Blockchain 4.0—The Evolutionary Transformation of Blockchain Technology. In: Blockchain Technology: Applications and Challenges, p. 300 (2021)
11. Wang, B., Luo, W., Zhang, A., Tian, Z., Li, Z.: Blockchain-enabled circular supply chain management: a system architecture for fast fashion. Comput. Ind. **123** (2020). https://doi.org/10.1016/j.compind.2020.103324
12. Kumar Agrawal, T., Sharma, A., Kumar, V.: Blockchain-based secured traceability system for textile and clothing supply chain. In: Springer Series in Fashion Business Artificial Intelligence for Fashion Industry in the Big Data Era, pp. 153–171 (2018)
13. Jain, G., Kamble, S.S., Ndubisi, N.O., Shrivastava, A., Belhadi, A., Venkatesh, M.: Antecedents of blockchain-enabled E-commerce platforms (BEEP) adoption by customers – a study of second-hand small and medium apparel retailers. J. Bus. Res. **149**, 576–588 (2022). https://doi.org/10.1016/j.jbusres.2022.05.041
14. Jabbar, S., Lloyd, H., Hammoudeh, M., Adebisi, B., Raza, U.: Blockchain-enabled supply chain: analysis, challenges, and future directions. Multimed. Syst. **27**(4), 787–806 (2021). https://doi.org/10.1007/s00530-020-00687-0
15. Oropallo, E., Secundo, G., Del Vecchio, P., Centobelli, P., Cerchione, R.: Blockchain technology for bridging trust, traceability and transparency in circular supply chain. Inf. Manag. (2021). https://doi.org/10.1016/j.im.2021.103508

Assessment System for a Large Container Management and Optimization Problem

Juan P. D'Amato[1,2](✉) and Pablo Lotito[1,2]

[1] Universidad Nacional del Centro de la Provincia de Buenos Aires, Facultad de Ciencias Exactas, PLADEMA, Tandil, Buenos Aires, Argentina
juan.damato@gmail.com
[2] National Scientific and Technical Research Council, Avenue Rivadavia 1917, Ciudad Autónoma de Buenos Aires, Argentina

Abstract. In logistics systems of medium and large enterprises, operators must analyze hundreds of thousands of variables to find potential improvements. This task becomes very complex in large scenarios and the taken decisions are generally based just on local changes, without regarding global factors.

In this work, we present a system for the analysis and optimization of a container transport system considering variations in demand over time and global conditions. We propose a coupled empty-full container reposition model that describes liners operations in a year. We also implement an assisted process that helps users to find and estimate missing data in order to describe a complete scenario. This system allows users to make predictions and risk analysis more accurately in a short time at the same time they can obtain an optimum benefit.

Keywords: Optimization · Container transport · logistic · expert system

1 Introduction

The liner shipping market is a very complex and variable environment, with seasonal sales fluctuations and extraordinary sale opportunities (e.g. harvest record in a country), which make enterprises to frequently re-evaluate their strategies. To reduce risks and identify opportunities, line operators must perform financial analysis at certain intervals of time, evaluating the economic benefit of incorporating or leaving a containers area. Likewise, the sea container industry is also confronted with the increasing container traffic demand. Liner companies must increase their capacities by investing in new containerships or reviewing their actual routes. Such problem is still an open challenge as [1] mention.

As it is known, freight transport generates a significant number of empty vehicle movements caused by the unbalanced directional flows between two specific points. This issue is an intrinsic element of vehicle fleet management and overall logistics scheduling process. If the inventory fails, then new loads must be rejected or containers have to be leased at high costs. Consequently, an important decision at the operational level is when to accept new loads that provide a real benefit, regarding empty repositioning and lease containers costs. Most container shipping companies assign several ships on a particular trade route, which is characterized by two end ports (i.e., start-service and end-service ports) and many intermediate calling ports. In a ship itinerary, all ports in a line from the start port to the end port are referred to as up-bound direction and then when called back are down-bound direction. In a single line, the up-bound carries the full containers; the down-bound the empty ones; what makes a predictable repositioning model. In global companies that operate with several lines, lines are inter-connected and the association of up-bound with full containers and down-bounds with empty containers is blurry and a key factor is to distribute the loads between lines, as studied in [7].

At the same time, global liner companies with hundreds of sales points all over the world have operations coordinators to perform forecast or what if analysis. These experts use a simplified transport model analysis (without stocks and with an inexact repositioning scheme), propose hypothetic scenarios where varies the ships capacity, alter the ports visiting sequence or propose new sales markets. This task demands a great human endeavor and it's common to commit many mistakes and to omit data to the model. Clearly, all the omitted information, leads to wrong further analysis and optimization. These issues force the coordinators to start using systems that helps them in decision-making and in creating and managing scenarios. This system should also help to maximize overall benefit which can often oppose to the individual thoughts. The whole information flow must be managed properly and presented in a way that users can understand and implement in the real business.

This paper presents an assessment system for analyzing and optimizing the benefits of container transportation, considering the demand on each port, transport capacity and empty containers repositioning. The implemented model includes of costs and benefits and market trends. As many of the analysis are forecasts, it's required administer a large volume of data which is partially unknown. It is proposed a methodology for estimating missing data which reduces user errors and helps to quickly detect faults in the scenario. The proposal is split in several phases. First, the transport model is proposed, considering full and empty containers and storage along time. Then, the second phase deals with complex scenario generation. An assisted process to guide users during scenario construction is presented, providing visual aids and strategies to recognize data that do not meet constrains. The third phase deals with transport model repositioning and profit optimization. The solving algorithm is presented. Finally, results and different application views are shown.

2   Literature Review

Several aspects of the work here presented has already been studied but separately. About containership routing problems, [12] try to find the most economical ship size and mix of fleet for a defined trade route with a known trade demand over a finite planning horizon. The work of [4] addresses a shipping model of minimizing the costs including opportunity costs such as penalty costs for cargo not shipped due to ship capacity constraints. In Rana and [14] it is discussed the optimal routing for a fleet of containerships operating on a trade route, to maximize the liner shipping company's profit. Besides the route, the optimal set of calling port sequence is also provided. They assume that non-profitable ports should not be selected as calling ports on the route. They formulate the problem as a mixed integer non-linear programming model and solve it by using Lagrangean relaxation techniques.

In [2], a Linear Programming is used to assign an existing fleet of containerships to a given set of routes based on detailed realistic models of operating costs. Considerable research has also been performed about empty container management. Gavish (Gavish 1981) developed a support system for vehicle fleet management. In his study, prior to planning on empty container relocation, owned and leased containers are assigned to the demand points based on the marginal cost criterion. In [5] treat leased container allocation, which is determined together with empty container relocation. Imai and [10] deal with fleet size planning for refrigerated containers where they determine the necessary number of containers required to meet predicted future transportation demand. The studies of [3] deal with empty container distribution in a relatively broad geographical area. In contrast, [7] and [8] focus on empty container repositioning in the hinterland of a specific port, despite the many similarities that exist in theory and in practice with repositioning in a board area. In [13] study the empty container allocation in a port with the aim to reduce redundant empty containers. They consider the problem as a non-standard inventory problem with simultaneous positive and negative demand under a general holding cost function.

Recent works, such as [6] and [11] deal with container management in an integrated approach to determine the optimal fleet composition with corresponding routing characteristics and empty container repositioning. Authors of [15] propose a heuristic based on genetic algorithms (GA), to find a set of calling ports, an associated port calling sequence, the number of ships by ship size category and the resulting cruising speed to be deployed in the service networks, with the objective of profit maximization for a liner shipping company but this proposal do not model operations in time. [9] propose a model for improving the operation efficiency of the seaports and to develop an analytical tool for yard operation planning. [16] presented a DSS quite similar to our proposal, dealing with full and empty containers tested with empirical data, trying to minimize system costs. In practice, these models generally fit well as small regions solutions, but when a world-broad solution is needed, it's best to let users manipulate some data (but not the whole analysis process) as they know best how container market behaves.

3   Proposal

One of the main purposes of this research work is to find the optimal movement of empty and full containers among ports to maximize profit. The proposed model has two decision variables, the number of full containers and the amount of empty containers, deployed on a transport network. The network is composed by several lines covering a region following a port itinerary. Each line has associated a group of ships of different characteristics: speed, capacity, length, fuel consumption, among others. As the system is used in strategic way, the time is discretized into months and some simplifications respect to an operational model are taken, e.g. Transport capacity per arc is the average capacities of the ships passing through a port in a month. All containers' operations are broken down into arcs operations, an intuitive decomposition for transportation network.

As the loading/unloading from vessels does not involve a real movement between ports, imaginary arcs are created with the corresponding cost. Leasing and leaving empty containers are modeled in the same way.

Full containers are loaded from an origin port, and they are sent via a *route* to a destination port and they are considered to be loaded and carried to destination in one month. There could be more than one alternative route for the same pair ports. A sale demands an empty container in the origin port and offers it in the end port. To make a sale from a port, there must be at least as many empty containers as full ones, so system keeps balanced. For simplification, in this work just one container size is considered.

The income for a full container is variable and depends on competitors and demands preferences. Generally, as the number of containers sold at a port change, also the price does. To model this behavior, the price is defined proportional to the number of containers sold at this port by a given elasticity coefficient, also known as *price elasticity of demand* (PED). The cost of a full container is composed by *direct costs* such as stowage, handling, days of use, leasing, stevedoring, electricity (for some kind of containers) costs among others incurring along a route, plus *indirect costs* arising from repositioning the necessary empty containers to do a sale. Ships costs are not considered for this problem and beyond the scope of this work. Costs are specific for full and empty containers, although they might have the same value.

Empty containers are offered at destination port in the month following the sale that generates it. If empty containers are not used, they can be reserved in a port, placed on other ports that demand them or released at a certain cost. If empty containers are required, they can be taken from stocks, require from other ports or leased at a certain cost.

3.1 Mathematical Formulation

We now introduce some definitions and notations. We consider the network as a graph $G(V, A)$, and each line is a sub-graph G^s, where S are available lines (or *services*) in the scenario and $s \in S$. Each port n has several attributes and roles and can belong for more than one service. Nodes are split out in artificial nodes and arcs between these nodes, representing operations. For loading or unloading containers from vessels, there are *unloading arcs* δ and *loading* ones λ. Arcs (i_s, j_s) between nodes i, j in a service s are called *transport arcs* τ. For instance, arcs $a(n_{s1}, n_{s2})$ that connect two lines at a port n are called *transshipment arcs* τ where $s1, s2 \in S$ and $n_{s1}, n_{s2} \in V$.

Full containers q transported between a pair-port v are sent via one of the alternative routes r and a route r_v is a succession of arcs $a_{is1,js2}$. The cost of an arc $a(i_s, j_s)$ is equal to the travel and waiting costs associated with each service $s1, s2$. Empty containers are grouped in transported y, stored s, leased x and released z arcs, each one with its associated cost. As they're calculated to fulfill a balance equation, they have not a predefined route.

Finally, the benefit function B can be formulated as:

$$B(q, y, x, z, s) = G(q) - CS(s) - CR(y) - CV(x, z)$$

where

- G is the obtained income $G(q) = \sum_{t,v}(p_v^t - c_v^t).q_v^t$,
- CS is the stock cost $CS(q) = \sum_{t,n}\sigma_n^t s_n^t$,
- CR is the empty repositioning cost $CR(y) = \sum_{\sigma,t}c_\sigma^t y_\sigma^t + \sum_{\tau,t}c_\tau^t y_\tau^t + \sum_{\delta,t}y_\delta^t c_\delta^t + \sum_{\lambda,t}y_\lambda^t c_\lambda^t$,
- CV is the leasing/leaving empty costs. $CV(y) = \sum_{t,n}\gamma_n^t x_n^t + \sum_{t,n}\delta_n^t z_n^t$.

  Indexes and variables refer to:

- t represents the evaluated period (in month)
- n is the interest port (always the origin of an arc)
- v is a port-to-port sale arc,
- r is the proposed route for a sale
- q_v^t number of containers sold for a pair port v in the period t
- σ, τ, δ, λ are the stock, transshipment, download and load arcs, respectively.
- p_v^t is the sale price $p_v^t = \bar{p}_v^t - \beta_v(q_v^t - \bar{q}_v^t)$ where $\bar{q}_v^t$, β_v^t and $\bar{p}_v^t$ are data.
- c_v^t is the full container cost for v in t, where $c_v = \sum_{r \in v}\lambda_r c_r$ and $c_r^t = \sum_{\sigma \in r}\sigma c_\sigma + \sum_{\tau \in r}\tau c_\tau + \delta c_\delta + \lambda c_\lambda$
- c_a^t, y_a^t is the cost and number of empty containers in arc a, respectively
- x_n^t, z_n^t is the number of empty containers leased/released at port n
- γ_n, δ_n are the leasing/releasing cost at port n
- s_n^t is the stock capacity at port n in period t

  The model fulfills the following constrains:

- Stock at time t and port n is equal to stock in $t - 1$ plus full (q) and empty (y) imbalance, $s_n^{t+1} = s_n^t - \sum_{v \leftarrow n}q_v^t + \sum_{v \rightarrow n}q_v^t + x_n^{t+1} - z_n^{t+1} + \sum_{\delta \rightarrow n}y_\delta^t - \sum_{\lambda \leftarrow n}y_\lambda^t$
- Stock at time t should be enough to fulfill sales with origin in n and inferior to stock capacity $\sum_{v \leftarrow n}q_v^t \leq s_n^t \leq CA_n$
- The amount of containers (full + empty) transported in arc σ should by fewer than volume vessel capacity for this arc $y_\sigma^t + \sum_{r \leftarrow \sigma}q_r^t \leq CapMax_\sigma$
- At the same time, the amount of containers are limited by a weight vessel capacity $\pi_0 y_\sigma^t + \sum_{r \leftarrow \sigma}\pi_v q_r^t \leq \pi_\sigma^{max}$
- The amount of empty containers bought or sold in a port are in a range: $0 \leq x_n^t \leq x_n^{max}$; $0 \leq z_n^t \leq z_n^{max}$
- The amount of full containers per route v with destination n are weighted by λ $q_r^t = \lambda_r^t q_v^t$, $\lambda_r^t \geq 0$, $\sum \lambda_n^t = 1$
- The amount of sales at each port are defined in a range $\pi_0 y_\sigma^t + \sum_{r \leftarrow \sigma}\pi_v q_r^t \leq \pi_\sigma^{max}$
- The amount of incoming containers at port n in period t are equal to the amount of outgoing containers $\sum_{\lambda \rightarrow w}y_\lambda^t + \sum_{\sigma \rightarrow w}y_\sigma^t = \sum_{\delta \leftarrow w}y_\delta^t + \sum_{\sigma \leftarrow w}y_\sigma^t$

3.2 Assisted Process of Generating Hypothetic Scenarios

As it was shown, the container transport model is quite complex, and the user has to manipulate carefully a huge amount of data to get a coherent scenario. Here we present a systematic and incremental process to reduce time in designing, analyzing and validating scenarios that are lately optimized. The process begins when the user generates an initial analysis scenario, taking one from their operations database, with their ships capacities configurations, routes and associates sales and costs.

Then, he makes some disturbances such as replacing vessels with different capacities, modify the routes, add sales areas, set future analysis periods, among others. At this point the system automatically validates the scenario, generating *Level 1* (L1) alerts, checking that all ports of interest are reachable with the defined routes and that all ship and port costs are defined. If some data is not set, the system arise a module for helping user to estimate it (this module is described in the next section). Once the user checks system suggestions and after making the necessary modifications, he gets what is called an *Alpha* scenario.

After this, the user enters or modifies the expected sales in the period of interest. The system checks that all sales are valid, and that the vessel capacity is enough for this freight. The system shows a series of indicators of "overload" and warnings such as "undefined paths" (routes that are not covered with the current structure) called *Level 2*(L2) warnings. Once the user attends to all suggestions, the system goes to a *Beta* stage. Both the L1 as L2 alerts do not allow the user to advance until they accomplish the proposals.

As a third step, the user calculates the empty repositioning movements needed to balance the system, using an algorithm described later. Upon completion of this process, the user already has a complete scenario version (Release 0), and the system generates a set of indicators and table views (Application Container, Relationship Full/Empty, among others) highlighting extreme values, called *Level 3* (L3) alerts, that is data that must be analyzed in depth. Once the user validates the results, he decides if the proposal is interesting watching the ships capabilities and the benefit obtained. Now, he can change data, restarting the process.

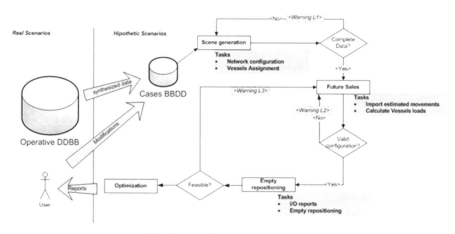

Fig. 1. Scenarios process generation

Finally, the user makes a profit optimization (using the algorithm later explained) and gets a *Release 1* scenario with the proposed improvements. As user always analyzes results in detail, it is necessary to have multiple data views that allow him to navigate data both in a spatial and temporal manner. The overall proposed process is shown in [Fig. 1]. The reposition and optimization model used is explained in the next section. If results are reliable, an optimal plan is approved, and it passes to management who imposed the changes aimed to fulfill the plan.

3.3 Data Conversion Step

Generally, user understands and manipulates data in an Origin-Transshipment-Destination Port format, without seeing intermediate transport arcs and ports. Since the transport model uses information with a different structure, it's necessary to incorporate a data conversion step which is performed in the DB engine. In [Fig. 2] we compare the different data models. When user imports data, the system converts from user format to expected one. When user calculates empty containers using the proposed algorithm, a new reverse conversion must be done hiding the intermediate data. All this conversion is transparent to the user and it's efficiently resolved.

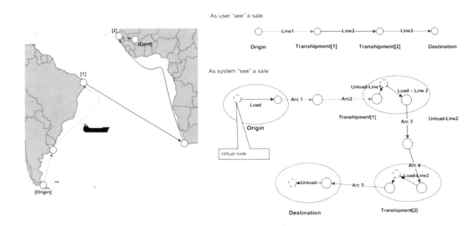

Fig. 2. Data conversion among modules

At the same time, when performing future scenarios analysis, it is common to propose new ports for evaluating how this impact on transportation and benefit. For these new markets, user generally does not have all the needed information to make the model work correctly. System can ask the user to manually enter it; but when there is a lot of missing data, the task becomes much more complex than the analysis itself.

For this problem, when system does not have all the necessary data, it was proposed some strategies or rules:

- Default values. The system set constant values instead of missing ones. For example: Stock capacity per port, container cost.
- Estimation based on historical data. Missing values are calculated based on historical data. For example, leasing cost.
- Estimation based on geographic data. System assigns a value according to a rule of geographic association; all ports in one country have the same cost. User enters just one per country. For example, handling cost, days at port.

- Distance weighting: System assigns a value using a weighting based measure. For example, elasticity coefficient, stevedoring cost.

The system detects missing data and then asks the user to select which of these rules wants to apply. This task is performed during the scenario generation stage and simplifies the gathering task. Generally, each strategy is related to a kind of variable.

3.4 Empty Repositioning and Benefit Optimization Algorithm

Both, empty reposition and benefit optimization problem are modeled in a quadratic function, coupled with two variables: q the quantities sold, and y the containers available to assign. The problem is mathematically generalized as follows:

$$\max_{q,y} B = \max_{q,y} \tfrac{1}{2} q^T A q + b^T q + c^T y$$
$$s..t.Dq + Fy \leq 0$$

Since the solution will be used in a interactive application, the solved method should be efficient. To achieve this and due to the problem is defined with linear constraints, we propose a variation of the Frank-Wolfe method [REF]. This method needs an initial feasible solution and solves a linearization of the problem, at each iteration k. The problem is rewritten as:

$$\max_{q,y} q_k^T A q + b^T q + c^T y - \left(\tfrac{1}{2} q^T A q k \right)$$
$$s.t.Dq + Fy \leq 0$$

Then, a descent direction is calculated in a step k and a linear search in that direction looking for a λ that minimizes the solution is done:

$$\max_{\lambda \in [0,1]} g(\lambda) = \frac{1}{2} \left(q_k + \lambda d_k^q \right)^T A \left(q_k + \lambda d_k^q \right) + b^T \left(q_k + \lambda d_k^q \right) + c^T \left(y_k + \lambda d_k^y \right)$$

This quadratic function has a minimum in the projection $[0.1]$: $\left(d_k^q \right)^T A q_k + \lambda \left(d_k^q \right)^T A d_k^q + b^T d_k^q + c^T d_k^y = 0$

From where: $\lambda_k = \min \left\{ \max \left\{ 0, -\frac{\left(d_k^q \right)^T A q_k + b^T d_k^q + c^T d_k^y}{\left(d_k^q \right)^T A d_k^q} \right\}, 1 \right\}$.

Finally, when λ_k is found, step is updated as: $\begin{cases} q_{k+1} = q_k + \lambda_k d_k^q \\ y_{k+1} = y_k + \lambda_k d_k^y \end{cases}$

To solve each step, it's used a linear solver. As termination criterion, the benefit is measured for each iteration and compared with the previous one. If less than an acceptable value, it is considered that the algorithm converged. For calculating empty containers required for a known sales configuration, only the first step is accomplished. It's important to note, that this task can only be accomplished if the scenario was previously validated.

4   Results

The proposed system is divided into two modules, a generation and analysis scenarios module and an optimization one. The user applies the process to generate scenarios, and get a first scenario of analysis using the first module. Then, he uses the second one to both calculate necessary empty containers and to optimize the overall sales profit. The proposed formulation let user to model complex situations and to get trustable forecasts. Generated scenarios could have from one service covering just a region, to more than twenty services with hundreds of ports in a world situation. We evaluated usability aspects of the system and the optimization model. The optimization algorithm used is iterative and has a quasi-linear computational cost with the amount of data. When the scenarios are complex (with more than 50k variables), the optimization can take only a few minutes in a multi-processor personal computer, being fully acceptable times. Average times that required the algorithm to solve an iteration are shown in [Fig. 3]. It's related to the number of variables $(y + q)$.

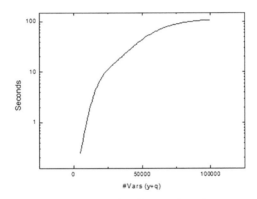

Fig. 3. Computing time per iteration

At the same time, it's discussed how the benefit evolve according to the number of iterations. As was expected, in the early iterations further improvements are observed, after then, the solution tends to stabilize. In practice, regardless of the number of variables, with 5 or 6 iterations the rate of improvement converges to less than 0.1%. The user can even choose to reduce the number of iterations to obtain a faster response.

Then, we evaluate the user experience with the system. It was taken the time in days that takes the user to generate a scenario and to perform a complete analysis considering the sales and empty variables in a period $(y + q)$ compared with times before counting with the system. The extracted results are shown in [Table 1]. Before using the system, the user could only make forecasting in a period (a year or a semester), without month discretization and without modeling some issues, like stocks.

Table 1. Required time in days for generating and testing global scenarios

Amount of variables	Without System	With System
5000	10	2
20000	21	3
40000	–	5
60000	–	10

All obtained information by the different modules are presented in several table views where important data is colored by topics shown in [Fig. 4].

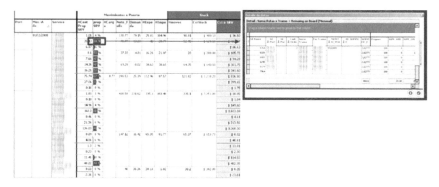

Fig. 4. Different table views for navigating results.

Formulas, data composition and filters are also provided to browse the resulting data (shown in a *red* being storage). For very experienced users who want to see everything in detail, this sometimes is not enough, so tables are exported to the commercial format of Microsoft Excel ® with the corresponding formulas. In this way the user can perform all the additional checks required (filters, sums, graphical), browsing the views exported from the system.

5   Conclusions

In this paper, the experiences of implementing an expert system for a container transport company using a robust optimization method were presented. The proposed scenario design process was efficiently applied in multiple scenarios and analysis, with different number of variables. The system allowed managing from a few thousand data in an annual review to a full scenario with hundreds of variables in a more detailed monthly step. According to the user experience, the same work was achieved far less time, even for more complex scenarios than the originals. The errors usually committed by the user, such as missing data, invalid routes definition or excessive ships occupation, were effectively detected and corrected. As future works, we propose to reduce computing time using a parallel linear solver and to extend the model incorporating ships replacement.

References

1. Zhou, C., Zhao, Q., Li, H.: Simulation optimization iteration approach on traffic integrated yard allocation problem in transshipment terminals. Flex. Serv. Manuf. J. **33**(3), 663 (2021)
2. Perakis, A.N., Jaramillo, D.I.: Fleet deployment optimization for liner shipping. Part 1: background, problem formulation and solution approaches. Marit. Policy Manag. **18**(3), 183–200 (1991)
3. Choong, S.T., Cole, M.H., Kutanoglu, E.: Empty container management for intermodal transportation networks. Transp. Res. **38**, 423–438 (2002)
4. Claessens, E.M.: Optimization procedure in maritime fleet management. Marit. Policy Manag. **14**(1), 27–48 (1987)
5. Crainic, T.G., Gendreau, M., Dejax, P.: Dynamic and stochastic models for the allocation of empty containers. Oper. Res. **41**, 102–126 (1993)
6. Ghorpade, T., Rangaraj, N.: Rolling Horizon Models for Inter-Depot Empty Container Repositioning, pp. 1825–1836 (2019). https://doi.org/10.1109/WSC40007.2019.9004884
7. Jula, H., Chassiakos, A., Ioannou, P.: Port dynamic empty container reuse. Transp. Res. Part E **42**, 43–60 (2006)
8. Epstein, R., et al.: A strategic empty container logistics optimization in a major shipping company. Interfaces **42**(1), 5–16 (2012)
9. Hajeeh, M., Behbehani, W.: Optimizing empty containers distribution among ports. J. Math. Stat. **7**(3), 216–221 (2011)
10. Jula, H., Chassikos, A., Ioannou, P.: Port dynamic empty container reuse. Transp. Res. Part E **42**(1), 43–60 (2006)
11. Kastner, M., Nellen, N., Schwientek, A., Jahn, C.: Integrated simulation-based optimization op. decisions at container terminals. Algorithms **14**(2), 42 (2021)
12. Lane, T., Heaver, T.D., Uyeno, D.: Planning and scheduling for efciency in liner shipping. Marit. Policy Manag. **14**(2), 109–125 (1987)
13. Li, J.-A., Liu, K., Leung, S.C.H., Lai, K.K.: Empty container management in a port with long-run average criterion. Math. Comput. Model **40**, 85–100 (2004)
14. Rana, K., Vickson, R.G.: Routing container ships using Lagrangean relaxation and decomposition. Transport. Sci. **25**(3), 201–214 (1991)
15. Shintani, K., Imai, A., Nishimura, E., Papadimitriou, S.: The container shipping network design problem with empty container repositioning. Transp. Res. Part E **43**, 39–59 (2007)
16. Bandeira, D.L., Becker, J.L., Borenstein, D.: A DSS for integrated distribution of empty and full containers. Decis. Supp. Syst. **47**, 383–397 (2009). https://doi.org/10.1016/j.dss.2009.04.003

Technologies for Biomedical Applications

Serious Games and the Cognitive Screening of Community-Dwelling Older Adults: A Systematic Review

Rute Bastardo[1] (ORCID), João Pavão[2] (ORCID), Bruno Gago[3] (ORCID), and Nelson Pacheco Rocha[4(✉)] (ORCID)

[1] UNIDCOM, Science and Technology School, University of Trás-Os-Montes and Alto Douro, Vila Real, Portugal
[2] INESC-TEC, Science and Technology School, University of Trás-Os-Montes and Alto Douro, Vila Real, Portugal
jpavao@utad.pt
[3] IBIMED, Department of Medical Sciences, University of Aveiro, Aveiro, Portugal
bmgago@ua.pt
[4] IEETA, Department of Medical Sciences, University of Aveiro, Aveiro, Portugal
npr@ua.pt

Abstract. This paper presents a systematic review of the literature aiming to analyze the state of the art of the use of serious games to detect age related cognitive impairments of community-dwelling older adults by identifying (i) the interaction paradigms being used, (ii) the cognitive domains being assessed, and (iii) how the proposed applications were evaluated. A systematic electronic search was performed, and 25 studies were included in the systematic review. The most used interaction paradigm was simulation of daily tasks, while memory, executive functions, and attention were the cognitive domains most referred. Six articles reported on diagnostic accuracy studies and their results indicate that serious games can discriminate between normal cognition and nonnormal cognition, since they present high sensitivity and specificity values.

Keywords: Serious games · Cognitive screening · Older adults · Systematic review

1 Introduction

Older adults experience concerns about their cognition, even if they do not consult a clinician [1]. Moreover, the early detection of cognitive impairment or other neurodegenerative disorders is associated with improved access to care and better outcomes [1–4].

The increase in life expectancy and population aging and the consequent increase in the prevalence of neurodegenerative diseases has promoted the interest in remote cognitive screening [1]. Therefore, multiple technologies are being used to support remote methods to detect mild cognitive impairment, including smart devices, machine learning algorithms, virtual reality, or serious games [3, 5–8].

This systematic review aimed to analyze the state of the art of applications using serious games to detect age related cognitive impairments of community-dwelling older adults. After this first section with the introduction, the rest of the paper is organized as follows: (i) the methods that were applied; (ii) the respective results; and (iii) a discussion and conclusion.

2 Methods

2.1 Study Design

This study consisted of a systematic review reported according to the Preferred Reporting Items for Systematic Reviews and Meta-Analyses (PRISMA) statement [9].

2.2 Search Strategy

The authors searched the PubMed, Scopus, and Web of Science for studies published before the end of April 2022. The expression used for the electronic search was: ('cognitive screening' OR 'cognitive test' OR 'memory screening' OR 'memory test' OR 'attention screening' OR 'attention test') AND ('computer' OR 'game' OR 'gaming' OR 'virtual' OR 'online' OR 'internet' OR 'mobile' OR 'app' OR 'digital'). EndNote and Microsoft Excel were used to record titles, authors, abstracts, publication date and inclusion/exclusion decisions.

2.3 Study Criteria

All English peer-reviewed studies published on scientific journals and conference proceedings and reporting on applications to remotely assess cognitive functions of community-dwelling older adults by means of serious games were considered.

The exclusion criteria were articles that: (i) were not published in English; (ii) did not have abstracts or authors' identification; (iii) their full texts were not available; (iv) reported on reviews or surveys; (v) were books, tutorials, editorials, and special issues announcements; (vi) were extended abstracts or posters; and (vii) reported on studies already covered by other included references (i.e., when two references reported on the same study in different venues, such as a scientific journal and a conference proceeding, the less mature one was excluded).

2.4 Selection Procedure and Data Extraction

The selection of the studies to be included in this systematic review was performed according to the following steps: (i) first step, the authors removed the duplicates, the references without abstract or authors, references reporting on reviews or surveys and references not written in English; (ii) second step, titles and abstracts of the retrieved studies were screened for inclusion; and (iii) third step, the authors assessed the full text of the remaining references against the outlined inclusion and exclusion criteria to achieve the list of the studies to be included in this systematic review.

Concerning data extraction, the following information was registered in a data sheet prepared by the authors for each one of the included studies: (i) the demographics of the study (i.e., authors and respective affiliations, year and source of publication); (ii) the scope of the study and the respective objectives; (iii) details of the technologies being used; (iv) the study research methods; (v) the findings of the study; and (vi) the limitations of the study.

2.5 Synthesis and Reporting

Considering the demographic data, the authors prepared a synthesis of the characteristics of the included studies, considering (i) the number of studies published in scientific journals and conference proceedings; (ii) the distribution of the studies by publication years; and (ii) the distribution of the studies by geographical areas, considering the affiliation of the first author.

In addition to general inclusion and exclusion criteria, the included studies were assessed against the following quality questions, which were adopted and adjusted from other studies [10, 11]: Q1 - are the objectives of the study clearly identified? Q2 - is the context of the study clearly stated? Q3 - does the research methods support the aims of the study? Q4 - has the study an adequate description of the technologies being used? Q5 - is there a clear statement of the findings? Q6 - are limitations of the study discussed explicitly? Each question was answered according to a binary scale (i.e., 1 for Yes or 0 for No).

Moreover, syntheses were prepared to systematize (i) the interaction paradigms being used, (ii) the cognitive domains being assessed and (iii) how the proposed applications were evaluated. For the latter purpose, the included studies were classified as (i) usability evaluation studies, (ii) feasibility studies (i.e., evaluation of the practicality of the proposed applications), (iii) validity and reliability studies (i.e., evaluation of the accuracy and consistence of the results of the proposed applications), and (iv) diagnostic accuracy studies (i.e., evaluation of how well the proposed applications discriminate between healthy and cognitive impaired individuals).

3 Results

3.1 Study Selection

Figure 1 presents the PRISMA flowchart of this systematic review. The search of the studies to be included was conducted in the first week of May 2022. A total of 8557 references was retrieved from the initial search: (i) 3580 references from PubMed; (ii) 3733 references from Scopus; and (iii) 1244 references from Web of Science.

The initial step of the screening phase yielded 4481 references by removing duplicates (n = 3452), references reporting on reviews or surveys (n = 311), references without abstracts or authors (n = 171), references not written in English (n = 141) and one retraction.

Based on titles and abstracts, 4436 references were removed since they did not meet the eligibility criteria. Finally, the full texts of the remaining 45 references were screened

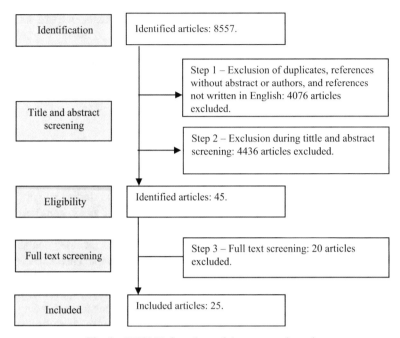

Fig. 1. PRISMA flowchart of the systematic review.

and 20 were excluded because they were not relevant for the specific objective of this systematic review. Therefore, 25 studies were considered eligible and included in the review [12–36].

3.2 Demographic Characteristics

In terms of publication types, 20 studies [13, 15, 17–21, 23, 25, 36] were published in scientific journals, being the remainder five studies [12, 14, 16, 22, 24] published in conference proceedings. The included studies were published between 2008 (i.e., one study [12]) and the first four months of 2022 (i.e., two studies [35, 36]). Nine studies [26–34] were published in 2021, the year with the highest number of publications.

Concerning the geographical distribution, Europa contributed with 13 studies (i.e., Greece [17, 26, 29, 35], Portugal [21, 23, 36], France [12, 15], Belgium [28], Finland [24], Switzerland [16], and United Kingdom [27]), North America contributed with six studies (i.e., Canada [18, 20, 30], and United States [13, 19, 32]), Asia contributed with five studies (i.e., China [22, 34], Korea [14], Taiwan [31], and Turkey [25]), and Argentina, South America, contributed with one study [33].

3.3 Quality Assurance

Table 1 presents the results of the quality assurance (i.e., the number of articles that conform with each one of the six quality assurance questions). The results varied from 17 (i.e., question 5) to 13 (i.e., question 2).

Table 1. Quality assurance.

Question	Yes	No
Q1. Are the objectives of the study clearly identified?	15	10
Q2. Is the context of the study clearly stated?	13	12
Q3. Does the research methods support the aims of the study?	15	10
Q4. Has the study an adequate description of the technologies being used?	14	11
Q5. Is there a clear statement of the findings?	17	8
Q6. Are limitations of the study discussed explicitly?	15	10

3.4 Interaction Paradigms and Cognitive Functions

As can be seen in Table 2, the most referred interaction paradigm was simulation of daily tasks. Some studies used specific tasks such as driving [12, 15], meal preparation [32], and shopping in a virtual supermarket [17, 25, 29, 34, 35], while others considered a set of diverse daily tasks (e.g., navigation, shopping, cooking, preparing a table, cleaning, taking the medication, turn off the alarm, or using the telephone [16, 22, 24]).

Table 2. Interaction paradigms and cognitive functions.

Interaction Paradigms		Articles	Cognitive functions
Daily Tasks			
	Set of diverse daily tasks		
		[16]	Not specified
		[22]	Attention, executive functions, language, and memory
		[24]	Executive functions, language, memory, and orientation
	Driving		
		[12]	Memory
		[15]	Memory
	Meal preparation		
		[32]	Memory
	Virtual supermarket		
		[17]	Attention, executive functions, memory, and spatial navigation
		[25]	Attention, executive functions, memory, and spatial navigation
		[29]	Executive functions, and spatial navigation
		[34]	Attention, executive functions, language, learning, memory, time orientation
		[35]	Executive functions, and spatial navigation

(continued)

Table 2. (*continued*)

Interaction Paradigms	Articles	Cognitive functions
Set of diverse abstract tasks		
	[13]	Attention, and memory
	[14]	Memory, perception, and reaction
	[21]	Attention, executive functions, memory, perception, and reasoning
	[23]	Attention, calculus, construction, executive functions, language, and memory
	[26]	Attention, executive functions, memory, perception, and orientation
	[31]	Attention, executive functions, and memory
	[36]	Attention, executive functions, memory, and spatial navigation
Start to goal spatial navigation tasks		
	[20]	Not specified
	[33]	Memory
Mimicry of traditional cognitive tests		
	[18]	Inhibitory control
	[27]	Attention, executive functions, memory, psychomotor
	[30]	Inhibitory control
Commercial games		
	[19]	Not specified
	[28]	Not specified

Moreover, the following interaction paradigms were also identified: (i) realization of diverse abstract tasks [13, 14, 21, 23, 26, 31, 36] (e.g., throw objects into portals according to their shape and color or memorize light patterns and report when the patterns are repeated [36]); (ii) start to goal spatial navigation tasks [20, 33]; (iii) mimicry of traditional cognitive tests [18, 27, 30]; and (iv) commercial games [19, 28].

Four studies [16, 19, 20, 28] did not specify the cognitive functions being assessed. Analyzing the remainder studies, memory was the most referred cognitive function and was assessed by 17 studies [12–15, 17, 21–27, 31–36], while executive functions were assessed by 13 studies [17, 21–27, 29, 31, 34–36] and attention was assessed by 11 studies [13, 17, 21–23, 25–27, 31, 34, 36]. In turn, multiple interaction paradigms were used to assess these functions. For instance, in terms of the assessment of memory, simulation of daily tasks was used by eight studies [12, 15, 17, 22, 24, 25, 32, 34] (five of them using the virtual supermarket metaphor [17, 25, 29, 34, 35]), while the execution of diverse abstract tasks was used by seven studies [13, 14, 21, 23, 26, 31, 36], the start to goal spatial navigation task was used by one study [33], commercial games were used by two studies [19, 28], and the mimicry of traditional cognitive tests was used by one study [27].

3.5   Assessment Characteristics

In terms of the assessment of the proposed applications, two articles [14, 36] present the design of the application, but did not report on the respective evaluation. Another article [20] presented the design of the application and reported on the test of the user interaction involving one participant. The assessments performed by the remainder 22 studies were classified in the following categories: (i) usability evaluation, five studies [16, 22, 24, 27, 35]; (ii) feasibility, six studies [12, 19, 26, 28, 32, 33]; (iii) validity and reliability, five studies [18, 21, 29–31]; and (iv) diagnostic accuracy, six studies [13, 15, 17, 23, 25, 34].

In terms of usability studies, the participants varied from eight [16] to 89 [27] healthy adults, while the evaluation instruments included the System Usability Scale [16, 24, 27, 35], the Simulated Sickness Questionnaire [24], the User Satisfaction Evaluation Questionnaire [24], the Igroup Presence Questionnaire [24], and the NASA Task Load Index [16]. Moreover, one study [22] reported on a focus group to evaluate the usability of the proposed solution.

Table 3 presents the characteristics of the feasibility studies.

Table 3. Feasibility studies.

Articles	Outcomes	Instruments	Participants
[12]	Comparison between the performance of young adults and older adults	Episodic Memory Classical Test, Trail Making Test A and B, and Wechsler Adult Intelligent Scale	113 students and 57 healthy older adults
[19]	Stratification of cognitive impairment	Mini-Mental State Examination	11 older adults
[26]	Comparison between normal condition and cognitive impairment	Montreal Cognitive Assessment	Ten older adults
[28]	Comparison between normal condition and cognitive impairment	Clinical diagnostic	23 health older adults and 23 cognitive impaired older adults
[32]	Comparison of game performance according to age groups	Wechsler Memory Scale and California Verbal Learning Test	41 students, 52 healthy older adults, and seven older adults with neuro-degenerative diagnosis
[33]	Analysis of performance	–	46 healthy adults

In turn, Table 4 presents the characteristics of the validity and reliability studies, and Table 5 presents the characteristics of the diagnostic accuracy studies. Concerning the diagnostic accuracy studies, the reported values on sensitivity (i.e., ability to identify an individual with cognitive impairment) varied from 73% [25] to 90% [23], while the reported values on specificity (i.e., ability to identify an individual without cognitive impairment) varied from 54% [23] to 100% [25].

Table 4. Validity and reliability studies.

Articles	Instruments	Participants
[18]	Mini-Mental State Examination, Confusion Assessment Method, Delirium Index, Richmond Agitation-Sedation Scale, Digit Vigilance Test, and a Choice Reaction Time Task	147 older adults
[21]	Montreal Cognitive Assessment	52 healthy older adults and 51 cognitive impaired older adults
[29]	Montreal Cognitive Assessment	43 healthy older adults and 33 cognitive impaired older adults
[30]	Go/No-Go discrimination task	30 young adults
[31]	Mini-Mental State Examination	120 older adults

Table 5. Diagnostic accuracy studies.

Articles	Instruments	Participants
[13]	Expert neuropsychologist's assessment	405 older adults (172 controls and 233 cognitive impaired)
[15]	Cognitive Difficulties Scale, Mill Hill Test, Trail-Making Test and Wechsler Adult Intelligence Scale	51 older adults (21 healthy controls, 15 cognitive impaired and 15 with dementia)
[17]	Mini-Mental State Examination, Ray Auditory Verbal Learning Test, Verbal Fluency Test, Rey-Osterrieth Complex Figure Test, Rivermead Behavioral Memory Test, Test of Everyday Attention, Trail Making Test	55 older adults (21 healthy controls and 34 cognitive impaired)
[23]	Mini-Mental State Examination and Montreal Cognitive Assessment	88 healthy older adults, 44 older adults with mild impairment, and 44 older adults with mild dementia
[25]	Mini-Mental State Examination	87 older adults (52 healthy controls and 35 cognitive impaired)
[34]	Mini-Mental State Examination and Montreal Cognitive Assessment	122 older adults (60 healthy controls and 62 cognitive impaired)

4 Discussion and Conclusion

This systematic review identified 25 studies related to applications supported on serious games to be used remotely as screening tools for age related cognitive impairment of community-dwelling older adults. Although the first study was published in 2008, a significant percentage of the studies were published in the last two years, which shows the current interest in the topic.

In terms of the geographical distribution of the studies, Europe had the highest contribution (i.e., 52% of the included studies), followed by North America (i.e., 24%). This means that more that 75% of the studies come from regions with a high percentage of older adults, that is, where there is a high prevalence of neurodegenerative diseases.

Most of the studies (i.e., 80% of the studies) where published in scientific journals. However, in terms of quality assurance, a significant number of articles present methodological drawbacks. The fifth quality assurance question (i.e., is there a clear statement of the findings?) obtained the highest number of positive results (i.e., 68% of the studies), while the second question (i.e., is the context of the study clearly stated?) obtained the highest number of negative results (i.e., 48% of the studies).

The simulation of daily tasks was the most used interaction paradigm since it was identified in 44% of the studies. Moreover, 28% of the studies considered as interaction paradigm the realization of diverse abstract tasks, while the remainder 28% of the studies considered the start to goal spatial navigation task, commercial games, and the mimicry of traditional cognitive tests.

In terms of cognitive functions being assessed, memory, executive functions, and attention where the most referred cognitive functions and were evaluated using different interaction paradigms. In addition to memory, executive functions, and attention, other ten cognitive domains were identified: (i) language (ii) orientation and navigation; (iii) perception; (iv) inhibitory control; (v) learning; (vi) reaction; (vii) reasoning; (viii) calculus; (ix) construction; and (x) psychomotor ability.

Additionally, concerning the assessment of the proposed applications, two studies did not report on the assessment of the applications and another study reported on the test of the user interaction involving one participant. The remainder 22 articles reported on usability evaluation studies (20% of the studies), feasibility studies (24% of the studies), validity and reliability studies (20% of the studies), and diagnostic accuracy studies (24% of the studies).

Considering the articles reporting on usability evaluation studies, five validated usability assessment instruments were identified: (i) the System Usability Scale; (ii) the Simulated Sickness Questionnaire; (iii) the User Satisfaction Evaluation Questionnaire; (iv) the Igroup Presence Questionnaire; (v) and the NASA Task Load Index [16].

The objectives of the feasibility studies included: (i) analysis of performance; (ii) comparison between normal condition and cognitive impairment; (ii) comparison between the performance of young adults and older adults; (iii) comparison of game performance according to age groups; and (v) stratification of cognitive impairment. In turn, the participants varied between 11 older adults [19], and 113 students and 55 healthy older adults [12].

The validity and reliability studies compared the results of the proposed applications with validated cognitive screening instruments being used in the clinical practice (e.g., Montreal Cognitive Assessment, Mini-Mental State Examination, Go/No-Go discrimination task, Confusion Assessment Method, Delirium Index, Richmond Agitation-Sedation Scale, Digit Vigilance Test, and a Choice Reaction Time Task). The number of participants varied between 30 young adults [30] and 147 older adults [18].

Finally, six articles [13, 15, 17, 23, 25, 34] reported on diagnostic accuracy studies. The discrimination between healthy controls and cognitive impaired was compared with the results of expert neuropsychologist's assessment [13] or different validated cognitive instruments (e.g., Mini-Mental State Examination or Montreal Cognitive Assessment). All the studies included healthy older adults and cognitive impaired older adults. The number of participants varied from 405 (i.e., the study reported by [13] included 172 healthy older adults and 233 cognitive impaired older adults) to 51 (i.e., the study reported by [15] included 21 healthy controls, 15 cognitive impaired and 15 with dementia).

The results of these diagnostic accuracy studies show that serious games applications can discriminate between normal cognition and nonnormal cognition. Therefore, further research should include clinical to collect robust scientific evidence to facilitate the translation for the clinical practice of serious games applications to detect cognitive impairments of community-dwelling older adults.

Acknowledgements. This work was supported by Programa Operacional Competitividade e Internacionalização (COMPETE 2020), Portugal 2020 and Lisboa 2020 of the Fundo Europeu de Desenvolvimento Regional (FEDER)/European Regional Development Fund (ERDF), under project project SH4ALL – Smart Health for ALL, POCI-01-0247-FEDER-046115.

References

1. Binng, D., Splonskowski, M., Jacova, C.: Distance assessment for detecting cognitive impairment in older adults: a systematic review of psychometric evidence. Dement. Geriatr. Cogn. Disord. **49**(5), 456–470 (2020)
2. Olivari, B.S., Baumgart, M., et al.: CDC grand rounds: promoting well-being and independence in older adults. MMWR Morb. Mortal Wkly Rep. **67**(37), 1036 (2018)
3. Snyder, P., Jackson, C., Petersen, R., et al.: Assessment of cognition in mild cognitive impairment: a comparative study. Alzheimers Dement. **7**(3), 338–355 (2011)
4. Gates, N., Kochan, N.: Computerized and on-line neuropsychological testing for late-life cognition and neurocognitive disorders: are we there yet? Curr. Opin. Psychiatry **28**(2), 165–172 (2015)
5. Pereira, C., Pereira, D., Weber, S., et al.: A survey on computer-assisted Parkinson's disease diagnosis. Artif. Intell. Med. **95**, 48–63 (2019)
6. Diaz-Orueta, U., Blanco-Campal, A., Lamar, M., et al.: Marrying past and present neuropsychology: is the future of the process-based approach technology-based? Front. Psychol. **11**, 361 (2020)
7. Lumsden, J., Edwards, E., Lawrence, N., et al.: Gamification of cognitive assessment and cognitive training: a systematic review of applications and efficacy. JMIR Serious Games **4**(2), e11 (2016)
8. Bastardo, R., Pavão, J., Martins, A.I., Silva, A.G., Rocha, N.P.: Cognitive screening instruments for community-dwelling older adults: a mapping review. In: The 2018 International Conference on Digital Science, pp. 533–544. Springer, Cham (2021)
9. Moher, D., Liberati, A., Tetzlaff, J., Altman, D.G.: Preferred reporting items for systematic reviews and meta-analyses: the PRISMA statement. BMJ **339**, b2535 (2009)
10. Shahin, M., Liang, P., Babar, M.A.: A systematic review of software architecture visualization techniques. J. Syst. Softw. **94**, 161–185 (2014)
11. Yang, L., et al.: Quality assessment in systematic literature reviews: a software engineering perspective. Inf. Softw. Technol. **130**, 106397 (2021)

12. Plancher, G., Nicolas, S., Piolino, P.: Virtual reality as a tool for assessing episodic memory. In Proceedings of the 2008 ACM Symposium on Virtual Reality Software and Technology, pp. 179–182. ACM, New York (2008)
13. Wright, D.W., Nevárez, H., Kilgo, P., LaPlaca, M., et al.: A novel technology to screen for cognitive impairment in the elderly. Am. J. Alzheimer's Disease Other Dementias® **26**(6), 484–491 (2011)
14. Byun, S., Park, C.: Serious game for cognitive testing of elderly. In: International Conference on Human-Computer Interaction, pp. 354–357. Springer, Heidelberg (2011)
15. Plancher, G., Tirard, A., Gyselinck, V., et al.: Using virtual reality to characterize episodic memory profiles in amnestic mild cognitive impairment and Alzheimer's disease: influence of active and passive encoding. Neuropsychologia **50**(5), 592–602 (2012)
16. Vallejo, V., Mitache, A.V., Tarnanas, I., Müri, et al.: Combining qualitative and quantitative methods to analyze serious games outcomes: a pilot study for a new cognitive screening tool. In: 2015 37th Annual International Conference of the IEEE Engineering in Medicine and Biology Society (EMBC), pp. 1327–1330. IEEE, Piscataway (2015)
17. Zygouris, S., Giakoumis, D., Votis, K., Doumpoulakis, S., et al.: Can a virtual reality cognitive training application fulfill a dual role? Using the virtual supermarket cognitive training application as a screening tool for mild cognitive impairment. J. Alzheimers Dis. **44**(4), 1333–1347 (2015)
18. Tong, T., Chignell, M., Tierney, M.C., Lee, J.: A serious game for clinical assessment of cognitive status: validation study. JMIR Ser. Games **4**(1), e5006 (2016)
19. House, G.P., Burdea, G., Polistico, K., Ross, J., Leibick, M.: A serious gaming alternative to pen-and-paper cognitive scoring: a pilot study of BrightScreener™. J. Pain Manag. **9**(3), 255–264 (2016)
20. Doucet, G., Gulli, R.A., Martinez-Trujillo, J.C.: Cross-species 3D virtual reality toolbox for visual and cognitive experiments. J. Neurosci. Methods **266**, 84–93 (2016)
21. Neto, H.S., Cerejeira, J., Roque, L.: Cognitive screening of older adults using serious games: an empirical study. Entertain. Comput. **28**, 11–20 (2018)
22. Zeng, Z., Fauvel, S., Hsiang, B.T.T., Wang, D., Qiu, Y., et al.: Towards long-term tracking and detection of early dementia: a computerized cognitive test battery with gamification. In: Proceedings of the 3rd International Conference on Crowd Science and Engineering, pp. 1–10. ACM, New York (2018)
23. Ruano, L., Severo, M., Sousa, A., Ruano, C., et al.: Tracking cognitive performance in the general population and in patients with mild cognitive impairment with a self-applied computerized test (brain on track). J. Alzheimers Dis. **71**(2), 541–548 (2019)
24. Dulau, E., Botha-Ravyse, C.R., Luimula, M., Markopoulos, P., Markopoulos, E., Tarkkanen, K.: A virtual reality game for cognitive impairment screening in the elderly: a user perspective. In: 2019 10th IEEE International Conference on Cognitive Infocommunications (CogInfoCom), pp. 403–410. IEEE, Piscataway (2019)
25. Eraslan Boz, H., Limoncu, H., Zygouris, S., Tsolaki, M., et al.: A new tool to assess amnestic mild cognitive impairment in Turkish older adults: virtual supermarket (VSM). Aging Neuropsychol. Cognit. **27**(5), 639–653 (2020)
26. Karapapas, C., Goumopoulos, C.: Mild cognitive impairment detection using machine learning models trained on data collected from serious games. Appl. Sci. **11**(17), 8184 (2021)
27. McWilliams, E.C., Barbey, F.M., Dyer, J.F., et al.: Feasibility of repeated assessment of cognitive function in older adults using a wireless, mobile, dry-EEG headset and tablet-based games. Front. Psych. **12**, 574482 (2021)
28. Gielis, K., Abeele, M.E.V., Verbert, K., et al.: Detecting mild cognitive impairment via digital biomarkers of cognitive performance found in klondike solitaire: a machine-learning study. Digit. Biomark. **5**(1), 44–52 (2021)

29. Iliadou, P., Paliokas, I., Zygouris, S., Lazarou, E., et al.: A comparison of traditional and serious game-based digital markers of cognition in older adults with mild cognitive impairment and healthy controls. J. Alzheimer's Dis. **79**(4), 1747–1759 (2021)
30. Tong, T., Chignell, M., et al.: Using a serious game to measure executive functioning: response inhibition ability. Appl. Neuropsychol. Adult **28**(6), 673–684 (2021)
31. Lin, C.W., Mao, T.Y., Huang, C.F.: A novel game-based intelligent test for detecting elderly cognitive function impairment. Comput. Math. Methods Med. **2021**, 1698406 (2021)
32. Barnett, M.D., Childers, L.G., Parsons, T.D.: A virtual kitchen protocol to measure everyday memory functioning for meal preparation. Brain Sci. **11**(5), 571 (2021)
33. Rodríguez, M.F., Ramirez Butavand, D., Cifuentes, M.V., Bekinschtein, P., Ballarini, F., García Bauza, C.: A virtual reality platform for memory evaluation: assessing effects of spatial strategies. Behav. Res. Methods **2021**, 1–13 (2021)
34. Yan, M., Yin, H., Meng, Q., Wang, S., Ding, Y., et al.: A virtual supermarket program for the screening of mild cognitive impairment in older adults: diagnostic accuracy study. JMIR Ser. Games **9**(4), e30919 (2021)
35. Zygouris, S., et al.: Usability of the virtual supermarket test for older adults with and without cognitive impairment. J. Alzheimer's Dis. Rep. **6**(1), 229–234 (2022)
36. Brugada-Ramentol, V., Bozorgzadeh, A., Jalali, H.: Enhance VR: a multisensory approach to cognitive training and monitoring. Front. Digit. Health **2022**(4), 916052 (2022)

The Fast Health Interoperability Resources (FHIR) and Clinical Research, a Scoping Review

João Pavão[1], Rute Bastardo[2], and Nelson Pacheco Rocha[3(✉)]

[1] INESC-TEC, Science and Technology School, University of Trás-Os-Montes and Alto Douro,
Vila Real, Portugal
jpavao@utad.pt
[2] UNIDCOM, Science and Technology School, University of Trás-Os-Montes and Alto Douro,
Vila Real, Portugal
[3] IEETA, Department of Medical Sciences, University of Aveiro, Aveiro, Portugal
npr@ua.pt

Abstract. The focus of this scoping review was the application of Fast Health Interoperability Resources (FHIR) to support clinical research. After the search and selection processes 36 studies were identified and included in the review. Analysing these studies, it is possible to conclude that FHIR is being used in the implementation of information systems to support (i) clinical trials data management, (ii) data integration for clinical research, and (iii) secondary use of clinical information. Moreover, the results also show the interest on the development of novel FHIR profiles to answer specific data requirements of clinical research.

Keywords: Clinical Research · Fast Health Interoperability Resources · FHIR · Scoping Review

1 Introduction

The flexibility of Fast Health Interoperability Resources (FHIR) has promoted its adoption to satisfy interoperability requirements of a broad range of healthcare information systems [1, 2].

The massive accumulation of large-scale molecular and clinical data in recent decades led to the conceptualization of personalized medicine seen as a practice of medicine that uses the individuals' genetic profile to guide decisions regarding the diagnosis and treatment of disease [3]. Therefore, it is possible to classify the patients into different subtypes, which can be helpful in the diagnosis by generating disease decision rules and personalized recommendations, especially for the rare diseases, or to develop predictive models that might be helpful to assess the disease progression states [3, 4]. This contributes to the prediction, prevention, personalization, and participation (P4) model [5] that is supported on the aggregation of clinical data, which is only possible if the interoperability of these data is assured.

Assuming the importance of FHIR to assure the interoperability of clinical data, the objective of this article was to analyse recent studies related to the application of FHIR in the context of clinical research.

A. Rocha et al. (Eds.): WorldCIST 2023, LNNS 801, pp. 409–418, 2024.
https://doi.org/10.1007/978-3-031-45648-0_40

2   Methods

A review protocol was defined with explicit descriptions of the methods to be used and the steps to be taken: (i) research questions; (ii) search strategies; (iii) eligibility criteria; (iv) screening procedures; (v) data extraction; and (vi) synthesis and reporting.

2.1   Research Questions

The research objective was informed by the following research questions:

- RQ1 - what type of clinical research applications benefit from the use of FHIR?
- RQ2 - what FHIR resources are being used to support clinical research?

2.2   Search Strategies

In terms of resources to be searched three databases were considered: PubMed, Web of Science and Scopus. PubMed was selected considering its importance among clinical researchers. In turn, Web of Science and Scopus are the two major existing multidisciplinary databases, have a high coverage of scientific journals and conference proceedings and contain a significant number of references indexed by other databases, such as ACM Digital Library, Science Direct, SpringerLink or IEEE Xplore.

The search queries were prepared to include articles that have in their titles, abstract or keywords the expression 'Fast Healthcare Interoperability Resources' or the term 'FHIR'.

2.3   Eligibility Criteria

Table 1 provides details of the eligibility criteria.

Table 1. Eligibility criteria.

Inclusion	Exclusion
References of full articles that deal with the application of FHIR in clinical research	References that did not deal with FHIR or related acronyms or did not address issues related to the defined research questions
English references published in peer reviewed proceedings or journals	Non-English references or references not published in peer reviewed proceedings or journals
References published before 31st March 2022	References published after 31st March 2022
References whose full texts were available	References whose full texts were not available
References reporting on primary studies	References not reporting on primary studies
References reporting on studies not covered by other included articles	References reporting on studies already covered by other included articles

2.4 Synthesis and Reporting

The included studies were analysed in terms of: (i) their demographic characteristics (i.e., number of studies published in scientific journals and conference proceedings, distribution of the studies according to their publication years, distribution of the studies by geographical areas, according to the affiliation of the first author); (ii) purposes of the applications being developed, and (iii) details of the FHIR resources being used.

3 Results

3.1 Study Selection

The search on the selected databases was conducted in April 2022 and identified 1343 references: (i) 350 from PubMed; (ii) 394 from Web of Science; and (iii) 599 from Scopus. After the selection process (Table 2) 36 studies were considered for this review [6–41].

3.2 Demographic Characteristics

Twenty-four studies [7, 10, 12, 13, 16–19, 21, 23, 24, 26, 30–41] were published in scientific journals and 12 studies [6, 8, 9, 11, 14, 15, 20, 22, 25, 27–29] were published in conference proceedings. The studies were published between 2016 and 2022: 2016, one study [6]; 2017, three studies [7–9]; 2018, four studies [10–13]; 2019, three studies [14–16]; 2020, eight studies [17–24]; 2021, 14 studies [25–38]; and 2022, three studies [39–41].

Concerning the geographical distribution, most studies were published by researchers from Germany (i.e., 16 studies [6, 11, 14, 17, 18, 20, 25, 27–30, 33, 36, 37, 40, 41]) and United States (i.e., 14 studies [8, 10, 12, 16, 19, 21, 23, 24, 26, 31, 32, 34, 38, 39]). The remainder six studies were published by researchers from Australia [7], Canada [35], Ireland [22], Pakistan [9], South Africa [15], and United Kingdom [13]. Moreover, three studies involved multinational research teams: Canada and United States [35]; Germany and Austria [14]; and Germany and Netherlands [36].

Table 2. Selection process.

Excluded	
	660 duplicates
	60 without abstract or authors
	29 secondary studies (reviews or surveys)
	516 excluded by the title and abstract analysis
	42 excluded by the full text analysis
Included	
	36 studies

3.3 Purposes of the Studies

From the analysis of the studies included in the review the following purposes were identified: (i) clinical trials data management, ten studies [11, 20, 21, 26, 27, 29, 33, 34, 37, 38]; (ii) data integration for clinical research, 17 studies [6–10, 16–18, 22, 25, 28, 30, 35, 36, 39–41]; and (iii) secondary use of clinical information, nine studies [12–15, 19, 23, 24, 31, 32].

Although most of the studies considered clinical research in general, some studies were focused on clinical research related to specific pathologies: (i) Covid-19, seven studies [17, 19, 25, 35, 36, 40, 41]; (ii) cancer, four studies [6, 24, 28, 34]; (iii) asthma, one study [16]; (iv) paediatrics, one study [23]; (v) obesity, one study [32]; (vi) obesity comorbidities, one study [31]; and (vii) haematology, one study [38].

Concerning the studies focused on the management of clinical trials data, (i) one study [33] developed a central trial registry, (ii) four studies [11, 20, 29, 37] developed applications for the identification of participants for clinical trials, and (iii) four studies [21, 26, 27, 34, 38] were focused on seamless data exchange between the electronic data capture (EDC) systems and Electronic Health Record (EHR).

Article [33] presented a prototypically implemented open-source central trial registry based on FHIR as a data storage and exchange format. The central trial registry contains records from university hospitals, which are automatically exported and updated by local clinical trial' management systems.

In turn, the four studies [11, 20, 29, 37] focused on the identification of suitable participants for specific clinical trials developed: (i) a FHIR-based representation of eligibility criteria in a machine-readable format to automate the identification of suitable participants [11]; (ii) FHIR patients' screening lists of potential participants for specific clinical trials [20]; and (iii) trials case number estimation for the planning of multicentre clinical trials [29, 37].

The identification and standardization of data elements used in clinical trials may control and reduce the cost and errors during the operational process. In this respect, five studies [21, 26, 27, 34, 38] proposed the use of FHIR to enable seamless data exchange between the EDC systems and EHR systems.

In terms of data integration for clinical research, there studies [7, 8, 16] were focused on clinical data interoperability, while 14 studies [6, 9, 10, 17, 18, 22, 25, 28, 30, 35, 36, 39–41] aimed to develop clinical data repositories to support clinical research.

Looking for the first group (i.e., focused on clinical data interoperability), (i) article [7] proposed two FHIR-based models to capture the metadata and data from clinical studies and facilitate their syntactic and semantic interoperability, (ii) article [8] reported the design and assessment a consensus-based approach for harmonizing the Observational Medical Outcomes Partnership (OMOM) common data model with FHIR resources, and (iii) other study [16] aimed to create an open-source application to transform clinical data to FHIR to support custom data management.

Concerning the articles focused on clinical data repositories to support clinical research:

- The study reported by [6] aimed to develop a hospital-integrated research data management system supporting biobank-based research, which included the translation of metadata information about data elements into FHIR resources.

- A framework for public health data analysis and mining was presented by [9].
- The transformation of primary FHIR data directly into data repositories is described by [10].
- The study reported by [17] collected, prioritized, and consolidated a COVID-19 compact core dataset using FHIR as data exchange store format.
- The integration of genomics and clinical data for statistical analysis was reported by [18].
- A FHIR-based ontology as a basis for managing demographic health care data in Ireland was proposed by [22].
- The study reported by [25] assessed the possibility of FHIR be used as the data store and exchange format to support a central search hub to gather COVID-19 research (e.g., studies, questionnaires, or documents).
- The study reported by [28] employed use cases from the German Cancer Consortium to present a harmonized FHIR-based data model.
- The study reported by [30] aimed to evaluate how FHIR data can be queried directly with a pre-processing service to be used for statistical analyses.
- The use of FHIR resources to represent clinical data for automated transformation to common model of the Patient-Centred Outcomes Research Network (PCORnet) and Observational Medical Outcomes Partnership (OMOP) was reported by [35].
- A consensus metadata model to facilitate structured searches of COVID-19 studies and resources was described by [36].
- The study reported by [39] assessed how a search ontology can be automatically generated using FHIR profiles and a terminology server.
- A cross-hospital query platform for researchers based on FHIR resources was proposed by [40].
- The design and implementation of the components involved in creating a cross-hospital feasibility query platform for researchers based on FHIR resources, which is part of a larger COVID-19 data exchange platform (CODEX), was described by [41].

Finally, in what concerns the studies [12–15, 19, 23, 24, 31, 32] focused on the secondary use of clinical information, the following objectives were identified:

- Article [12] reported a study that developed a representation of patients' information based on FHIR resources to demonstrate that deep learning methods using this representation are capable of accurately predicting multiple medical events from multiple centres without site-specific data harmonization.
- A general-purpose semantic search system to surface semantic data from clinical notes was presented by [13].
- The study reported by [14] aimed to demonstrate how the output of a commercial clinical text-mining tool can be integrated with FHIR resources to generate instances from clinical narratives.
- The study reported by [15] aimed to propose FHIR resources to represent unstructured nicotine use data in a structured format for secondary use.
- A machine-readable dataset to allow the extraction of relevant associations for COVID-19 and other coronavirus infectious diseases was described by [19].

- The study reported by [23] aimed to demonstrate the use of random forest and multilayer perceptron neural network techniques to predict multi-centre paediatric readmissions, using a data platform supported on FHIR.
- The study reported by [24] aimed to identify novel associations between genotypes and clinical phenotypes, which were represented by FHIR resources.
- The study reported by [31] aimed to develop natural language processing algorithms represented by FHIR resources.
- The study reported by [32] demonstrated how FHIR-based representation of unstructured clinical data can be ported to deep learning models for text classification in clinical phenotyping.

3.4 FHIR Resources

A broad range of normative FHIR resources to support clinical research were referred by the included studies. The most referred FHIR resources were: (i) Observation, 11 studies [7, 10, 14, 16, 17, 24, 26, 28, 30, 34, 40]; (ii) Condition, ten studies [14, 16, 17, 24, 26, 28, 30, 31, 37, 40]; (iii) Patient, nine studies [7, 10, 11, 16, 17, 26, 28, 30, 40]; (iv) Procedure, seven studies [16, 17, 28, 30, 31, 37, 40]; (v) Encounter, six studies [7, 10, 16, 17, 28, 30]; (vi) Medication, five studies [14, 17, 28, 31, 40]; and (vii) DiagnosticReport, three studies [10, 14, 34]. Moreover, (i) AllergyIntollerance, CarePlan, ClinicalImpression, Consent, EpisodeOfCare, MedicationRequest, Organization, Practitioner, Questionnaire, QuestionnaireResponse and ResearchStudy were referred by at least two studies, and (ii) Binary, Composition, DocumentReference, FamilyMemberStory, Group, ImagingStudy, Immunization, PractitionerRole, Provenance, RiskAssesment, Specimen, and Subscription were referred by at least one study.

Additionally, three studies [7, 24, 39] proposed FHIR extensions: (i) the ClinicalStudy Plan resource to capture clinical studies' metadata and the ClinicalStudyData to describe the data of clinical studies [7]; (ii) Cohort, Report, Count and Phecod resources to explore new associations between genotypes and clinical phenotypes [24]; and (iii) the Researcher and Participant resources to respectively characterise the researchers and participants of clinical studies, as well as the Credential resource to provide security access control and the Tag resource to integrate extensibility and backward capabilities [39].

4   Discussion and Conclusion

This scoping review identified 36 studies. Considering the types of clinical research applications that benefit from the use of FHIR (i.e., the first research question) three different types of applications were identified: (i) clinical trials data management; (ii) data integration for clinical research; and (iii) secondary use of clinical information.

Concerning the second research question (i.e., what FHIR resources are being used to support clinical research?), the normative FHIR resource most referred by the included studies was Observation. Other normative FHIR resources with significant number of references in included studies were Condition, Patient, Procedure, Encounter and Medication. In turn, three studies [7, 24, 39] propose extensions to FHIR resources, namely (i) ClinicalStudy Plan to capture clinical studies' metadata, (ii) ClinicalStudyData to describe the data of clinical studies, (iii) Credential to provide security access control, (iv) Tag to integrate extensibility and backward capabilities, and (v) Cohort, (vi) Report, (vii) Count, and (vii) Phecod for exploring new associations between genotypes and clinical phenotypes.

The results indicate that FHIR seems to answer to the interoperability requirements of clinical research, although some extensions to normative resources should be considered.

The oldest study [6] included in this scoping review was published in 2016. In turn, only 11 studies [6–16] were published before 2020. This recent interest in FHIR to support clinical research is a natural consequence of the fact that the first FHIR publication, the FHIR Draft Standard for Trial Use 1, was published in 2013. After this first publication, the Draft Standard for Trial Use 2 was published in 2015, the Standard for Trial Use 3 was published in 2017, and the Release 4, the first version with normative content, was published in 2019.

The limitations of this scoping review are related to the dependency on its keywords and the databases selected. Despite these limitations, the authors followed rigorous methodological steps and the systematically collected evidence might be useful to understand the current trends of the application of FHIR in the context of clinical research.

Acknowledgements. This work was supported by Programa Operacional Competitividade e Internacionalização (COMPETE 2020), Portugal 2020 and Lisboa 2020 of the Fundo Europeu de Desenvolvimento Regional (FEDER) / European Regional Development Fund (ERDF), under project project ACTIVAS - Ambientes Construídos para uma Vida Ativa, Segura e Saudável, POCI-01–0247-FEDER-046101.

References

1. Yogesh, M.J., Karthikeyan, J.: Health informatics: engaging modern healthcare units: a brief overview. Front. Public Health **10**, 854688 (2022)
2. Staff, M.: FHIR: reducing friction in the exchange of healthcare data. Commun. ACM **65**(12), 34–41 (2022)
3. Catlow, J., Bray, B., Morris, E., Rutter, M.: Power of big data to improve patient care in gastroenterology. Frontline Gastroenterol. **13**(3), 237–244 (2022)
4. Hassan, M., Awan, F.M., Naz, A., de Andrés-Galiana, E.J., et al.: Innovations in genomics and big data analytics for personalized medicine and health care: a review. Int. J. Mol. Sci. **23**(9), 4645 (2022)
5. Flores, M., Glusman, G., Brogaard, K., et al.: P4 medicine: how systems medicine will transform the healthcare sector and society. Pers. Med. **10**(6), 565–576 (2013)
6. Ulrich, H., Kock, A.K., Duhm-Harbeck, P., et al.: Metadata repository for improved data sharing and reuse based on HL7 FHIR. In: MIE, pp. 162–166. IOS Press, Amsterdam (2016)
7. Leroux, H., Metke-Jimenez, A., Lawley, M.J.: Towards achieving semantic interoperability of clinical study data with FHIR. J. Biomed. Semant. **8**(1), 1–14 (2017)
8. Jiang, G., Kiefer, R.C., Sharma, D.K., Prud'hommeaux, E., Solbrig, H.R.: A consensus-based approach for harmonizing the OHDSI common data model with HL7 FHIR. Stud. Health Technol. Inf. **245**, 887 (2017)

9. Khalique, F., Khan, S.A.: An FHIR-based framework for consolidation of augmented EHR from hospitals for public health analysis. In: 11th International Conference on Application of Information and Communication Technologies (AICT), pp. 1–4. IEEE, Piscataway (2017)

10. Solbrig, H.R., Hong, N., Murphy, S.N. and Jiang, G.: Automated population of an i2b2 clinical data warehouse using FHIR. In: AMIA Annual Symposium Proceedings, vol. 2018, p. 979. American Medical Informatics Association, Bethesda (2018)

11. Kraus, S.: Investigating the capabilities of FHIR search for clinical trial phenotyping. In: Proceedings of the 63rd Annual Meeting of the German Association of Medical Informatics, Biometry and Epidemiology, p. 3. IOS Press, Amsterdam (2018)

12. Ajkomar, A., Oren, E., Chen, K., Dai, A.M., et al.: Scalable and accurate deep learning with electronic health records. NPJ Dig. Med. **1**(1), 1–10 (2018)

13. Wu, H., Toti, G., Morley, K.I., et al.: SemEHR: A general-purpose semantic search system to surface semantic data from clinical notes for tailored care, trial recruitment, and clinical research. J. Am. Med. Inform. Assoc. **25**(5), 530–537 (2018)

14. Daumke, P., Heitmann, K.U., Heckmann, S., Martínez-Costa, C., Schulz, S.: Clinical text mining on FHIR. In: MedInfo, pp. 83–87. IOS Press, Amsterdam (2019)

15. Ngwenya, M. and Bankole, F.: Mining and representing unstructured nicotine use data in a structured format for secondary use. In: Proceedings of the 52nd Hawaii International Conference on System Sciences. University of Hawai, Hawai (2019)

16. Pfaff, E.R., Champion, J., Bradford, R.L., et al.: Fast healthcare interoperability resources (FHIR) as a meta model to integrate common data models: development of a tool and quantitative validation study. JMIR Med. Inf. **7**(4), e15199 (2019)

17. Sass, J., Bartschke, A., Lehne, M., Essenwanger, A., et al.: The German Corona Consensus Dataset (GECCO): a standardized dataset for COVID-19 research in university medicine and beyond. BMC Med. Inf. Decis. Mak. **20**(1), 1–7 (2020)

18. Gruendner, J., Wolf, N., Tögel, L., Haller, F., Prokosch, H.U., Christoph, J.: Integrating genomics and clinical data for statistical analysis by using GEnome MINIng (GEMINI) and Fast Healthcare Interoperability Resources (FHIR): System design and implementation. J. Med. Internet Res. **22**(10), e19879 (2020)

19. Oniani, D., Jiang, G., Liu, H., Shen, F.: Constructing co-occurrence network embeddings to assist association extraction for COVID-19 and other coronavirus infectious diseases. J. Am. Med. Inf. Assoc. **27**(8), 1259–1267 (2020)

20. Reinecke, I., Gulden, C., Kümmel, M., Nassirian, A., Blasini, R., Sedlmayr, M.: Design for a modular clinical trial recruitment support system based on FHIR and OMOP. In: Digital Personalized Health and Medicine, pp. 158–162. IOS Press, Amsterdam (2020)

21. Garza, M.Y., Rutherford, M., Myneni, S., Fenton, S., et al.: Evaluating the coverage of the hl7® fhir® standard to support esource data exchange implementations for use in multi-site clinical research studies. In: AMIA Annual Symposium Proceedings, vol. 2020, p. 472. American Medical Informatics Association, Bethesda (2020)

22. McGlinn, Kris, Hussey, Pamela: An analysis of demographic data in irish healthcare domain to support semantic uplift. In: Krzhizhanovskaya, V.V., Závodszky, G., Lees, M.H., Dongarra, J.J., Sloot, P.M.A., Brissos, S., Teixeira, J. (eds.) ICCS 2020. LNCS, vol. 12140, pp. 456–467. Springer, Cham (2020). https://doi.org/10.1007/978-3-030-50423-6_34

23. Ehwerhemuepha, L., Gasperino, G., et al.: HealtheDataLab–a cloud computing solution for data science and advanced analytics in healthcare with application to predicting multi-center pediatric readmissions. BMC Med. Inf. Decis. Mak. **20**(1), 1–12 (2020)

24. Zong, N., Sharma, D.K., Yu, Y., Egan, J.B., et al.: Developing a FHIR-based framework for phenome wide association studies: a case study with a pan-cancer cohort. In AMIA Summits on Translational Science Proceedings 2020, vol. 750. AMIA, Bethesda (2020)

25. Klofenstein, S.A.I., Vorisek, C.N., Shutsko, A., et al.: Fast healthcare interoperability resources (FHIR) in a FAIR metadata registry for COVID-19 research. In: Applying the FAIR Principles to Accelerate Health Research in Europe in the Post COVID-19 Era, pp. 73–77. IOS Press, Amsterdam (2021)
26. Cheng, A.C., Duda, S.N., Taylor, R., Delacqua, F., et al.: REDCap on FHIR: clinical data interoperability services. J. Biomed. Inf. **121**, 103871 (2021)
27. Riepenhausen, S., Mertens, C., Dugas, M.: Comparing SDTM and FHIR® for real world data from electronic health records for clinical trial submissions. In: MIE, pp. 585–589. IOS Press, Amsterdam (2021)
28. Lambarki, M., Kern, J., Croft, D., et al.: Oncology on FHIR: A data model for distributed cancer research. Stud. Health Technol. Inf. **278**, 203–210 (2021)
29. Wettstein, R., Hund, H., et al.: Feasibility queries in distributed architectures-concept and implementation in HiGHmed. Stud Health Technol. Inf. **278**, 134–141 (2021)
30. Gruendner, J., Gulden, C., Kampf, M., et al.: A framework for criteria-based selection and processing of fast healthcare interoperability resources (FHIR) data for statistical analysis: design and implementation study. JMIR Med. Inf. **9**(4), e25645 (2021)
31. Wen, A., Rasmussen, L.V., Stone, D., Liu, S., et al.: CQL4NLP: development and integration of FHIR NLP extensions in clinical quality language for EHR-driven phenotyping. In: AMIA Annual Symposium Proceedings, vol. 2021, p. 624. American Medical Informatics Association (2021)
32. Liu, S., Luo, Y., Stone, D., Zong, N., Wen, A., Yu, Y., et al.: Integration of NLP2FHIR representation with deep learning models for EHR phenotyping: a pilot study on obesity datasets. In: AMIA Annual Symposium Proceedings, vol. 2021, p. 410. American Medical Informatics Association, Bethesda (2021)
33. Gulden, C., Blasini, R., Nassirian, A., Stein, A., Altun, F.B., et al.: Prototypical clinical trial registry based on fast healthcare interoperability resources (FHIR): design and implementation study. JMIR Med. Inf. **9**(1), e20470 (2021)
34. Zong, N., Stone, D.J., Sharma, D.K., Wen, A., et al.: Modeling cancer clinical trials using HL7 FHIR to support downstream applications: a case study with colorectal cancer data. Int. J. Med. Inf. **145**, 104308 (2021)
35. Lenert, L.A., Ilatovskiy, A.V., Agnew, J., Rudisill, P., et al.: Automated production of research data marts from a canonical fast healthcare interoperability resource data repository: applications to COVID-19 research. J. Am. Med. Inf. Assoc. **28**(8), 1605–1611 (2021)
36. Schmidt, C.O., Darms, J., Shutsko, A., Löbe, M., et al.: Facilitating study and item level browsing for clinical and epidemiological COVID-19 studies. Stud. Health Technol. Inf. **281**, 794–798 (2021)
37. Banach, A., Ulrich, H., Kroll, B., Kiel, A., et al.: APERITIF–automatic patient recruiting for clinical trials based on HL7 FHIR. In: Public Health and Informatics, pp. 58–62. IOS Press, Amsterdam (2021)
38. Wood, W.A., Marks, P., Plovnick, R.M., Hewitt, K., et al.: ASH research collaborative: a real-world data infrastructure to support real-world evidence development and learning healthcare systems in hematology. Blood Adv. **5**(23), 5429–5438 (2021)
39. Vaidyam, A., Halamka, J., Torous, J.: Enabling research and clinical use of patient-generated health data (the mindLAMP platform): digital phenotyping study. JMIR Mhealth Uhealth **10**(1), e30557 (2022)

40. Rosenau, L., Majeed, R.W., Ingenerf, J., Kiel, A., et al.: Generation of a fast healthcare inter-operability resources (FHIR)-based ontology for federated feasibility queries in the context of COVID-19: feasibility study. JMIR Med. Inf. **10**(4), e35789 (2022)

41. Gruendner, J., Deppenwiese, N., Folz, M., Köhler, T., et al.: The architecture of a feasibility query portal for distributed COVID-19 fast healthcare interoperability resources (FHIR) patient data repositories: design and implementation study. JMIR Med. Inf. **10**(5), e36709 (2022)

Scoping Review: Application of Machine Learning Techniques in Genetic Diagnosis

Beatriz Faria[1], Mariana Ribeiro[1], Raquel Simões[1], Susana Valente[1], and Nelson Pacheco Rocha[2]([✉]) [ID]

[1] Department of Medical Sciences, University of Aveiro, Aveiro, Portugal
{beatriz.faria,mrpr,raqueladsimoes,susanamvalente}@ua.pt
[2] IEETA, Department of Medical Sciences, University of Aveiro, Aveiro, Portugal
npr@ua.pt

Abstract. Machine learning techniques have been an important support for disease prediction and diagnosis utilizing genetic variants classification. This scoping review aimed to synthetize the application of machine learning techniques in processing of genetic variants to support clinical diagnosis. Scopus, Web of Science and PubMed databases were used for retrieving the studies to be included in this scoping review. A total of 522 records were retrieved and after the selection process 11 studies were included. Several machine learning techniques have been applied to processing genetic variants related to rare genetic diseases, cancer, autism spectrum disorder and hereditary hearing loss. However, there is no evidence of the translation of these machine learning techniques to the clinical practice.

Keywords: Machine Learning · Genetic Variants · Genetic Diagnosis

1 Introduction

Deoxyribonucleic Acid (DNA) sequencing technologies have led to an increase in the number of individual-specific genotype data, which might allow the identification of genetic variants that contribute to specific diseases [1]. Therefore, the study of genetic variants may help the diagnosis of various types of diseases, such as cancer, neurological disorders, or rare diseases.

Machine learning techniques have been an important support for disease prediction and diagnosis utilizing genetic variants' classification. Through sets of statistical and computational algorithms, machine learning techniques can make predictions based on associations between genetic variants and complex diseases' phenotypes [1, 2].

Machine learning techniques need to be evaluated for their performance. This evaluation allows comparison of different implementation to select the most suitable one. Accuracy, sensitivity, specificity, precision, F1-score, Mathew's correlation coefficient and the area under the receiver operating characteristic curve are common performance measures, where higher values indicate a better performance. Also, false discovery rate is used to evaluate the classifiers, in which the method with the lower value has better performance.

A. Rocha et al. (Eds.): WorldCIST 2023, LNNS 801, pp. 419–428, 2024.
https://doi.org/10.1007/978-3-031-45648-0_41

This scoping review aimed to synthetize the current state of the application of machine learning techniques to process genetic variants to support clinical diagnosis.

2 Methods

2.1 Research Questions

The objective of this scoping review was decomposed in the following research questions.

- RQ1 - What diseases can be diagnosed using genetic variants?
- RQ2 - What machine learning techniques are currently being used to process genetic variants?
- RQ3 - What are the purposes of genetic variants processing?
- RQ4 - Are the reported solutions being translated to the clinical practice?

2.2 Information Sources and Search Strategy

To retrieve the studies to be included in this scoping review, three scientific databases were considered.

First, Scopus, the largest peer-review scientific literature database, was considered. The following query was used for the Scopus database: TITLE-ABS-KEY (genomics AND variant AND ((artificial AND intelligence) OR (machine AND learning))), thus looking for literature with the terms 'genomics', 'variant' and 'artificial intelligence' or 'machine learning' in either the title, abstract or keywords.

Web of Science, a platform that allows access to multiple databases of various academic disciplines, was also considered. For this database the following query was prepared: TS = (Variant AND genomics AND ("machine learning" OR "artificial intelligence")) to look for literature that had the words 'variant', 'genomics' and "machine learning" or "artificial intelligence" in their whole text.

Finally, PubMed was also considered. PubMed is a free search engine accessing primarily the MEDLINE database on life sciences and biomedical topics, and the following query was used: genomics [Title/Abstract] AND variant [Title/Abstract] AND ((artificial [Title/Abstract] AND intelligence [Title/Abstract]) OR (machine [Title/Abstract] AND learning [Title/Abstract])).

2.3 Eligibility Criteria

In terms of inclusion criteria, the authors aimed to include full articles written in English and published in scientific journals reporting studies with experimental results of the application of machine learning techniques to support clinical diagnosis based on genetic variants.

In terms of exclusion criteria, articles were excluded if they were not published in English, reported on reviews or surveys, were published in conference proceedings, were books, tutorials, or editorials, were not relevant for the purpose of this scoping review or their full texts were not available.

2.4 Selection of the Studies

The selection of the studies to be included in this scoping review was performed according to three steps. In the first step the authors removed the duplicates. In the second step, the titles and abstracts of retrieved articles were screened for inclusion according to the predefined eligibility criteria. Finally, in the third step, the authors assessed the full text of the remaining references against the outlined eligibility criteria to achieve the list of the studies to be included in this scoping review.

2.5 Synthesis and Reporting

After the characterization of the publication years of the included studies, the authors performed a synthesis of the target diseases. Moreover, a narrative synthesis was prepared to describe the included articles in terms of target disease, purpose (e.g., diagnosis, gene prioritization or variant prioritization), machine learning techniques being used, and if the reported solutions are being translated to clinical practice.

3 Results

3.1 Selection Process

The selection process is illustrated in the Preferred Reporting Items for Systematic Reviews and Meta-Analyses (PRISMA) [3] flowchart presented in Fig. 1. The Scopus, Web of Science and PubMed databases were accessed on May 11th, 2022, and a total of 522 records were identified (313 from Scopus, 168 from Web of Science and 47 from PubMed).

In the first step of the selection process 366 articles were left after the removal of duplicates. In the second step of the selection process 353 articles were excluded due to the following reasons: one article does not have abstract; 15 articles reported on reviews; and 337 articles were not relevant for the purpose of this scoping review (e.g., did not report on the application of machine learning techniques, or reported on studies focused on animals and not on humans). In the third step of the selection process, the full texts of the remainder 13 articles were analyzed. Of these, 11 studies were included in this scoping review [4–14].

3.2 Publication Years

The studies were published between 2017 and 2021: 2017, one study [4]; 2018, one study [5]; 2019, one study [6]; 2020, three studies [7–9]; and 2021, five studies [10–14].

3.3 Target Diseases

In terms of target diseases, the included studies were focused on the following diseases: rare genetic diseases [6, 8, 9, 13, 14], namely Mendelian diseases; cancer [4, 5, 7, 11]; autism spectrum disorder [10]; and hereditary hearing loss [12].

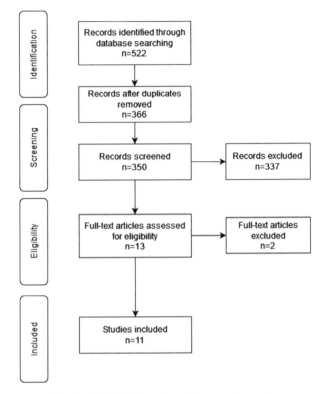

Fig. 1. PRISMA flowchart of the scoping review.

3.4 Rare Genetic Diseases

In terms of rare diseases, three studies [6, 8, 9] aimed to contribute to the research and diagnosis of Mendelian diseases, one study [14] was focused on genetic rare diseases in general and another study [13] was focused on the identification of digenic disease genes of undiagnosed diseases (i.e., rare diseases whose cause cannot be identified).

Mukherjee et al. [13] studied the use of a Random Forest classifier for identifying digenic disease genes in undiagnosed diseases. The aim of the study was to find the genetic causes of rare non-monogenic diseases, which consists of gene pairs' evaluation for the potential to cause digenic disease. Several realistic scenarios showed the accuracy and robustness of their approach [13]. Although the proposed algorithm is available online, it is not clear if it is already being used in clinical practice.

In turn, De La Vega et al. [14] proposed Fabric GEM, a computational tool using the Naive Bayes machine learning technique to support the diagnosis of rare genetic diseases, including Mendelian diseases. The performance of the proposed system was assessed and compared to other approaches using a cohort from the Rady Children's Institute for Genomic Medicine. A limitation of this study is related to the fact of the referred cohort still have a significant number of unsolved cases. Moreover, the authors referred the availability of a demo of the proposed tool, but it is not clear if it was already used in clinical practice [14].

Considering the studies targeting Mendelian diseases, Birgmeier et al. [8] developed a novel machine learning tool, the Automatic VAriant evidence DAtabase (AVADA), supported on Gradient Boosting Decision Tree that automatically identifies pathogenic genetic variant evidence in full-text primary literature about Mendelian diseases and converts it to genomic coordinates using natural language processing.

According to the authors [8], by combining AVADA-based rapid variant retrieval with validation it will be possible the creation and upkeep of cheaper, better, or faster updating variant databases, which will ultimately contribute to both rapid diagnosis and reanalysis [8]. AVADA presents an estimated 73% sensitivity and 49% precision over relevant papers. Besides, the length and complexity of biomedical texts also resulted in AVADA's 61% precision in mapping mentioned variants to their correct genomic coordinates [8]. The authors did not present evidence of the use of AVADA in clinical practice.

Furthermore, Li et al. [9] tried to overcome the current challenge for Mendelian diseases research and diagnosis by presenting the Consequence-Agnostic Pathogenicity Interpretation of Clinical Exome (CAPICE) platform. This platform supports an ensemble of machine learning techniques, including Random Forest and Gradient Boosting Decision Tree for prioritizing pathogenic variants [9].

Still in terms of Mendelian diseases, Li et al. [6] developed a phenotype similarity score called Emission-Reception Information Content (ERIC), and the Xrare system based on the machine learning technique Gradient Boosting Decision Tree for prioritizing pathogenic variants explaining the disease phenotypes, which, according to the authors, is crucial for genetic diagnosis of rare Mendelian diseases. ERIC score ranked consistently higher for disease genes than other phenotypic similarity scores in the presence of imprecise and noisy phenotypes. Extensive simulations and real clinical data demonstrated that Xrare outperforms existing alternative methods by 10–40% at various genetic diagnosis scenarios [6]. However, the authors highlight a few limitations such as the need to further extend the model when more comprehensive phenotype annotations are available and the fact that the followed approach is mainly useful for prioritization of variants near coding regions, thus being a less powerful tool to handle noncoding regulatory variants [6]. As future work, the authors expressed the intention to expand the model to increase its performance to handle noncoding regulatory variants but did not mention if they intend to translate it to clinical practice.

3.5 Cancer

Kumar et al. [7] presented the Supervised Machine Learning Framework (SVFX) supported on a Random Forest machine learning-based workflow to identify and prioritize structural variants. This integrative framework was applied to structural variants in six different cancer cohorts (i.e., breast, ovarian, liver, esophageal, stomach, and skin cancer) and the authors concluded that, overall, the framework presents a high accuracy in identifying pathogenic somatic and germline structural variants from benign variants. Likewise, the cancer germline deletion model was highly accurate across the six cancer cohorts [7]. However, there is no evidence in the literature or in the referred article that this method is being used in clinical practice.

The study of Özcan Şimşek et al. [11] aimed to diagnose and differentiate various cancer types (e.g., breast cancer, lung adenocarcinoma and kidney renal clear cell carcinoma) by utilizing and transferring the information of existing mutations for a novel gene selection method for gene expression data. Logistic regression was the used machine learning technique for the gene selection. According to the experiment's findings, the proposed gene selection method presents better performance when compared to other feature selection methods. However, once again, there is no proof that this procedure is employed in clinical practice in the literature or the cited paper [11].

The authors [11] considered that one of the major challenges of using gene expression data is the small sample size with high dimensionality, in other words, the fact that there may be thousands of genes in each sample but only a few of them are effective on the target disease, and most of them are irrelevant. To overcome this high dimensionality problem, gene selection techniques are applied prior to classification. However, it turns into another problem, the feature selection step may exclude genes that in general have minimal influence on disease development while still being significant for the diagnosis of some cancer types [11].

The study presented by Soh et al. [4] aimed to develop a genetic alteration-based diagnosis tool for patients with cancer, such as when metastases are found, but from whom the primary site remains unknown. The study employed three machine learning techniques, namely linear Support Vector Machine with recursive feature selection, Logistic Regression and Random Forest, to select a small subset of gene alterations that are most informative for cancer-type prediction [4]. The authors [4] used those genes as biomarkers to classify sequenced tumor samples that span 28 different cancer types. They found Support Vector Machine with recursive feature selection to be the most predictive model of cancer type from gene alterations. By using only 100 somatic point-mutated genes for prediction, the overall accuracy was 49%. From a biological point of view, their results showed that "the most discriminatory power comes from copy number alterations", meaning that somatic point mutations may be more similar across cancer types, however, copy number alterations are more specific for each cancer type [4]. However, the authors did not mention the intention to translate the proposed model to clinical practice.

By using a machine learning technique Support Vector Machine to select gene subsets Lu et al. [5] created an integrated framework for identifying and prioritizing subtype-specific driver genes is critical to guide the diagnosis of breast cancer.

Overall, the experimental results demonstrated that the power of integrative analysis can generate a short list of biologically significant genes that can facilitate the process of biomarkers discovery [5]. Given that the breast cancer data is very heterogeneous, the authors [5] tried to overcome this limitation by using their approach. Despite the good results obtained, this method presents some limitations. It is computationally expensive due to the iteration in training the model and searching for the best model. Also, it needs a high number of samples to perform well since it is a method based on statistical machine learning. Possible use of the proposed framework in clinical practice was not mentioned by the authors.

3.6 Autism Spectrum Disorder

The study reported by Wang and Avillach [10] aimed to identify important mutations and to create a diagnosis system for autism spectrum disorder, using 100 significant common variants related to this disease and Convolutional Neural Networks.

Based on experimental results, the authors concluded that the proposed method was a reliable method for distinguishing the diseased group from the control group [10]. Although the system achieved high accuracy, it has limitations for selection of common variants and for the Convolutional Neural Networks proposed, which was a straightforward solution. However more advanced machine learning techniques could be applied to this problem [10]. In turn, there is no evidence in the referred article that this method is being used in clinical practice [10].

3.7 Hereditary Hearing Loss

Luo et al. [12] studied the diagnosis of hereditary hearing loss based on genetic variants and using six different machine learning techniques: Support Vector Machine, Random Forest; Decision Tree; K-Nearest Neighbor; Adaptive Boosting; and Multilayer Perceptron Algorithms.

The authors compared the performance of all the six machine learning techniques to predict hereditary hearing loss by analyzing genes related to hereditary hearing loss with variants. They concluded that among the six machine learning techniques, the Support Vector Machine showed the best performance and has great potential to be an efficient and effective tool for hereditary hearing loss prediction when high throughput sequencing data are available [12]. However, there is no evidence concerning its translation to clinical practice.

4 Discussion

This scoping review identified 11 studies related to the application of machine learning techniques in the study of genetic variants to support the clinical diagnosis. The included studies were published between 2017 and 2021, which shows the current interest in the topic of this scoping review.

Considering the first research question (i.e., what diseases can be diagnosed using genetic variants?) a significant percentage of the studies were focused on rare genetic diseases (i.e., five studies [6, 8, 9, 13, 14]), specifically in Mendelian diseases (i.e., three studies [6, 8, 9]), and cancer (i.e., four studies [4, 5, 7, 11]). Moreover, the remainder two studies were focused on autism spectrum disorder [10] and hereditary hearing loss [12].

Concerning the machine learning techniques currently being used to process genetic variants (i.e., the second research question), the following methods were used by the included studies: Random Forest [4, 7, 9, 12, 13]; Gradient Boosting Decision Tree [6, 8, 9]; Support Vector Machine [4, 5, 12]; Logistic Regression [4, 11]; Naïve Bayes [14]; Convolution Neural Network [10]; Decision Tree [12]; K-Nearest Neighbor [12]; Adaptive Boosting [12]; and Multilayer Perceptron Algorithms [12].

Random Forest was used in [13] to identify digenic diseases genes of undiagnosed diseases, and in [7] to identify and prioritize structural variants in six different cancer cohorts. Moreover, Random Forest was used together with Gradient Boosting Decision Tree for prioritizing pathogenic variants related to Mendelian diseases [9] and together with Support Vector Machine and Logistic Regression to develop a genetic alteration-based diagnosis tool for patients with cancer [4]. Moreover, Random Forest was compared with other five machine learning techniques to support the diagnosis of hereditary hearing loss [12].

Gradient Boosting Decision Tree was applied in the study reported by [8] to identify pathogenic genetic variant evidence in full-text primary literature about Mendelian diseases and in the study reported by [6] for prioritizing pathogenic variants explaining the Mendelian diseases phenotypes. Still in terms of Mendelian diseases, Gradient Boosting Decision Tree was used together with Random Forest for prioritizing pathogenic variants [9].

One study [5] assessed the Support Vector Machine method for identifying and prioritizing subtype-specific driver genes that are critical to guide the diagnosis of breast cancer. Moreover, as already mentioned, Support Vector Machine was with Random Forest and Logistic Regression to develop a genetic alteration-based diagnosis tool for patients with cancer [4]. Finally, Support Vector Machine was also compared with other five machine learning techniques to support the diagnosis of hereditary hearing loss [12].

Logistic Regression was used in conjunction with Support Vector Machine to implement a diagnosis tool for patients with cancer [4]. Moreover, Logistic Regression was also used in [11] to diagnose and differentiate various cancer types.

The use of the Naive Bayes machine learning technique is referred by [14] to support the diagnosis of rare genetic diseases, including Mendelian diseases, while [10] reported the use Convolutional Neural Networks to create a diagnosis system for autism spectrum disorder.

Finally, more four other methods were also referred [12]: Decision Tree; K-Nearest Neighbor; Adaptive Boosting; and Multilayer Perceptron Algorithms. These methods are part of the six-methods set, which also includes Random Forest and Support Vector Machine that were compared by Luo et al. [12] to support the diagnosis of hereditary hearing loss.

Considering the third research question (i.e., what are the purposes of genetic variants processing?), the following purposes were identified: variant retrieval from the literature; gene prioritization; variant prioritization; and clinical diagnosis.

One study [8] proposed the AVADA system to automate the process of variant evidence retrieval from the primary literature about Mendelian diseases. In turn, two studies [5, 13] were focused on gene prioritization: one article [5] described an integrated framework for identifying and prioritizing subtype-specific driver genes to guide the diagnosis of breast cancer, and a second article [13] reported a study aiming to identify digenic disease genes in undiagnosed diseases via machine learning techniques. Moreover, since prioritizing variants is a commonly used procedure to reduce the number of genetic variants that need to be manually evaluated, three other studies [6, 7, 9] were focused on variant prioritization for: the identification of digenic diseases genes of undiagnosed diseases [13]; the identification and prioritization of structural variants in six different cancer cohorts [7]; and the prioritizing of pathogenic variants explaining the Mendelian diseases phenotypes [6]. Finally, five studies [4, 10–14] were focused on the clinical diagnosis: a diagnosis system for autism spectrum disorder [10]; a method to diagnose and differentiate various cancer types (e.g., breast cancer, lung adenocarcinoma and kidney renal clear cell carcinoma) [11]; a genetic alteration-based diagnosis tool for patients with cancer [4]; a system to support the diagnosis of hereditary hearing loss [12]; and a computational tool to support the diagnosis of rare genetic diseases, including Mendelian diseases [14].

In what concerns the possibility of the reported solutions being translated to the clinical practice (i.e., the fourth research question), it is possible to conclude that the included studies are research-oriented, and the author failed to demonstrate how their results might be used in clinical practice.

Like all reviews, this scoping review presents some limitations, namely both the chosen keywords and the databases that were used in the research and even the judgement of the authors when screening the articles. However, the authors followed rigorous procedures in the selection of the studies and the extraction of data, which means that this scoping review is a contribution to the understanding of the current state of the art of using machine learning techniques in genetic diagnosis.

5   Conclusion

In conclusion, this scoping review included 11 studies and synthetize the current state of the application of machine learning techniques to process genetic variants to support clinical diagnosis. According to the included studies, this machine learning techniques application is still restricted to few diseases, namely rare genetic diseases, including Mendelian diseases, cancer, autism spectrum disorder and hereditary hearing loss. According to the publication years of the included studies, the specific topic of this scoping review is an emerging topic, and, therefore, it is expected that the number of studies will increase in the future.

Although, in general, the proposed solutions present good performance indicators, a major drawback of the current research is the lack of translation to the clinical practice.

Acknowledgments. This review was carried out within the scope of the course unit Clinical Information Management of the Master's in Clinical Bioinformatics at the University of Aveiro.

References

1. Ho, D.S.W., Schierding, W., Wake, M., Saffery, R., O'Sullivan, J.: Machine learning SNP based prediction for precision medicine. Front. Genet. **10**, 267 (2019)
2. Dasgupta, A., Sun, Y.V., König, I.R., Bailey-Wilson, J.E., Malley, J.D.: Brief review of regression-based and machine learning methods in genetic epidemiology: the Genetic Analysis Workshop 17 experience. Genet. Epidemiol. **35**(S1), S5–S11 (2011)
3. Moher, D., Liberati, A., Tetzlaff, J., Altman, D.G.: Preferred reporting items for systematic reviews and meta-analyses: the PRISMA statement. BMJ **339**, b2535 (2009)

4. Soh, K.P., Szczurek, E., Sakoparnig, T., Beerenwinkel, N.: Predicting cancer type from tumour DNA signatures. Genome Med. **9**(1), 1–11 (2017)
5. Lu, X., Li, X., Liu, P., Qian, X., Miao, Q., Peng, S.: The integrative method based on the module-network for identifying driver genes in cancer subtypes. Molecules **23**(2), 183 (2018)
6. Li, Q., Zhao, K., Bustamante, C.D., Ma, X., Wong, W.H.: Xrare: a machine learning method jointly modeling phenotypes and genetic evidence for rare disease diagnosis. Genet. Med. **21**(9), 2126–2134 (2019)
7. Kumar, S., Harmanci, A., Vytheeswaran, J., Gerstein, M.B.: SVFX: a machine learning framework to quantify the pathogenicity of structural variants. Genome Biol. **21**(1), 1–21 (2020)
8. Birgmeier, J., Deisseroth, C.A., Hayward, L.E., Galhardo, L.M., Tierno, A.P., et al.: AVADA: toward automated pathogenic variant evidence retrieval directly from the full-text literature. Genet. Med. **22**(2), 362–370 (2020)
9. Li, S., van der Velde, K.J., De Ridder, D., Van Dijk, A.D., Soudis, D., Zwerwer, L.R., et al.: CAPICE: a computational method for consequence-agnostic pathogenicity interpretation of clinical exome variations. Genome Med. **12**(1), 1–11 (2020)
10. Wang, H., Avillach, P.: Diagnostic classification and prognostic prediction using common genetic variants in autism spectrum disorder: genotype-based deep learning. JMIR Med. Inf. **9**(4), e24754 (2021)
11. Özcan Şimşek, N.Ö., Özgür, A., Gürgen, F.: A novel gene selection method for gene expression data for the task of cancer type classification. Biol. Direct **16**(1), 1–5 (2021)
12. Luo, X., Li, F., Xu, W., Hong, K., Yang, T., Chen, J., et al.: Machine learning-based genetic diagnosis models for hereditary hearing loss by the GJB2, SLC26A4 and MT-RNR1 variants. EBioMedicine **69**, 103322 (2021)
13. Mukherjee, S., Cogan, J.D., Newman, J.H., Phillips, J.A., III., Hamid, R., Network, U.D., et al.: Identifying digenic disease genes via machine learning in the Undiagnosed Diseases Network. Am. J. Human Genet. **108**(10), 1946–1963 (2021)
14. De La Vega, F.M., Chowdhury, S., Moore, B., Frise, E., McCarthy, J., Hernandez, E.J., et al.: Artificial intelligence enables comprehensive genome interpretation and nomination of candidate diagnoses for rare genetic diseases. Genome Med. **13**(1), 1–19 (2021)

Radiomics and Radiogenomics Platforms Integrating Machine Learning Techniques: A Review

Rafael Oliveira[1], Beatriz Martinho[1], Ana Vieira[1], and Nelson Pacheco Rocha[2]([✉]) [iD]

[1] Department of Medical Sciences, University of Aveiro, Aveiro, Portugal
{rafael.slo,beatrizmartinho,ana.vieira19}@ua.pt
[2] IEETA, Department of Medical Sciences, University of Aveiro, Aveiro, Portugal
npr@ua.pt

Abstract. Radiomics and radiogenomics are still new fields to be explored in oncology, although there are several platforms and tools already developed, or in development. This review aimed to identify the current state of the art of radiomics and radiogenomics in terms of platforms and tools integrating machine learning techniques to support the analysis of medical imaging studies of oncological patients. A systematic literature search was performed, and 19 studies were included in the review. These studies were published between 2016 and 2022 and considered several platforms and tools integrating machine learning techniques for the analysis of medical imaging studies of oncological patients together with molecular and clinical endpoints. However, the included articles present scarce evidence related to the translation of the reported platforms and tools to the clinical practice.

Keywords: Radiomics · Radiogenomics · Machine Learning · Oncology · Medical Imaging · Precision Medicine

1 Introduction

Machine learning techniques have revolutionized many fields and have been extensively used in medical sciences. The applications of these techniques in health care are immense and range from patients' diagnosis and prognosis to drug discovery and development, e-Health applications, or even classification of clinical documentation using natural language processing techniques [1].

Cancer is characterized as a set of genetic abnormalities, which could be caused by hereditary factors of environmental factors, where epigenetics plays a major role with histone modifications or Deoxyribonucleic Acid (DNA) methylation. Replication errors can originate mutations, as well the activation of oncogenes and the inactivation of tumor suppressors, resulting in oncogenesis [2]. To acquire genetic information regarding a patient, molecular and clinical techniques may involve biopsies for further genetic analysis. This method, however, is quite invasive and may not even be possible depending on the tumor location. The inaccessibility of the tumors poses a challenge in the diagnosis, prognosis, and stratification of oncological patients. Not only that, but biopsy samples lose key environmental, spatial, and temporal factors that can help identify a subtype of cancer, and gene expression profiles may be incomplete precisely due to this lost. Besides, a single biopsy is not enough to collect all the information regarding a tumor given their inter- and intra-heterogeneity. These features, however, can be obtained by using specific algorithms to process medical imaging studies to infer predictive and prognostic information [3].

In terms of diagnosis, particularly the diagnosis of oncological patients, machine learning techniques have been used for predictions and automation, improvement of clinical practice, and even to achieve a better understanding of tumor molecular biology [4]. Segal et al. [5] were the first to see the relation between gene expression in hepatocellular carcinoma and imaging features using Computed Tomography (CT).

The extraction of such features has defined a new field called radiomics. Radiomics main goal is to extract features from medical imaging and convert these images into mineable data to predict molecular characteristics. In essence, radiomics can be used to create patients' "silhouettes" from a molecular standpoint and then confirm these same features with the integration of other molecular and clinical endpoints, also known as radiogenomics or genetic imaging [6].

Radiogenomic's main goal is then to discover radiomic features that can reflect gene expression or mutations to optimize the understanding of the molecular basis of diseases, the prediction of mutations, thus aiding the clinical decision processes. Radiomics and radiogenomics have been associated with gene expression and molecular subtypes of tumors [7] and have proven to be a thriving field to help understand molecular profiles of patients and, also, to stratify patients given the inter- and intra-heterogeneity of tumors allowing a precise treatment for each patient accordingly to their risk [8].

Radiomics and radiogenomics can push precision medicine in oncology to the next level to predict the correct treatment for a given patient, since an early evaluation of response to a treatment assumes a great importance in terms of treatment decisions. For instance, Arshad et al. [9] used data from Fluorine-fluoro-D-glucose Positron Emission Tomography-Computed Tomography to verify if these data in association with data from Positron Emission Tomography (PET) could optimize survival prediction after chemoradiotherapy in 358 non-small cell lung cancer patients. Features such as histogram, shape and texture were obtained in combination with dimensionality reduction techniques and could predict a 14-month survival difference in the validation cohort [9].

Although radiomics and radiogenomics have high potential in the context of precision medicine, their application is not an easy task. Therefore, there is a need for platforms and tools incorporating efficient algorithms, namely using machine learning and deep learning techniques, to facilitate the clinical application of radiomics and radiogenomics. In this context, it is important to know how machine learning techniques are being used by radiomics and radiogenomics platforms and tools to support diseases' diagnosis, namely the diagnosis of oncological diseases, which was the objective of this review.

2 Methods

The main goal of this review was to systematize the state of the art of radiomics and radiogenomics in terms of platforms and tools supported on machine learning techniques for the analysis of medical imaging studies of oncological patients.

To perform the review, a research protocol was prepared to define: (1) the information sources and search strategies; (2) the eligibility criteria; (3) the studies' selection and data extraction processes; and (4) the synthesis and reporting of the results.

The search of the relevant studies was performed in four databases: PubMed, Scopus, Web of Science (WoS) and Science Direct. For that, databases queries were prepared combining the following terms: ("genomic*" OR "omic*" OR "molecular" OR "radiomic" OR "radiogenomic") AND ("medical imaging") AND ("machine learning" OR "artificial intelligence" OR "deep learning") AND ("oncology" OR "cancer").

The inclusion criteria considered for this review were any study that: (1) was published in English; (2) was published in the last ten years; (3) was published in a peer-reviewed scientific journal; (4) included a platform or tool that integrates genetic information in the analysis of medical imaging studies of oncological patients; and (5) included the application of machine learning techniques.

In turn, the exclusion criteria were any study that: (1) was not a primary study, but a review, a meta-analysis, or a book; (2) was published in a conference proceeding or as a books chapter; (3) was not relevant for the objective of this review; (4) its full text was not available; and (5) was published in other language than English.

After retrieving the articles through the database searching, the authors removed the duplicates and articles without abstract or authors. Then, the titles and abstracts of remaining articles were screened for inclusion according to the predefined eligibility criteria. Finally, in the third step, the authors assessed the full text of the eligible references against the inclusion and exclusion criteria to achieve the studies to be included in this review. In all the selection steps, the assessment of the articles was performed independently by two authors and the disagreements were solved by a third author.

Finally, in terms of synthesis and reporting of the results, data from the included studies were extracted to an Excel file with the descriptive details collected and summarized, namely, authors, publication year, title, abstract, study type, and as well the characteristics of the platforms and tools and their purposes. Moreover, the included studies were analyzed in terms of publication years and a narrative description was prepared to describe the identified platforms and tools and how they were used.

3 Results

3.1 Literature Search

The database search was performed in May 2022 and 610 articles were retrieved.

Figure 1 presents the Preferred Reporting Items for Systematic Reviews and Meta-Analyses (PRISMA) flowchart [10] describing the selection process of the articles included in the review.

In the first step of the selection process 109 articles were removed because they were duplicated and other 62 were removed because they were reviews or surveys, or they did

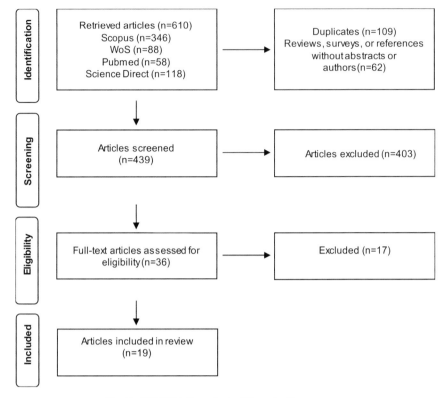

Fig. 1. PRISMA flowchart of the selection process.

not have abstracts or authors. According to the outline inclusion and exclusion criteria, 403 articles were excluded during the title and abstract screening, leading 36 articles for eligibility. From these 36 articles, 17 were excluded during the full-text analysis. Therefore, 19 studies were included in this review [11–29].

All the included articles were published between 2016 and 2022: (1) 2017, one study [11]; (2) 2017, one study [12]; (3) 2019, four studies [13–16]; (4) 2020, five studies [17–21]; (5) 2021, four studies [22–25], and (6) 2022, four studies [26–29].

3.2 Platforms and Tools

Table 1 presents the platforms and tools identified in the included studies.

PyRadiomics [12] is an open-source platform implemented in Python aiming to support the radiomic analysis and radiomic quantification, namely for processing and extraction of radiomics features using a large panel of engineered hard-coded features algorithms. PyRadiomics can analyze and process (e.g., 3D and 2D segmentations) medical imaging studies, including CT, PET, MRI, mammography, or ultrasounds. Specifically, data extracted from CT, PET, and MRI images are linked to various clinical outcomes to support clinical decision in oncology. The authors demonstrate the use of the platform to characterize benign and malignant lung lesions [12]. However, this platform is not intended for clinical use, since it was created with the goal of setting up a reference standard for radiomic analyses, as a resource to address the needs in cancer research and bring awareness to the potential use of radiomics technologies.

Table 1. Platforms and tools identified in the included studies.

#	Platform and Tools
[11]	An OsiriX Python plugin
[12]	A platform aiming to support the radiomic analysis and radiomic quantification
[13]	A platform to handle the radiomics process (e.g., extracting features, analysis, selection, and generation of predictive models)
[14]	A machine learning tool to perform texture analysis of multi-energy virtual monochromatic images from dual-energy CT
[15]	A toolkit, the Medical Imaging Interaction Toolkit (MITK) Phenotyping, to widespread the application of radiomics
[16]	A platform for individualizing radiotherapy dose
[17]	A platform to support the analysis of medical imaging studies
[18]	A platform to support the treatment of pediatric cancers
[19]	A quantitative image feature pipeline to evaluate and create new medical imaging biomarkers
[20]	A 3D slicer radiomics extension for nodule detection and prediction of invasiveness
[21]	A visualization tool to detect and classify the polyps identified by CT colonography
[22]	A tool to support preoperative prediction of the stage, size, grade, and necrosis score in clear cell renal cell carcinoma from Magnetic Resonance Imaging (MRI) studies
[23]	A platform to optimize the accuracy of diagnostics, prognostics, and prediction
[24]	A tool for the selection, analysis, and comparison of regions of interest of breast medical imaging studies
[25]	A platform to interpret and predict the results MRI studies, namely for detecting prostate cancer without the need of biopsies
[26]	A platform for omics data management to be used in clinical studies
[27]	A platform able to link clinical factors, imaging metadata, and extracted radiomics features
[28]	An OsiriX plugin to support single-modality and multimodality radiomics
[29]	A platform to predict and mine the genotype of a gene responsible for the lung cancer

Two studies [11, 28] used OsiriX as the basis of new radiomics platforms. Blackledge et al. [11] presented the pyOsiriX, and OsiriX Python plugin. In turn, Amini et al. [28] used OsiriX as the basis of new radiomics platforms that implements different combinations of machine learning techniques to analyze single-modality (i.e., CT and

PET) and multimodality radiomics models towards overall survival prediction in head and neck squamous cell carcinoma and non-small cell lung carcinoma patients.

In the study conducted by Gatta et al. [13] an agent-based decision support system was proposed. The proposed system can extract a wide range of features from a set of medical imaging studies, assess and select them in accordance with the outcome to be predicted, and produce optimized prediction models. Prediction models can be optimized for a particular center and then traded to compare the differences. It is also possible to combine datasets to produce broad prediction models or to create predictive models using more general methods [13]. The system was assessed using two independent datasets with rectal cancer data and the results showed that the system can generate predictive models with good performance [13].

Seidler et al. [14] proposed a machine learning assisted-texture analysis of multi-energy virtual monochromatic image datasets from dual-energy CT to differentiate metastatic head and neck squamous cell carcinoma lymph nodes from lymphoma, inflammatory, or normal lymph nodes.

One study [15] proposes the MITK Phenotyping framework for the widespread application of radiomics. It is an open-source platform, offering the option of storing data from images and meshes, while also extending other public platforms. MITK Phenotyping also provides shared libraries containing multiple algorithms to use in general radiomics studies [15].

Lou e al. [16] proposed an image-based deep learning platform for individualizing radiotherapy dose in the context of the treatment of lung cancer. In this platform, a multi-tasking deep neural network predicts treatment outcomes from medical imaging studies and the authors referred a treatment failure rate below 5% [16]. Also considering the lung cancer, Wang et al. [29] proposed a fully automated system to predict and mine the genotype for a gene responsible for the cancer and its progression. The authors used whole-lung imaging and automated lung segmentation, and the results showed that it is a viable alternative to whole tumor analysis without a significant cost difference. The system also identifies patients with a considerable risk of having resistance to tyrosine kinase inhibitors, which promotes tumor growth and disease progression [29].

Another platform of interest is the Joint Imaging Platform (JIP) of the German Cancer Consortium [17]. JIP is a flexible digital infrastructure open for the community focused on the use of machine learning techniques to process medical imaging studies. It aims to support multicenter medical imaging trials on large cohorts, by facilitating collaborative imaging projects addressing ethical and legal constraints as well technical and organizational issues. The platform provides data integration and harmonization mechanisms, interactive and automatic analysis, as well federated machine learning applications and was used by ten hospitals in Germany [17].

The PRIMAGE platform resulted from a four-year project founded by the European Commission and aims to process imaging to support the treatment of pediatric cancers [18]. It uses high-quality anonymized datasets about imaging, molecular, clinical, and genetic data for training and validating machine learning algorithms and offers clinical decision making for diagnosis, predictions, and prognosis. The developed prototype was focused on two specific pediatric cancers (i.e., neuroblastoma and diffuse intrinsic pontine glioma), although the authors referred the intention to translate it into other malignant tumors [18].

Mattonen et al. [19] proposed the Quantitative Image Feature Pipeline (QIFP) [19]. The QIFP enables researchers to link publicly accessible datasets (e.g., the Cancer Imaging Archive), upload repositories of linked medical imaging studies, including segmentations and clinical data, and even to upload their own algorithms. Researchers can use the tools and infrastructure provided by the QIFP to evaluate and create new imaging biomarkers, which can then be used in physical, virtual, and multicenter clinical trials. QIFP aims to surpass the lack of sufficient evaluation and validation as well translation into the clinical workflow of radiomics tools, mostly because of accessible shared software algorithms and architectures for comparison and evaluation are scare [19].

Xu et al. [20] used PyRadiomics to implement a 3D slicer radiomics extension. This extension might be used to support an interactive visualization of volumetric voxel images, polygonal meshes and volume renderings, focused on clinical and biomedical applications. In the included study [20], plain CT images were loaded into the 3D Slicer for nodule detection and segmentation, and identification of regions of interest to support the prediction of invasiveness of lung adenocarcinoma [20]. Still, in terms of regions of interest, Marinov et al. [24] presented a tool for the selection, analysis, and comparison of regions of interest of breast medical imaging studies. Although the tool was developed in Matlab, it provides a library of commonly applied algorithms and a friendly graphical user interface [24].

Moreover, Kotecha et al. [21] propose the use of image processing, deep learning, and convolutional neural network (i.e., Google-Net architecture) to process and analyze CT colonoscopies and CT DICOM images to detect polyps.

Tree-based Pipeline Optimization Tool (TPOT) is an open-source genetic programming-based software package, developed for the data science community and is one of the first automated machine learning methods. In the study of Choi et al. [22], this tool was used for comparison of performance, using an extra trees classifier for prognostic for patients with clear cell renal cell carcinoma. A comparison was made between TPOT and a manually optimized radiomics model using Random Forest technique, and the results showed that TPOT presents a good performance [22].

Computer pathology, ComPath for short, is a recent field aiming to promote the accuracy of diagnostics, prognostics, and prediction, and it uses data from thousands of patients and different image processing algorithms. Corvo et al. [23] proposed the IIComPath, which is a visual analytics approach that might support cohort construction, hypothesis generation, data analysis and data collection available to researchers. This platform relies on Python-supported servers and its main aim is to surpass several gaps related to the need to consider thousands of patients and to implement image algorithms based on machine learning techniques to guarantee the accuracy of diagnostics, prognostics, and prediction [23].

Most radiomics platforms were developed considering CT and PET scans. However, MRI images support histopathologic clinical procedure to detect certain types of cancer, like prostate cancer. Therefore, Shao et al. [25] proposed a platform to interpret and predict the results MRI studies, mostly using deep learning methods. This platform, the ProsRegNet, was proven to be about 20 times faster than other state-of-art algorithms and it is seen as useful for clinical use, detecting early-stage prostate cancer without invasive methods, like biopsies, while also improving the accuracy of reading MRI by the radiologists [25].

The need for information systems capable of handling heterogeneous digital assets is growing as technological advancements in omics and biomedical imaging boost the throughput of data generation in the life sciences. A system based the data management principles of findability, accessibility, interoperability, and reusability (FAIR), the qPortal, was proposed by Kuhn Cuellar et al. [26]. This portal is a web-based platform for omics data management to be used in clinical studies, namely in clinical studies related to cancer research [26]. Also, considering the FAIR principles, Jha et al. [27] proposed a pipeline for big medical imaging data processing able to link clinical factors, metadata of medical imaging studies, and extract radiomics features. This pipeline was implemented in the Tata Memorial Hospital in Mumbai, India, as part of the BIONIC Indo-Dutch research collaboration. The aim of this collaboration was to collect and aggregate data (i.e., using domain semantic ontologies) for large-scale multicenter collaboration and federated machine learning [27].

4   Conclusion

According to the distribution of the studies by the respective publication years, it possible to infer the importance of the current interest in radiomics and radiogenomics solutions integrating machine-learning techniques to support the analysis of medical imaging studies of oncological patients. This review shows that the role of machine learning in radiomics is significant in terms of available platforms and tools to support clinical decision process in oncology.

Most of the platforms and tools referred by the included studies are in-house developed, although some research groups made them available as open sources to the scientific and clinical communities. This might contribute to speed up the development of skill-based competences in radiomics and to facilitate the reproducibility and comparability of results and standardize feature definitions and computation methods for an improved reliability of the results [30].

The proposed platforms and tools might be extremely useful to support clinical decision, by promoting the diagnosis accuracy and helping to determine which is the best treatment option for a specific patient. Some limitations remain, however, as the identified platforms and tools are still in a very premature state of implementation, which means that they must assessed by clinical studies before released for widespread use in investigation and translated to clinical practice.

Acknowledgments. This review was carried out within the scope of the course unit Clinical Information Management of the Master's in Clinical Bioinformatics at the University of Aveiro.

References

1. Cook, G.J.R., Goh, V.: What can artificial intelligence teach us about the molecular mechanisms underlying disease? Eur. J. Nucl. Med. Mol. Imaging **46**(13), 2715–2721 (2019)
2. Ding, L., Bailey, M.H., Porta-Pardo, E., Thorsson, V., Colaprico, A., Bertrand, D., et al.: Perspective on oncogenic processes at the end of the beginning of cancer genomics. Cell **173**(2), 305-320.e10 (2018)
3. Panayides, A.S., Pattichis, M.S., Leandrou, S., Pitris, C., Constantinidou, A., Pattichis, C.S.: Radiogenomics for precision medicine with a big data analytics perspective. IEEE J. Biomed. Health Inf. **23**(5), 2063–2079 (2019)
4. Greenspan, H., van Ginneken, B., Summers, R.M.: Guest editorial deep learning in medical imaging: overview and future promise of an exciting new technique. IEEE Trans. Med. Imaging **35**(5), 1153–1159 (2016)
5. Segal, E., Sirlin, C.B., Ooi, C., Adler, A.S., Gollub, J., Chen, X., et al.: Decoding global gene expression programs in liver cancer by noninvasive imaging. Nat. Biotechnol. **25**(6), 675–680 (2007)
6. Aerts, H.J.W.L.: The potential of radiomic-based phenotyping in precision medicine: a review. JAMA Oncol. **2**(12), 1636–1642 (2016)
7. Skoulidis, F., Heymach, V.: Co-occurring genomic alterations in non-small-cell lung cancer biology and therapy. Nat. Rev. Cancer **19**(9), 495–509 (2019)
8. Beig, N., Bera, K., Prasanna, P., Antunes, J., Correa, R., Singh, S., et al.: Radiogenomic-based survival risk stratification of tumor habitat on Gd-T1w MRI Is associated with biological processes in glioblastoma. Clin. Cancer Res. Off. J. Am. Assoc. Cancer Res. **26**(8), 1866–1876 (2020)
9. Arshad, M.A., Thornton, A., Lu, H., Tam, H., Wallitt, K., Rodgers, N., et al.: Discovery of pre-therapy 2-deoxy-2-18F-fluoro-D-glucose positron emission tomography-based radiomics classifiers of survival outcome in non-small-cell lung cancer patients. Eur. J. Nucl. Med. Mol. Imaging **46**(2), 455–466 (2019)
10. Moher, D., Liberati, A., Tetzlaff, J., Altman, D.G.: Preferred reporting items for systematic reviews and meta-analyses: the PRISMA statement. BMJ **339**, b2535 (2009)
11. Blackledge, M.D., Collins, D.J., Koh, D.M., Leach, M.O.: Rapid development of image analysis research tools: bridging the gap between researcher and clinician with pyOsiriX. Comput. Biol. Med. **69**, 203–212 (2016)
12. Van Griethuysen, J.J., Fedorov, A., Parmar, C., Hosny, A., Aucoin, N., Narayan, V., et al.: Computational radiomics system to decode the radiographic phenotype. Can. Res. **77**(21), e104–e107 (2017)
13. Gatta, R., Vallati, M., Dinapoli, N., Masciocchi, C., Lenkowicz, J., Cusumano, D., et al.: Towards a modular decision support system for radiomics: a case study on rectal cancer. Artif. Intell. Med. **96**, 145–153 (2019)
14. Seidler, M., Forghani, B., Reinhold, C., Pérez-Lara, A., Romero-Sanchez, G., Muthukrishnan, N., et al.: Dual-energy CT texture analysis with machine learning for the evaluation and characterization of cervical lymphadenopathy. Comput. Struct. Biotechnol. J. **17**, 1009–1015 (2019)
15. Götz, M., Nolden, M., Maier-Hein, K.: MITK Phenotyping: An open-source toolchain for image-based personalized medicine with radiomics. Radiother. Oncol. J. Eur Society for Therapeutic Radiology and Oncology **131**, 108–111 (2019)
16. Lou, B., Doken, S., Zhuang, T., Wingerter, D., Gidwani, M., Mistry, N., et al.: An image-based deep learning framework for individualising radiotherapy dose: a retrospective analysis of outcome prediction. Lancet Dig. Health **1**(3), e136–e147 (2019)

17. Scherer, J., Nolden, M., Kleesiek, J., Metzger, J., Kades, K., Schneider, V., et al.: Joint imaging platform for federated clinical data analytics. JCO Clin. Cancer Inf. **4**, 1027–1038 (2020)
18. Martí-Bonmatí, L., Alberich-Bayarri, Á., Ladenstein, R., Blanquer, I., Segrelles, J.D., Cerdá-Alberich, L., et al.: PRIMAGE project: predictive in silico multiscale analytics to support childhood cancer personalised evaluation empowered by imaging biomarkers. Eur. Radiol. Exp. **4**(1), 22 (2020)
19. Mattonen, S.A., Gude, D., Echegaray, S., Bakr, S.H., Rubin, D.L., Napel, S.: Quantitative imaging feature pipeline: a web-based tool for utilizing, sharing, and building image-processing pipelines. J. Med. Imaging (Bellingham, Wash.) **7**(4), 42803 (2020)
20. Xu, F., Zhu, W., Shen, Y., Wang, J., Xu, R., Qutesh, C., et al.: Radiomic-based quantitative CT analysis of pure ground-glass nodules to predict the invasiveness of lung adenocarcinoma. Front. Oncol. **10**, 872 (2020)
21. Kotecha, S., Vasudevan, A., Holla, V.K., Kumar, S., Pruthviraja, D., Latte, M.V.: 3D visualization cloud-based model to detect and classify the polyps according to their sizes for CT colonography. J. King Saud Univ. Comput. Inf. Sci. **34**(8), 4943–4955 (2020)
22. Choi, J.W., Hu, R., Zhao, Y., Purkayastha, S., Wu, J., McGirr, A.J., et al.: Preoperative prediction of the stage, size, grade, and necrosis score in clear cell renal cell carcinoma using MRI-based radiomics. Abdominal Radiol. (New York) **46**(6), 2656–2664 (2021)
23. Corvo, A., Caballero, H.G., Westenberg, M.A., van Driel, M.A., van Wijk, J.J.: Visual analytics for hypothesis-driven exploration in computational pathology. IEEE Trans. Visual Comput. Graph. **27**(10), 3851–3866 (2021)
24. Marinov, S., Buliev, I., Cockmartin, L., Bosmans, H., Bliznakov, Z., Mettivier, G., et al.: Radiomics software for breast imaging optimization and simulation studies. Physica Med. **89**, 114–128 (2021)
25. Shao, W., Banh, L., Kunder, C.A., Fan, R.E., Soerensen, S.J., Wang, J.B., et al.: ProsRegNet: a deep learning framework for registration of MRI and histopathology images of the prostate. Med. Image Anal. **68**, 101919 (2021)
26. Kuhn Cuellar, L., et al.: A data management infrastructure for the integration of imaging and omics data in life sciences. BMC Bioinf. **23**(1), 61 (2022)
27. Jha, A.K., Mithun, S., Sherkhane, U.B., Jaiswar, V., Shi, Z., Kalendralis, P., et al.: Implementation of big imaging data pipeline adhering to FAIR principles for federated machine learning in oncology. IEEE Trans. Radiat. Plasma Med. Sci. **6**(2), 207–213 (2022)
28. Amini, M., et al.: Overall survival prognostic modelling of non-small cell lung cancer patients using positron emission tomography/computed tomography harmonised radiomics features: the quest for the optimal machine learning algorithm. Clin. Oncol. (Royal College of Radiologists (Great Britain)) **34**(2), 114–127 (2022)
29. Wang, S., Yu, H., Gan, Y., Wu, Z., Li, E., Li, X., et al.: Mining whole-lung information by artificial intelligence for predicting EGFR genotype and targeted therapy response in lung cancer: a multicohort study. Lancet Dig. Health **4**(5), e309–e319 (2022)
30. Avanzo, M., Wei, L., Stancanello, J., Vallieres, M., Rao, A., Morin, O., et al.: Machine and deep learning methods for radiomics. Med. Phys. **47**(5), e185–e202 (2020)

Development of a Biomechanical System for Rehabilitation Purposes of Bedridden Patients with Prolonged Immobility – Prototype and Preliminary Results

Luis Roseiro[1,2(✉)], Tomás Ribeiro[1], Marco Silva[1], Frederico Santos[1],
Alexandra André[3], Ruben Durães[4], William Xavier[5], Arménio Cruz[6],
and Cândida Malça[1,7]

[1] Polytechnic Institute Coimbra, Coimbra Institute of Engineering, Rua Pedro Nunes – Quinta da Nora, 3030-199 Coimbra, Portugal
{lroseiro,a21270456,msilva,fred,candida}@isec.pt

[2] CEMMPRE, Centre for Mechanical Engineering, Materials and Processes, University of Coimbra, Pinhal de Marrocos, 3045-043 Coimbra, Portugal

[3] Polytechnic Institute Coimbra, Coimbra Health School, Rua 5 de Outubro S. Martinho do Bispo, Ap. 7006, 3046-854 Coimbra, Portugal
alexandra.andre@estesc.ipc.pt

[4] Orthos XXI, Rua Santa Leocádia 2735, 4089-012 Guimarães, Portugal
desenvolvimento.or5@orthosxxi.com

[5] Wiseware, Z. Ind. da Mota, R. 12 Lt. 51-E, 3830-527 Gafanha da Encarnação, Portugal
william@wisewaresolutions.com

[6] The Health Sciences Research Unit, Nursing School of Coimbra (ESEnfC), 3000 Coimbra, Portugal
acruz@esenfc.pt

[7] Centre for Rapid and Sustainable Product Development, Polytechnic Institute of Leiria, Rua de Portugal, 2430-028 Marinha Grande, Portugal

Abstract. Citizens with physical limitations, mainly bedridden patients, cannot perform physical activity, which can translate into long periods of immobilization, with severe consequences for their health. This type of patient usually stays in bed for long periods, leading to getting several motor problems due to immobility. Thus, it is vital to develop biomechanical systems to implement physical rehabilitation activities for this type of patient. This work presents a prototype designed as Ablefit, created explicitly for bedridden patients, aiming to prevent complications associated with their immobility for long periods. The developed equipment is based on structural support and positioning unit, an active/passive linear module, and an active/passive rotary module, allowing them to perform different types of physical movements by the user. The work presented describes system requirements and the developed prototype's concept, emphasizing the equipment's mechanical system and its interface with the user. The first tests on the prototype of the system made it possible to identify the adjustments to be implemented and showed the possibility that it could represent an added value in physical motor rehabilitation procedures for bedridden patients.

© The Author(s), under exclusive license to Springer Nature Switzerland AG 2024
A. Rocha et al. (Eds.): WorldCIST 2023, LNNS 801, pp. 439–446, 2024.
https://doi.org/10.1007/978-3-031-45648-0_43

440   L. Roseiro et al.

Keywords: Biomechanics of Rehabilitation · Bedridden Mobilization · Mechanical Interface with Patient

1 Introduction

The World Health Organization (WHO) has set clear guidelines concerning sedentary behavior, recommending that all citizens have regular physical activity. However, in citizens with physical limitations, particularly in the case of bedridden patients, the capability to perform physical activity is limited, which can translate into long periods of immobilization, with severe consequences for their health. This type of patient usually stays in bed for long periods, leading to getting several motor problems due to immobility. Reductions in muscle mass, bone mineral density, and physical impairment are the first evidence associated with others that can appear, like muscular atrophy, muscular weakness, respiratory complications, blood circulation complications, and bone demineralization [1–3].

The absence of muscular stimulation will affect the skeletal system [4], and early mobilization is a key to increasing functional capacity and muscle strength in this type of patient, leading to significant outcomes [5, 6]. Thus, it is vital to develop biomechanical systems to implement physical rehabilitation activities specifically for this type of patient. An adequately defined physical rehabilitation program is necessary to create and implement this device correctly. Some authors have prescribed several programs, specifically for bedridden patients, like the works of Akar et al. [7] and Maimaiti et al. [8].

The use of assistive technologies, like cycle ergometry and electric muscle stimulation have been growing in rehabilitation plans for bedridden patients [1]. These technologies have been showing important results in the recovery of motor functions [9].

The literature and the market present some solutions for bedridden patients, such as the motored systems, essentially based on cycle ergometers [10]. There are additional solutions that use elastic bands combined with designs that promote rotational movement for use by the hands, as is the case of workout and recovery system [11]. In this line-up, there is also a set of equipment called Rocher cages, in which various accessories are coupled, allowing the implementation of rehabilitation procedures for different body parts. This type of equipment is based on the principle of cables and rubber bands and therefore contains the same disadvantages associated with these accessories.

There are also types of equipment capable of rehabilitating bedridden people, placing them on an inclined plane with variable adjustment, and stimulating the lower limbs by promoting the movement of walking. An example of this is the Hocoma Erigo Robot equipment [12]. This equipment also enables electrical stimuli in the muscles of the lower limbs. This equipment does not allow for an answer regarding the rehabilitation of the upper limbs, nor does it have the ability to assess vital signs and record them.

Based on the collected data and according to the research carried out, although there are several solutions on the market, none fully responds to the specific needs of control and evaluation of the rehabilitation process, nor does it include a well-defined rehabilitation plan.

As the physical rehabilitation programs implemented for bedridden patients are limited [2], recently, a scoping review guided by the Joanna Briggs Institute (JBI) methodology has been published [2], defining challenges and providing information to develop new programs suitable in clinical and organizational contexts.

One of the challenges concerning physical rehabilitation programs is the suitability to be implemented, which depends on the biomechanical devices that can be developed and used by the patients. Then, new Biomechanical systems must be designed based on rehabilitation programs. This leads to an essential task for the engineers, focusing on the interface human–device, and follows the consequence of having multidisciplinary teams working on these developments.

The work presented has its context in the Ablefit project, which aims to contribute to the development of methodologies and systems that ensure physical activity for this type of patient. This project focuses on the development of physical rehabilitation equipment for bedridden patients that can contribute to the prevention of complications associated with their immobility for long periods. The developed equipment involves a structural support and positioning unit, a set of actuators, and a control, monitoring, and gamification unit with a user interface.

The work presented was in line with a new rehabilitation program [1] and described the first development of the mechanical system of the equipment and its interface with the user. The Ablefit also.

2 The Ablefit System

Taking into account the intended objectives, one of the methodologies considered relevant in developing the biomechanical system involved discussion and multidisciplinary work. This work highlights this component for its importance in the entire process that led to the prototype.

2.1 System Requirements

The requirements to be considered for the development of the new solution represented an essential phase of the work, having been defined as an objective of a modular system that can be simple to use and correspond to the needs of each patient, with practical characteristics both for the user and for the possible caregivers. The equipment should also have an advanced control system to monitor various parameters related to the patient and the exercise performed, with a particular account of the speed and force realized or imposed by the equipment. This will allow recording and evaluation of the patient's progress and performance. This registration and the implementation of a new future gamification solution are two significant factors when it comes to patient motivation.

Gamification serves as a stimulus for the user to practice specific exercises. The existence of a record that allows users to understand their evolution throughout the rehabilitation period encourages them not to give up on the process, which can be an essential factor in the efficiency/effectiveness of the exercises. Real-time biofeedback systems are crucial to the point of rehabilitation plans, as there is a progressive growth in people's motivation and emotional involvement [13, 14].

Regarding the physical-motor rehabilitation component, more highlighted in work presented, the equipment must have the ability to perform a set of exercises that guarantee results for patients. Table 1 shows the survey of the main movements defined by the project team, which it was intended to take into account in the work. The base system was thus defined for mobilization requirements with flexion, extension, abduction, and adduction of both upper and lower limbs.

Table 1. Main movements to be implemented with the Ablefit System.

Movement	Upper Limb/ Joint	Lower Limb/Joint
Flexion	Shoulder	Hip
Extension		
Abduction		
Adduction		
Internal rotation		
External Rotation		
Flexion	Elbow	Knee
Extension		
Flexion	Fist, hand, finger	Ankle
Extension		

2.2 Prototype Design and 3D Modeling

3D modeling was fundamental in the design of the prototype. The Solidworks software was used to execute all component drawings and their virtual assembly. The structural calculations of the system, to ensure adequate mechanical strength and rigidity, were supported by the use of technical profiles from the Minitec brand and the structural analysis of the mechanical components using the finite element method, also with Solidworks software. This approach allowed the optimization of the geometry of the parts and the machining carried out in a computer numerical control machining center (CNC).

The concept of the prototype designated by Ablefit is based on two complementary modules, a linear module for the implementation of linear movements with a curvilinear trajectory of the upper and lower limbs and a rotating module for the performance of circular motions of the upper and lower limbs. For the intended modularity, a support structure was developed, a built-in technical profile, with a fitting system for replacing the modules and vertical tuning, which allows adjustment to the adequate and safe interface with the patient. The C-shaped system allows adjustment to the bed where the patient is. Figure 1 shows a view of the system and its interface and fits into a hospital bed.

Figure 2 presents a 3D representation of the system with the integrated modules and the visualization of the main components. This visualization allows an understanding of the concept developed and raised in the prototype. The support structure, in addition to guaranteeing the stability of the modules and the visualization interface, allows adjustment and fitting in any bed where a patient is.

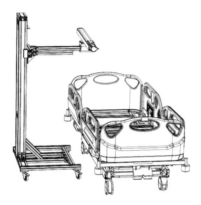

Fig. 1. 3D system visualization of Ablefit system prototype next to an hospital bed.

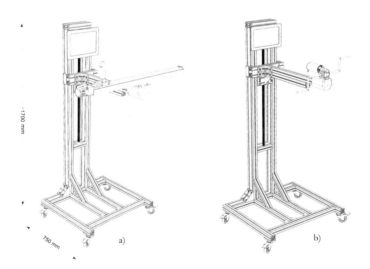

Fig. 2. 3D system visualization of the prototype: a) with linear component; b) with rotary component.

2.3   Linear and Rotary Modules - Interface with the User

The end of the structure has a mechanical component that allows the coupling and uncoupling of the two modules developed, thus guaranteeing the ease of its quick replacement and making the system modular. Figure 3 presents the two modules and the user interface designed to ensure the ergonomics and safety required for the system. In the case of the linear module, the system is based on a linear guide unit (Oriental Motor EZSM6E080AZAK), which guarantees the required level of force, speed, and range of motion. This type of actuator allows its control in terms of speed and position, ensuring the implementation of the system both in an active system line and in a passive approach. The user interface is positioned on the sliding base of the linear guide (Fig. 3). This interface includes a support and connection base, a vertical beam with a flexural load cell (Any load Model 108BA), which allows quantifying the force applied in both directions of linear movement, and an ergonomic external structure for gripping the hand. At the top of the beam, with a snap-on/disengage system, there is a clip-on pedal fitting system, which allows the coupling of a safety boot (walker), thus ensuring the use of the linear design with the lower limbs.

The rotary module involves using a rotary motor fixed to a mechanical protection component, which is attached to the end of the structure. An S-type load cell is attached to each of the lateral rotation shafts, which allows the quantification of the force exerted by each of the arms or feet, depending on whether the upper or lower limbs are used. The rotary motor guarantees the use of the system in both an active and passive system line. The hand rotation handle is shown in Fig. 3 and can also be used with the foot, as it allows rotation around the shaft. This handle can be replaced by a pedal loop, as described for the linear system, where a boot and loop are fitted.

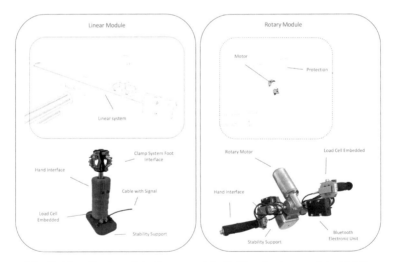

Fig. 3. Visualization of the linear module (left) and rotary module (right).

2.4   First Experimental Tests

The first tests of the prototype were implemented to identify the functionality of the prototype, having been carried out by health professionals. Figure 4 shows an example of performing exercises with the upper limbs and Fig. 5 shows an example of performing exercises with the lower limbs. The preliminary tests showed that extension and flexion movements were full performed in active and passive modes. However, in some movements, especially abduction and adduction, limitations in the fullness of the movement are identified, which will be the target of adjustment and improvement.

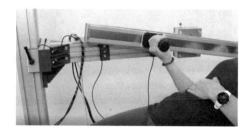

Fig. 4. Example of an experimental test with the upper limbs.

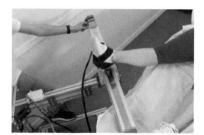
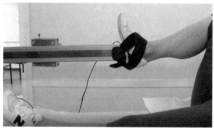

Fig. 5. Example of an experimental test with the lower limbs.

3 Discussion

This work presented the framework of the prototype of biomechanical equipment for accomplishing physical exercise by bedridden patients. The developed system allows the performance of a set of movements considered relevant for this type of patient. The design incorporates a linear module based on a linear guide with rotation in the support structure and a rotation module based on a rotary motor. Both systems can be used in active and passive modes, having developed an interface with the patient that guarantees the performance of movements in a safe way.

Preliminary results performed by health professionals showed the mechanical robustness of the initial prototype, evidencing an excellent adjustment and positioning capacity, fulfilling the function of framing and interfacing with the patient in hospital beds. As it is ongoing work, the results allowed us to observe the existing limitations in the performance of some movements, particularly abduction/adduction, and the need for implementing adjustments.

It must be pointed out that the system provides complete biofeedback to the clinics, allowing the quantification of several parameters relevant to the rehabilitation programs, like the force applied by the user. Also, the control unit will guarantee the user's safety, and a set of vital signals, like heart rate, blood pressure, body temperature, respiratory frequency, and oxygen saturation, are considered. A tablet interface connected to the system is used for interaction with the patient and the clinic. This interface is under implementation and can be used for gamification, allowing the patient to realize the movements defined by the rehabilitation program.

Aknowledgments. This research was co-financed by the European Regional Development Fund (ERDF) through the partnership agreement Portugal 2020—Operational Programme for Competitiveness and Internationalization (COMPETE2020) under the project POCI-01–0247-FEDER-047087 ABLE-FIT: Desenvolvimento de um Sistema avançado para Reabilitação.

Authors acknowledge the support of the CEMMPRE-UC, sponsored by FCT under the projects UIDB/00285/2020, LA/P/0112/2020 and the Mais Centro Program.

References

1. Parry, S.M., Puthucheary, Z.A.: The impact of extended bed rest on the musculoskeletal system in the critical care environment. Extrem. Physiol. Med **4**, 16 (2015)
2. Parola, V., et al.: Rehabilitation programs for bedridden patients with prolonged immobility: a scoping review protocol. Int. J. Environ. Res. PublicHealth **18**, 12033 (2021)
3. Campos, A., Cortés, E., Martins, D., Ferre, M., Contreras, A.: Development of a flexible rehabilitation system for bedridden patients. J. Brazilian Soc. Mech. Sci. Eng. **43**, 361 (2021)
4. Eimori, K., Endo, N., Uchiyama, S., Takahashi, Y., Kawashima, H., Watanabe, K.: Disrupted bone metabolism in long-term bedridden patients. PLoS ONE **11**, 1–11 (2016)
5. Rocha, A.M., Martinez, B.P., da Silva, V.M., Junior, L.F.: Early mobilization: why, what for and how? Medicina Intensiva **41**(7), 429–436 (2017)
6. Arias-Fernández, P., Romero-Martin, M., Gómez-Salgado, J., Fernández-García, D.: Rehabilitation and early mobilization in the critical patient: systematic review. J. Phys. Ther. Sci. **30**, 1193–1201 (2018)
7. Akar, O., et al.: Efficacy of neuromuscular electrical stimulation in patients with COPD followed in intensive care unit. Clin. Respir. J. **11**, 743–750 (2017)
8. Maimaiti, P., Sen, L.F., Aisilahong, G., Maimaiti, R., Yun, W.Y.: Statistical analysis with Kruskal Wallis test for patients with joint contracture. Future Gener. Comput. Syst. **92**, 419–423 (2019)
9. Phyo, S.T., Kheng, L.K., Kumar, S.: Design and development of robotic rehabilitation device for post stroke therapy. Int. J. Pharma Med. Biol. Sci. **5**, 31–37 (2016). https://doi.org/10.18178/ijpmbs.5.1.31-16-37
10. Motomed. https://www.motomed.com. Accessed 2 Jan 2023
11. Workoutandrecovery. https://www.workoutandrecovery.com. Accessed 2 Jan 2023
12. Hocoma. https://www.hocoma.com/solutions/erigo. Accessed 2 January 2023
13. Barandas, M., Gamboa, H., Fonseca, J.: A real time biofeedback system using visual user interface for physical rehabilitation. Procedia Manuf. **3**, 823–838 (2015)
14. Condino, S., Turini, G., Viglialoro, R., Gesi, M., Ferrari, V.: Wearable augmented reality application for shoulder rehabilitation. Electronics **8**, 1178 (2019)

Variational Inference Driven Drug Protein Binding Prediction

Neeraj Kumar[1] and Anish Narang[2(✉)]

[1] IIT Delhi, New Delhi, India
[2] AAAI, Washington, D.C., USA
anishnarang06@gmail.com

Abstract. The identification of drug-protein interactions (DPIs) is a key task in drug discovery, where drugs are chemical compounds and targets are proteins. Traditional DPI prediction methods are either time consuming (simulation-based methods) or heavily dependent on domain expertise (similarity-based and feature-based methods). Recent explorations involving deep learning either exploit 3D structure of the proteins and/or use GNNs (graph convolutions) to capture neighbour relationships.

In this paper, we present a novel end-end deep learning architecture, VED-BI, that leverages variational inference based encoder and decoder along with GraphSage based approach to predict drug-protein interaction. Due to better generalization, our architecture is able to deliver better results on the drug protein binding prediction. Detailed experimental comparative analysis using Precision, Recall and AUC metrics, has been performed across all the approaches on metador and bindingDB datasets. Our proposed novel architecture performs better on bindingDB dataset as compared to key state-of-the-art results.

Keywords: variational architecture · drug-protein interactions · transductive inference

1 Introduction

Identifying interactions between drug and proteins is important in the discovery and development of safe and effective drugs. However, identifying DPIs (Drug protein interactions) experimentally is time consuming and expensive. The DPIs can be expressed in the form of a bipartite network, with drugs and proteins as nodes. This means we can split the vertices V into two disjoint sets, V_{left} and V_{right}, such that edges can only exist between vertex v_{left} and v_{right} and they represent interactions between the corresponding drug-protein pairs. Recently, approaches based on Bipartite networks have made significant progress in the research of drug re-positioning, drug disease association prediction.

To identify DPIs, various types of drug and protein features and algorithms to predict DPIs have been investigated. The similarity based methods are considered to be the simplest link prediction techniques, which measure a score for each pair of unlinked nodes that defines the similarity between the nodes. For

A. Rocha et al. (Eds.): WorldCIST 2023, LNNS 801, pp. 447–456, 2024.
https://doi.org/10.1007/978-3-031-45648-0_44

example, The Common Neighbourhood(CN), Jaccard coefficient(JC), Preferential Attachment(PA) and Adamic Adar(AA) use neighbourhood based features between drugs and proteins. We study bipartite graph based learning methods that map drug and proteins to a common feature vector space and minimize the Euclidean distances between vectors linked by known interactions.

Recently, deep neural networks (DNNs) have achieved excellent performance for various applications, such as natural language understanding, computer vision tasks, speech recognition and synthesis and many more. End-to-end training using deep learning has produced superior results as it helps to learn the right representation as well. While some prior techniques use fixed input representation in a deep learning approach and hence have lower accuracy, [12] uses GNNs and CNNs along with neural attention mechanism in an end-end representation learning approach. We propose a novel variational encoder-decoder based architecture along with GraphSage based approach **VED-BI** *(Variational Encoder Decoder based Binding Inference)* to predict the drug protein interaction which leads to better generalization and hence relatively better results on well-known datasets. In this paper we make the following contributions:

- We present a novel variational inference based encoder decoder architecture (VED-BI) to predict the drug protein interaction (Sect. 3.3). The embedded representations learnt, via variational architecture, from the drug molecular graph and protein sequence respectively are fused in series and parallel fashion. We also used GraphSage [3] based approach to train drug protein bipartite graph transductively and capture the neighbourhood information, leading to better generalization.
- Experimental results on metador and binding DB [2] dataset on various evaluation metrics such as precision, recall and AUC score have been performed on all the methods including similarity based metrics, spectral curve fitting and VED-BI. The AUC score in most of the methods comes to more than 80 % which shows that a new link that does occur will be given a good score compared to the score of a potential link that does not occur. Further, VED-BI delivers superior AUC and Recall on the Binding DB dataset, as compared to [12].

2   Related Work

Methods based on drug similarity and protein similarity for DPI prediction, require characteristic information of drugs, proteins, and DPI, such as chemical structure [13], genomic sequence, type of binding, reason for interaction etc. When the above characteristic information is not available, these methods cannot be effectively executed. DTI predictions based on similarities between protein sequences or drug structures have limitations due to the underlying assumption that, similar drugs share similar targets, which is not necessarily true.

Bipartite network based approaches have made significant achievements in the research of drug re-positioning, drug-disease association analysis, drug-protein interaction prediction, and gene-disease association prediction [14,15].

Recent works such as OnionNet [16] use the 3D-structure of protein-ligand complexes as input to 3D convolutions (3D-CNNs) to predict binding affinity. [6] uses structure-aware interactive graph neural network to improve prediction performance. This can capture 3D spatial information through polar-inspired graph attention layers and global long-range interactions through pairwise interactive pooling. Our approach does not require 3D structure information and hence has wider applicability.

With the increasing popularity of deep learning, researchers are adopting deep neural models to predict DTIs. Instead of using RBM and DBN, [8] used stacked Autoencoder for representation learning and SVM for classification. [1] explored graph convolutional network to model chemical structures and leveraged recurrent neural network (RNN) on protein sequences. [9,11] also used RNN to model SMILES strings, which are sequential encoding of chemical structures. GraphDTA [7] adopts GNN models to learn drug presentation and combines the protein representation from 1D convolutions to predict binding affinity. [12] used GNN to extract the drug features along with protein embedding with late fusion to predict the drug protein interaction. Our approach generalizes the above GNN based approaches using variational encoder decoder and demonstrates superior results.

3   Materials and Methods

3.1   Similarity Metrics

In the bipartite graph, for given nodes (x, y) on opposite sides, let $S(x)$ and $S(y)$ respectively denote the 2-hop neighbours of x and neighbors of y. We used the following distance metrics for predicting the links between drug and protein:

- The Common Neighborhood (CN) method defines the number of co-neighbors of drug-protein pairs as a drug-protein similarity. If two nodes share many common neighbors, there may be a link between the two nodes. The more neighbors of drugs and proteins, the greater the possibility of drug-protein interaction. The formulae for calculating the CN is given by

$$CN = |S(x) \cap S(y)| \qquad (1)$$

- Jaccard coefficient method considers the influence of nodes in the network and is basically a normalized version of CN.

$$JC = \frac{|S(x) \cap S(y)|}{|S(x) \cup S(y)|} \qquad (2)$$

- Preferential Attachment (PA) method defines that the probability of connecting edges between any two pairs of nodes in the network is proportional to the product of the degrees of these two nodes.
- Adamic Adar looks at the common neighbors' features (common neighbors' neighbors) and gives weight to the feature by means of the reciprocal of the log of the feature.

$$AA = \sum_{z \in S(x) \cap S(y)} \frac{1}{\log \tau(z)} \qquad (3)$$

3.2 Spectral Curve Fitting

Consider performing an eigenvalue decomposition on the adjacency matrix A of graph G [5]. We'll then obtain the eigenvector matrix U and diagonal eigenvalue matrix Λ. [5] proposed that one can transform a graph with kernel function F by either applying it directly to the adjacency matrix F(A) or to its' eigenvalue matrix F(Λ). The assumption here is that the eigenvectors stay the same, because we assume that the original and transformed graph are not vastly different. This is reasonable for our case, where the original graph is our train set while the transformed graph is our test set.

$$A = U \wedge U^T, A' = F(A) = U F(\wedge) U^T \tag{4}$$

Now consider our link prediction task, where we have our training graph G (adjacency matrix A) and testing graph G' (adjacency matrix A'). The idea here is to find a function F that can map our original A to A'. The objective function is given by

$$min_F |F(A) - A'| = |U F(\wedge) U^T - A'| = |F(\wedge) - U^T A' U| \tag{5}$$

3.3 Variational Inference Based Deep Learning Approach

We have proposed variational encoder decoder based architecture to predict the drug protein interaction. We have proposed the two approaches as given in Figs. 1 and 2.

3.3.1 Variational Autoencoder

Let us consider some dataset $X = \{x^{(i)}\}_{i=1}^N$ consisting of N i.i.d. samples of some continuous or discrete variable x. We assume that the data is generated by some random process, involving an unobserved continuous random variable z. The process consists of two steps: (1) a value $z^{(i)}$ is generated from some prior distribution $p(z)$; (2) a value $x^{(i)}$ is generated from some conditional distribution $p(x|z)$. Given N data points $\mathbf{X} = \{\mathbf{x}_1, \dots, \mathbf{x}_N\}$ we typically aim at maximizing the marginal log-likelihood:

$$\ln p(\mathbf{X}) = \sum_{i=1}^N \ln p(\mathbf{x}_i), \tag{6}$$

with respect to parameters. This task could be troublesome because the integral of the marginal likelihood $p(x) = \int p(z)p(x|z)\,dz$ is intractable (so we cannot evaluate or differentiate the marginal likelihood). These intractabilities are quite common and appear in cases of moderately complicated likelihood functions $p(x|z)$, e.g. a neural network with a nonlinear hidden layer. To overcome this issue one can introduce an *inference model* (an *encoder*) $q(\mathbf{z}|\mathbf{x})$ and optimize the variational lower bound:

$$\ln p(\mathbf{x}) \geq \mathbb{E}_{q(\mathbf{z}|\mathbf{x})}[\ln p(\mathbf{x}|\mathbf{z})] - \mathrm{KL}\big(q(\mathbf{z}|\mathbf{x})||p(\mathbf{z})\big), \tag{7}$$

where $p(\mathbf{x}|\mathbf{z})$ is called a *decoder* and $p(\mathbf{z}) = \mathcal{N}(\mathbf{z}|\mathbf{0}, \mathbf{I})$ is the *prior*. There are various ways of optimizing this lower bound but for continuous $\mathbf{z}$ this could be done efficiently through a re-parameterization of $q(\mathbf{z}|\mathbf{x})$ [4,10]. Then the architecture is called a *variational auto-encoder* (VAE).

Typically, a diagonal covariance matrix of the encoder is assumed, *i.e.*, $q(\mathbf{z}|\mathbf{x}) = \mathcal{N}\big(\mathbf{z}|\boldsymbol{\mu}(\mathbf{x}), \mathrm{diag}(\boldsymbol{\sigma}^2(\mathbf{x}))\big)$, where $\boldsymbol{\mu}(\mathbf{x})$ and $\boldsymbol{\sigma}^2(\mathbf{x})$ are parameterized by the NN. However, this assumption can be insufficient and not flexible enough to match the true posterior.

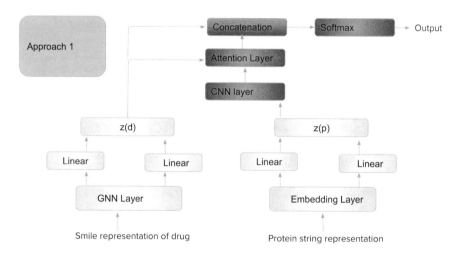

Fig. 1. Parallel Approach: Drug & protein encoders are in parallel followed by decoder

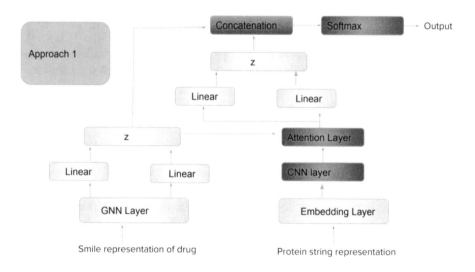

Fig. 2. Series Approach: Drug & protein encoders are in series followed by decoder

3.3.2   Drug Encoder

A drug is represented by a SMILES (?) sequence, which essentially encodes a chemical structure graph $G = (V, E)$, where V is a set of atoms and E is a set of chemical bonds that bind two atoms as undirected edges. We use RDKit[1] to transform SMILES string to chemical structure graph. We have used the GNN layer to generate the drug embedding . We have used the two linear layer to generate $\mu(x)$ and $\sigma(x)$ and used reparametrization trick to get z(d) = $\mu(x) + \sigma(x) \cdot x$.

3.3.3   Protein Encoder

There are 20 types of amino acids in the protein sequence and we have used n-gram to create overlapping words from protein sequences. The total number of possible n-grams is 20^n, and we have kept the n = 3. We then create the word embeddings using the n-gram amino acids and used two linear layers to predict the $\mu(x)$ and $\sigma(x)$. We then use the reparametrization trick to get z(p) = $\mu(x) + \sigma(x) \cdot x$.

3.3.4   Decoder

In the parallel approach, the protein embedding, $z(p)$, is fed to the CNN layer. Cross modal attention layer then takes the output of CNN layer and drug embedding, $z(d)$, as inputs. Further, concatenation is applied to the output of cross modal attention layer with $z(d)$. The output of this layer is then fed to a linear layer followed by softmax to predict the output $\in \{0, 1\}$. In the series approach, $z(p)$ is concatenated with $z(d)$, followed by linear layer and softmax to predict the output.

3.3.5   Losses

The optimisation function for variational inference is given in Eq. 8:

$$\ln p(\mathbf{x}) \geq \mathbb{E}_{q(\mathbf{z}|\mathbf{x})}[\ln p(\mathbf{x}|\mathbf{z})] - \mathrm{KL}\big(q(\mathbf{z}|\mathbf{x})||p(\mathbf{z})\big), \tag{8}$$

Cross Entropy Loss: In multi class classification, the class label (y^s) follows the multinomial distribution given in Eq. 9. So the first part of Eq. 8 i.e. $\ln p(\mathbf{x}|\mathbf{z}^{(T)})$ is the cross entropy loss. The cross entropy loss(CE) is given by Eq. 9 . In our experiment, it is a 2 class problem, hence it will follow binomial distribution.

$$CE(\hat{y}, y) = \sum_{j=1}^{N} y^{(j)} \log \hat{y}^{(j)}, \sum_{j=1}^{N} y^{(j)} = 1 \tag{9}$$

where $y^{(j)}$ is the true label i.e. 0 or 1 and $\hat{y}^{(j)}$ is the predicted probability of a class coming from softmax operation.

Kl Loss: The KL loss acts as the regularizer which prevents the model from overfitting and helps in learning a flexible posterior distribution. β value is used to control the effect of KL loss. The value of β decreases linearly with every epoch.

$$KL = \sum_{j=1}^{N} \sigma_i^2 + \mu_i^2 - \log(\sigma_i) - 1 \tag{10}$$

---

[1] http://www.rdkit.org/.

3.4   GraphSage Based Deep Learning Approach

We have transductively trained the bipartite graph using GraphSage ([3]) architecture to predict the link between drug and protein. We have used two layers of GrapghSage network with two linear layers on top of it to predict the link. Equations 11 and 12 show the GraphSage equations which transductively captures the neighbourhood information of the node.

$$\mathbf{h}^k_{\mathcal{N}(v)} \leftarrow \text{AGGREGATE}_k(\{\mathbf{h}^{k-1}_u, \forall u \in \mathcal{N}(v)\}) \tag{11}$$

$$\mathbf{h}^k_v \leftarrow \sigma\left(\mathbf{W}^k \cdot \text{CONCAT}(\mathbf{h}^{k-1}_v, \mathbf{h}^k_{\mathcal{N}(v)})\right) \tag{12}$$

4   Experiments

4.1   Datasets

- The Manually Annotated Target and Drug Online Resource (MATADOR) database is a free online database of DTIs and contains a total of 8936 DTI entries. The Chemical and Protein IDs in each entry of the database are used to form the DTI network.
- BindingDB [Gilson et al., 2016] is a public, web-accessible database of measured binding affinities, focusing chiefly on the interactions of small molecules. We construct a binary classification dataset with 12592 positive examples and 53852 negative examples.

4.2   Evaluation Metrics

- AUC-ROC curve: AUC-ROC curve is a performance measurement for the classification problems at various threshold settings. ROC is a probability curve and AUC (area under the curve) represents the degree or measure of separability. It tells how much the model is capable of distinguishing between classes. Higher the AUC, the better the model is at predicting 0 classes as 0 and 1 classes as 1.
- Precision: It recommends the quality of algorithm as it shows how accurate we are when we are suggesting the a particular drug to the protein. It is the fraction of relevant instances among the retrieved instances.
- Recall: Recall (also known as sensitivity) is the fraction of relevant instances that were retrieved.

4.3   Results - Similarity Based Metrics

We have calculated precision, recall and AUC-ROC for similarity based techniques: Adamic Adar and preferential attachment. Figures 3 and 4 show the graphical representation of ROC curve along with other metrics for metador and binding DB respectively. We can see lower precision values from similarity metrics because the number of predicted edges between drug and protein is very

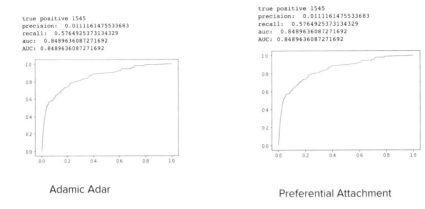

Fig. 3. Similarity based metrics results on metador dataset

high as the probability threshold for predicting the edge taken is low. The recall values obtained from all similarity based metrics is around 57% which shows that around 57% this percentage of true edges are being predicted. AUC metric is around 85% which shows that new link that does occur will be given a good score compared to the score of a potential link that does not occur.

In case of BindingDB, we have a low recall and precision, but AUC metric varies range from 81% in case of Adamic Adar to 94% in preferential attachment as can be seen in Fig. 4.

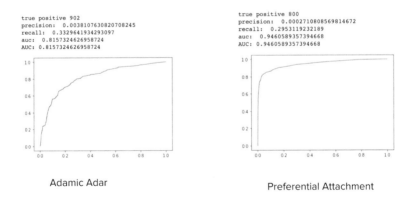

Fig. 4. Similarity based metrics results on BindingDB dataset

Table 1. Comparison of VED-BI with previous work for Binding DB dataset

Method	AUC	Precision	Recall
Tsubaki-et al	91.19	**77.42**	49.61
VED-BI, Approach-1	91.13	72.04	57.61
VED-BI, Approach-2	**92.46**	76.40	**59.41**

Table 2. Comparison of GraphSage approach for Binding DB and Metador datasets

Method	AUC	Precision	Recall
BindingDB	87.90	67.99	98.67
Metador	87.23	64.88	98.56

4.4 Deep Learning Approach Analysis

We have used 12592 positive interactions and 53852 negative interactions for training and have validated the proposed method on bindingDB dataset. In Table 1, we show results using our variational inference method (VED-BI). Approach 2 (Series) clearly shows better performance against prior work on AUC and Recall, while it has comparable precision metric. The superior performance of the Series Architecture (Approach 2), demonstrates better compositionality of features generated across protein sequence and drug molecule leading to better identification of links. In Table 2, we show the results of incorporating the bipartite graph driven features transductively in deep learning architecture using Graph sage approach. This delivers AUC of around 88 % and high Recall of around 98%, demonstrating the power of bipartite network based features from GraphSage.

Strengths and Weaknesses
Our approach performs well on above datasets and metrics but it can have better performance by leveraging 3D structure of the proteins and drugs. Further, we need to leverage inductive link prediction to target new proteins and drugs while also considering sparse data scenarios.

5 Conclusions and Future Work

In this paper, we did a comparative study of various techniques including similarity based methods such as common neighbourhood, Jaccard coefficient, Preferential Attachment(PA) and Adamic Adar on bipartite drug protein network for link prediction. We proposed a novel variational encoder decoder architecture (VED-BI) along with GraphSage based approach for transductive link prediction to model drug and protein features with better generalization in drug protein interaction prediction. Detailed analysis on metdaor and bindingDB datasets demonstrates improvements in AUC and Recall metrics as compared to well known prior deep learning approaches. In future, we aim to study variational inference architectures for few shot link prediction and new approaches such as stable diffusion based inference.

References

1. Altae-Tran, H., Ramsundar, B., Pappu, A.S., Pande, V.S.: Low data drug discovery with one-shot learning. ACS Cent. Sci. **3**, 283–293 (2017)
2. Chen, X., Lin, Y., Gilson, M.K.: The binding database: overview and user's guide. Biopolymers **61**(2), 127–41 (2001)
3. Hamilton, W.L., Ying, Z., Leskovec, J.: Inductive representation learning on large graphs. In: NIPS (2017)
4. Kingma, D.P., Welling, M.: Auto-encoding variational bayes. CoRR arxiv:1312.6114 (2014)
5. Kunegis, J.: On the spectral evolution of large networks (2011)
6. Li, S., et l.: Structure-aware interactive graph neural networks for the prediction of protein-ligand binding affinity. In: Proceedings of the 27th ACM SIGKDD Conference on Knowledge Discovery & Data Mining, pp. 975-985 (2021)
7. Nguyen, T., Le, H., Quinn, T.P., Nguyen, T., Le, T.D., Venkatesh, S.: GraphDTA: predicting drug-target binding affinity with graph neural networks. Bioinformatics **37**(8), 1140–1147 (2020)
8. Peng-Wei, Chan, K.C.C., You, Z.H.: Large-scale prediction of drug-target interactions from deep representations. In: 2016 International Joint Conference on Neural Networks (IJCNN), pp. 1236–1243 (2016)
9. Pollastri, G., Przybylski, D., Rost, B., Baldi, P.: Improving the prediction of protein secondary structure in three and eight classes using recurrent neural networks and profiles. Proteins Struct. **47**, 228–235 (2002)
10. Rezende, D.J., Mohamed, S., Wierstra, D.: Stochastic backpropagation and approximate inference in deep generative models. In: ICML (2014)
11. Schwaller, P., Gaudin, T., Lanyi, D., Bekas, C., Laino, T.: "found in translation": predicting outcomes of complex organic chemistry reactions using neural sequence-to-sequence models. CoRR arxiv:1711.04810 (2017)
12. Tsubaki, M., Tomii, K., Sese, J.: Compound-protein interaction prediction with end-to-end learning of neural networks for graphs and sequences. Bioinformatics **35**, 309–318 (2019)
13. Weininger, D.: Smiles, a chemical language and information system. 1. introduction to methodology and encoding rules. J. Chem. Inf. Comput. Sci. **28**, 31–36 (1988)
14. Yue, X., et al.: Graph embedding on biomedical networks: methods, applications and evaluations. Bioinformatics **36**, 1241–1251 (2020)
15. Zhang, W., Chen, Y., Liu, F., Luo, F., Tian, G., Li, X.: Predicting potential drug-drug interactions by integrating chemical, biological, phenotypic and network data. BMC Bioinf. **18**, 1–12 (2016)
16. Zheng, L., Fan, J., Mu, Y.: OnionNet: a multiple-layer intermolecular-contact-based convolutional neural network for protein–ligand binding affinity prediction. ACS Omega **4**(14), 15956–15965 (2019)

A Smart-Wear Device for Postural Real-Time Feedback in Industrial 4.0 Settings Using DREAM Approach

Augusto de Sousa Coelho[1,2] , Paulo Matos[3], Abílio Borges[3], Richardson Lacerda[1], Susana Lopes[4,5] , Mário Rodrigues[4,5] , Mário Lopes[4,5] , and Joaquim Alvarelhão[4(✉)]

[1] Aveiro North School, University of Aveiro, 3720-511 Oliveira de Azeméis, Portugal
{augustodesousacoelho,richardson.lacerda}@ua.pt
[2] Research Institute for Design, Media and Culture [ID+], 3720-511 Oliveira de Azeméis, Portugal
[3] Atena - Industrial Automation, 3770-355 Palhaça, Portugal
{paulo.matos,aborges}@atena-ai.pt
[4] School of Health Sciences, University of Aveiro, 3810-193 Aveiro, Portugal
{pslopes,mmpr,mariolopes77,jalvarelhao}@ua.pt
[5] Institute of Biomedicine, University of Aveiro, 3810-193 Aveiro, Portugal

Abstract. Concepts underlying the so-called *Industry 4.0* are an opportunity to develop strategies based on the merging of the physical, digital, and biological worlds, enabled by new technologies. This work reports the design proposal for an upside body wear involving industry workers to capture body signs, including real-time postural assessment, that could indicate risk for musculoskeletal disorders. Discover, Research, Engage, Approve, Make methodology allows us to consider factors such as situations, contexts, company employees, and workers (pe, gender, age, or type of task) to have a complete representation of the system. A series of participatory observations were performed in the industrial context which contributed to establishing a deeper relationship between the stakeholders involved around the theme with the analytical portion of the model enabling work in tandem with industrial partners. From participatory observation and worker feedback, the team developed a soft wearable to monitor the body conditions to prevent work-related injuries, instead of an exoskeleton to enhance the body. Nevertheless, the design team started to create concepts of soft wearables, always paying attention to the limitation of the components being used by the upper body device, designing where the circuits would be located around the body, resorting to a life-sized dummy as a model, and testing its arm movements. Future developments include the integration of several partial prototypes that will generate a final device that will integrate all the functions.

Keywords: smart-wear · postural assessment · industry 4.0

A. Rocha et al. (Eds.): WorldCIST 2023, LNNS 801, pp. 457–463, 2024.
https://doi.org/10.1007/978-3-031-45648-0_45

1 Introduction

Concepts underlying the so-called *Industry 4.0* are an opportunity to develop strategies based on the merging of the physical, digital, and biological worlds, enabled by new technologies [1]. The (r)evolution could be implemented by boosting the efficiency of direct and indirect work with tools for better and healthier workers in industrial environments. Many *Industry 4.0* initiatives have been developed over the recent years but few have developed a human-centric approach [2], which means not only incorporating new possibilities technologically driven but looking at the impact on and role of society and workers.

The health of workers has increasing concerns in the industrial sector since numerous work-related diseases compromise their professional performance and personal life. Current epidemiological evidence demonstrates that tasks that require excessive force, poor posture, high repetition, long duration, and are stressful have an increased risk of musculoskeletal disorders [3]. Due to the high prevalence of musculoskeletal complaints and diseases, a significant effort has been made in the prevention of work-related disorders, considering the high burden on health care systems, the economy, the individual, and co-workers and families affected directly and indirectly by the injured workers [4].

Current scientific literature is still to clarify if the regular use of exoskeletons can prevent work-related musculoskeletal complaints and musculoskeletal diseases (primary prevention), improve musculoskeletal complaints, or prevent the aggravation of established musculoskeletal diseases (secondary prevention), or mitigate the progression of musculoskeletal diseases or aid workers with musculoskeletal diseases in work reintegration (tertiary prevention). Consequently, the impact on workers' health with the use of an exoskeleton during occupational tasks, is still unknown, mostly due to the absence of long-term studies under real working conditions [5].

However, particularly for repetitive tasks injury prevention due to inadequate posture may be addressed with smart alerts and information delivered to the operator in real-time conditions or in strategically selected intervals of time during the work shift. A comfortable "smart wear", in opposition to a hard device, with posture assessment, could be an adequate solution to overcome wearing a heavy assistive option.

In this context, this work reports the design proposal for an upside body wear involving industry workers to capture body signs, including real-time postural assessment, that could indicate risk for musculoskeletal disorders.

In the following sections, the previous work will be described, and details about the methodology used for gathering information from potential users of the solution. In the final sections, preliminary results will be presented, some conclusions are drawn, and the continuity of the research work is envisaged.

2   Previous Work

As mentioned, repetitive activities are allied to motion disorders which are also associated with fatigue. Local fatigue of a determined muscle group and global fatigue produced by repetitive activities may lead to changes in both local and global characteristics of that activity. To prolong adequate motor performance, the body goes through complex adaptations and compensations that may lead to injury. It is of interest to acknowledge possible pattern alterations that can be detected, to prevent musculoskeletal disorders [6].

Measurement of external workload can be quantified by speed, accelera-tion/deceleration, and changes in movement direction during physical activity. For these indicators, the use of Inertial Measurement Units sensors could provide sufficient infor-mation if triangulated with the internal workload. However, for estimating internal work-load, a greater diversity of measurements is needed, needing a holistic approach, where physiological and psychological stimulation may be considered for measurement. For example, heart rate (HR), body temperature (BT), and electromyography (EMG) can contribute to this purpose. Current progress has been made in the field of the assessment of training loads in sports, including sporting garments with embedded EMG sensors. These garments can monitor players and can be worn during sporting activities allowing the quantification of muscle activity [7]. The internal load can reflect how an individual's physical capacity responds to an activity, and the external load is quantified indepen-dently of the physical capacity of the individual [7]. Based on these requirements, a draft of the architecture was prepared, which is presented in Fig. 1.

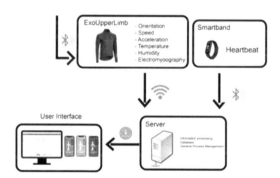

Fig. 1. Architecture for upper body monitoring device

The IMU sensor for this application is the BNO055 Smart Sensor is a System in Package (SiP) solution that integrates a triaxial 14-bit accelerometer, an accurate close-loop triaxial 16-bit gyroscope, a triaxial geomagnetic sensor and a 32-bit microcontroller running the BSX3.0 FusionLib software.

This smart sensor is significantly smaller than comparable solutions. By integrat-ing sensors and sensor fusion in a single device, the BNO055 makes integration easy, avoids complex multivendor solutions, and thus simplifies innovations [8]. For EMG the AD8226 sensor was added to the system [9], and a bracelet for monitoring heart rate [10].

3   Methodology

The need to mediate the stakeholder's expectations, especially for industry workers who perform repetitive tasks, for product design is a challenge. A technology-centric approach, not considering all elements can lead to a failure in identifying and achieving a proper design. Based on previous experiences the methodology – Discover, Research, Engage, Approve, Make (DREAM) was chosen [11]. Following this methodology implies deep research, first on *sensoring* systems, how it's used, and what defines them. After the initial research phase, engagement with the company and workers was performed to collect their feedback, needs, and expectations – Fig. 2.

The objective was to consider, as much as possible, factors such as situations, contexts, company employees, and workers (gender, age, type of task, etc.) to have a complete representation of the system. This approach contributes to establishing a deeper relationship between the stakeholders involved around the theme with the analytical portion of the model enabling work in tandem with industrial partners.

For the development of this activity, a series of participatory observations were performed in the industrial context. The first observation of a participant equipped with the soft wearables in the real factory setting was performed to validate the product architecture, and the interactions between them, as well as to obtain specific information about the worker tasks. With this participatory observation, a detailed description of the context, treatment, and analysis of the collected data, namely information on the location of the components of the monitoring device was obtained. Finally, the interface of system interaction for the worker was also tested through a participatory observation to obtain feedback on the colors and functionality of the interface.

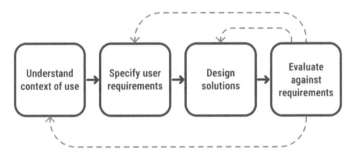

Fig. 2. Interaction model for data collection in the industrial context

4   Preliminary Results

To carry out the participatory observation, the concepts were developed and prototyped in a laboratory, using a mannequin as anthropomorphic support – Fig. 3.

The main goal of this task was to observe the worker in the real context, test the planned locations of the sensor circuits, and test if they can maintain in place during the activities.

With the valuable information collected, some crucial questions were answered regarding the design and development of a device of this nature, such as: what is the maximum size? How heavy should it be? Where should it be placed? How should it communicate with the user?

The worker identified the zones of discomfort caused by the repeated tasks and accordingly, with the worker's feedback, the controller should be placed on the upper back, below the back of the neck, and between the shoulders – Fig. 3. These locations allow the connection through a dock to the circuitry for the user´s freedom of movement, as well as for the visual feedback for the user's co-workers.

After the participatory observation and worker feedback, the team developed a soft wearable to monitor the body conditions to prevent work-related injuries, instead of an exoskeleton to enhance the body. Nevertheless, the design team started to create concepts of soft wearables, always paying attention to the limitation of the components being used by the upper body device, designing where the circuits would be located around the body, resorting to a life-sized dummy as a model, and testing its arm movements.

It was necessary to experiment with how the volume of the controller (the main component of the product) would be placed on the body. The team's concern was related to the weight of the garment, the freedom of movement of its user, and the comfort of using it for 8 h in a row in a hot environment. After these experiments and further analysis, the team concluded it would be to develop a soft wearable in the thinnest possible fabric, providing skin ventilation.

Fig. 3. Possible places to fix the components

Finally, a few volumetric models for the controller were made using blue foam. The aim was to fit all the system's components in a very compact body, so it does not cause discomfort to the user. The final shape of the body is still to be defined since the team is still experimenting with different sizes, shapes, and the arrangement of the components in the garment.

5   Conclusion and Future Work

The chosen user-centric approach, supported by a project-based learning methodology, was essential to integrate all the stake holder's expectations in defining the system interactions and outgoing communication.

To facilitate adherence and improve smart wear acceptability among industry workers, technology should (a) focus its use on improving workplace safety, (b) advance a positive safety climate, and (c) ensure sufficient evidence to support employees' beliefs that the wearable will meet its objective, and (d) involve and inform employees in the process of selecting and implementing wearable technology [12]. Therefore, the results of this task are aligned with current scientific reports.

The main goal of this activity was to develop a support system for the person incorporating the ergonomic features of the system to prevent health risks. The system aims to provide the worker with feedback on the best postures to prevent work-related injuries. Future developments include the integration of several partial prototypes that will generate a final device that will integrate all the functions.

Acknowledgments. The present study was also developed in the scope of the Project Augmented Humanity [POCI-01-0247-FEDER-046103 and LISBOA-01-0247-FEDER-046103], financed by Portugal 2020, under the Competitiveness and Internationalization Operational Program, the Lisbon Regional Operational Program, and by the European Regional Development Fund.

References

1. Malomane, R., Musonda, I., Okoro, C.S.: The opportunities and challenges associated with the implementation of fourth industrial revolution technologies to manage health and safety. Int. J. Environ. Res. Public Health **19**(2) (2022)
2. Cunha L, Silva D, Maggioli S. Exploring the status of the human operator in Industry 4.0: A systematic review. Front Psychol. **13** (2022)
3. Bridger, R.: Introduction to Human Factors and Ergonomics. CRC Press, Boca Raton (2017)
4. Steinhilber, B., Luger, T., Schwenkreis, P., Middeldorf, S., Bork, H., Mann, B., et al.: The use of exoskeletons in the occupational context for primary, secondary, and tertiary prevention of work-related musculoskeletal complaints. IISE Trans. Occup. Ergon. Hum. Factors **8**(3), 132–144 (2020)
5. Bär, M., Steinhilber, B., Rieger, M.A., Luger, T.: The influence of using exoskeletons during occupational tasks on acute physical stress and strain compared to no exoskeleton - a systematic review and meta-analysis. Appl. Ergon. **94** (2021). https://pubmed.ncbi.nlm.nih.gov/336 76059/. Accessed 7 Dec 2022
6. Côté, J.N., Feldman, A.G., Mathieu, P.A., Levin, M.F.: Effects of fatigue on intermuscular coordination during repetitive hammering. Motor Control **12**(2), 79–92 (2008). https://pub med.ncbi.nlm.nih.gov/18483444/. Accessed 7 Dec 2022
7. Cardinale, M., Varley, M.C.: Wearable training-monitoring technology: applications, challenges, and opportunities. Int. J. Sports Physiol. Perform. **12**(Suppl 2), 55–62 (2017). https:// pubmed.ncbi.nlm.nih.gov/27834559/. Accessed 7 Dec 2022
8. Bosch Sensortec. Smart Sensor BNO055 | Bosch Sensortec. https://www.bosch-sensortec. com/products/smart-sensors/bno055/. Accessed 7 June 2022

9. WTWH Media. Designing an Arduino-based EMG monitor. https://www.engineersgarage. com/arduino-based-emg-monitor-ad8226/. Accessed 8 June 2022

10. LILYGO. LILYGO® TTGO T-Wristband DIY Programmable Smart Bracelet ESP32-PICO-D4 Main Chip 0.96 Inch IPS Screen Silicone Bracelet Strap(1)(1)(1) - New Product - Shenzhen Xin Yuan Electronic Technology Co., Ltd. http://www.lilygo.cn/claprod_view.aspx?TypeId= 21&Id=1282&FId=t28:21:28. Accessed 7 Dec 2022

11. de Sousa Coelho, A., Branco, V.: Design research for the development of a medical emergency ambulance. Design as a symbolic qualifier in the design of complex systems/products. **20**(sup1), S2135–57 (2017). https://doi.org/10.1080/14606925.2017.1352731, https://www. tandfonline.com/doi/abs/10.1080/14606925.2017.1352731. Accessed 7 Dec 2022

12. Jacobs, J.V., et al.: Employee acceptance of wearable technology in the workplace. Appl. Ergon. **78**, 148–56 (2019)

Author Index

Printed in the United States
by Baker & Taylor Publisher Services